Groundwater Geophysics in Hard Rock

Groundwater Geophysics in Hard Rock

P.C. Chandra

Consultant, The World Bank, New Delhi, India
Advisor, WAPCOS Ltd., Ministry of Water Resources,
River Development & Ganga Rejuvenation, Government of India

CRC Press
Taylor & Francis Group
Boca Raton London New York

CRC Press is an imprint of the
Taylor & Francis Group, an **informa** business

A BALKEMA BOOK

Cover photo: 'Groundwater impounded in abandoned quarry in Chhotanagpur
Gneiss-Granulite Complex, Jharkhand State, India'
Courtesy: Dr Dipankar Saha, Central Ground Water Board, India

CRC Press
Taylor & Francis Group
6000 Broken Sound Parkway NW, Suite 300
Boca Raton, FL 33487-2742

First issued in paperback 2020

ISBN-13: 978-0-415-66463-9 (hbk)
ISBN-13: 978-0-367-78335-8 (pbk)

Library of Congress Cataloging-in-Publication Data

Chandra, P. C.
  Groundwater geophysics in hard rock / P.C. Chandra,
    pages cm
  Includes bibliographical references and index.
  ISBN 978-0-415-66463-9 (hardcover) – ISBN 978-0-203-09367-2 (ebook)   1. Water well drilling.
2. Hydrogeology.   3. Boring.   I. Title.
  TD412.C425   2015
  553.7′9–dc23
                                                                    2015031737

Visit the Taylor & Francis Web site at
http://www.taylorandfrancis.com

and the CRC Press Web site at
http://www.crcpress.com

Dedicated to the memory of my parents, maternal uncles, mother in law and friend Ashok Acharjya

Table of contents

Preface

Hard rock hydrogeology is quite complex and the application of geophysics is challenging. Though there are abundant success-stories worldwide on the applications of conventional and modern geophysical techniques, uncertainties still exist in providing the desired information. Obviously, techniques have their respective inherent limitations and dissatisfaction may arise due to over-expectations. But most of the uncertainties are either due to nature's complexities, mismatched and ill-equipped applications or inadequacies in hydrogeological transformation of geophysical information and vice versa. To ensure success, geophysical applications have to start with hydrogeology and end in hydrogeology.

There has been a long-standing demand for a book on groundwater geophysics in hard rock presenting a comprehensive view of problem-specific geophysical methodologies, appropriate techniques with combinations, titbits of local and regional scale field-level approaches and constraints, the art of interpretation, the imperatives of limitations and achievements in the form of case studies. The book *Groundwater Geophysics in Hard Rock*, written for field geophysicists, hydrogeologists and engineers, attempts to bring these aspects together. It does not provide a detailed explanation of the theory or interpretation of geophysical methods, but presents the utilities, practical aspects and applications through case studies, including drilling results, in understanding their usefulness and limitations in hard rock.

It is a humble attempt to put together the relevant views of eminent researchers in hard rock hydrogeology and exploration geophysics including conventional and modern approaches in one place and also to share my experience. The objective is to present a broad, practical review of geophysical methods for groundwater exploration in hard rocks, establish an understanding of the conceptual framework and basic principles involved and the physical properties measured and to provide a practical guide for making field measurements and interpret the data. It also attempts to put forward the status of groundwater geophysics in hard rock and a broad outlook of prospective research activities to overcome various uncertainties.

Chapter 1 is on emerging groundwater issues and problems in hard rocks and applications of geophysics, followed by Chapter 2 on the hydrogeological characterization of hard rock aquifers to understand and appreciate the issues. Chapter 3 on the geophysical conceptualization of hard rock aquifers gives an introductory account of the multi-scale applications of different geophysical investigation methods and techniques used worldwide, their convergence with other disciplines and experiences. It also incorporates the past developments, present status of knowledge and future potentials so that the reader enters into the environment of groundwater geophysics with perceptive indulgence. As hard rock groundwater investigations are mostly demand-controlled, near habitations, with a variety of constraints, this is followed by Chapter 4 on how to plan and select sites for geophysical investigations. Chapters 5 to 12 methodically present the effective geophysical methods and techniques, data acquisition, processing, interpretation techniques and limitations embedded with case studies on magnetic and aeromagnetic, electrical resistivity, self-potential, mise-a-la-masse, frequency domain electromagnetic, very low frequency electromagnetic, time-domain electromagnetic and borehole logging. In many cases a single geophysical technique may not yield the desired information and a combination of techniques is required. Chapter 13 presents a few modern methods and explains how the techniques can be prudently integrated for convergence of geophysical, hydrogeological and hydrochemical thematic investigations with case studies for a holistic approach reaching the best contextual commonality and concludes the subject. Finally Chapter 14 is devoted to artificial recharge and contamination – the inevitable contemporary hard rock aquifer development and management issues and an integral part of sustainable groundwater management. The chapter renders an insight into the potential applications of geophysics in the selection of artificial recharge sites, impact assessment and the identification of contamination and waste disposal sites. The conventional resistivity method, being the most popular, is discussed in detail. There are repetitions on high resolution airborne surveys to stress its essentiality in fast coverage, cost effective, large scale, high data-density aquifer characterization.

I hope the book will be useful for practising geologists, hydrogeologists, geophysicists and civil and environmental engineers who plan to enter the challenging domain of enigmatic hard rock-fractured zone-aquifer delineation and subsurface characterization. It may help increase awareness in groundwater geophysics and develop interest in and appreciation of its applications in hard rock. Also, it may help stimulate scientific research activities to prevail over non-uniqueness and poor resolution in complex hard rock heterogeneities.

P.C. Chandra
June, 2015
Varanasi, India

Acknowledgements

I am grateful to the Central Ground Water Board, Government of India and the CSIR-National Geophysical Research Institute, Government of India for providing me unlimited opportunities and freedom to work and experiment in groundwater geophysics in various parts of India, work which forms the basis of the book. I take this opportunity to thank all the hydrogeologists, geophysicists and engineers of Central Ground Water Board and CSIR-NGRI with whom I had the privilege to work, contribute and gain experience.

The case studies presented were carried out with K.C.B. Raju, K.R. Srinivasan, G.K. Dev Burman, A.N. Bhowmick, Hira Singh, P.C. Chaturvedi, A.D. Joseph, P.K. Das, B.B. Trivedi, Subhash Singh, P.H.P. Reddy, A.K. Ghosh, A. Ramakrishna, M.M. Srivastava, S. Tata, Mohd. Adil, S. Haq, Vikas Ranjan, M.K. Bhowmic, B.K. Oraon, Subroto Das and B.S. Tewari of CGWB. I thank them and gratefully acknowledge their contributions. I am thankful to the Director General, Geological Survey of India for providing the aeromagnetic maps. Also, I gratefully acknowledge the research work in the form of scientific papers and books, manuals and brochures of the authors and institutions and personal communications referred in this book.

I am grateful to Dipankar Saha of CGWB, and K. Mallick, R.L. Dhar, Shakeel Ahmed, S.K. Verma, S.S. Rai, D. Muralidharan, Subash Chandra and N.C. Mondal of NGRI for valuable discussions. Jiri Krasny of Charles University Prague and Ms Germaine Seijger of Taylor and Francis Group whom I met at the IAH conference at Hyderabad in 2009 encouraged me to write this book. I am thankful to them, and also to Alistair Bright, Acquisition editor, of Taylor and Francis Group for his persisting encouragement and the required copyright permissions, to anonymous reviewers, to Jean Roy, IGP, Montreal, Canada for very useful suggestions in writing and to E. Auken, Aarhus University and J.W. Lane, USGS for relevant interactions I had from time to time. I am thankful to Ms Anju Gaur, World Bank and J.K. Rai, WAPCOS Ltd. and S. Kunar, RITES for valuable discussions on geophysical applications in aquifer mapping.

I held several fruitful discussions with G.S. Yadav and M. Banerjee of B.H.U.; Jai Krishan Tandon, Lokendra Kumar and Shashi Kant Singh of CGWB helped drafting the figures; Dipankar Saha and Subash Chandra provided the photograph; Ms Pallavi Chattopadhyay, Tarun Gaur, Pradeep Maurya and Deepak Jaiswal of CSIR-NGRI; and A.K. Agrawal, S.N. Dwivedi, R.K. Singh and S.S. Vittala of CGWB provided various technical supports; James Rainbird, Language Editor, Lukas Goosen, Production Manager and Ms José van der Veer, Production Editor and the supporting staff of

Taylor and Francis Group did the painstaking job of producing the book. I remain grateful to all.

I am indebted to my wife Durga, son Arnav and daughter Aditi for their love and constant support and encouragement during this writing project which has taken a disproportionate amount of time. Aditi and Arnav helped in the critical review of parts of the book. While acknowledging it I feel proud of them. I am thankful to my brothers and sisters, daughter-in-law Priyanka, and sisters- and brothers-in-law who constantly encouraged and helped in various ways during these years, facilitating the writing.

I remain indebted to Late Amalendu Roy of CSIR-NGRI and to my teacher Late C.L. Singh of B.H.U. for their enduring guidance and motivation throughout my career.

About the Author

 Dr. Prabhat C. Chandra, a professional groundwater geophysicist, was born in Varanasi, India in 1950. He received B.Sc. and M.Sc degrees in Geology and Geophysics from Banaras Hindu University (BHU) in 1970 and 1972 and was awarded the N.L. Sharma Gold Medal in Geology and first rank in Geophysics.

Soon after, Dr. Chandra started his career as a groundwater geophysicist at the CSIR-National Geophysical Research Institute, Hyderabad (NGRI) India and in 1978 joined the Central Ground Water Board (CGWB), Govt. of India. His doctoral thesis was on groundwater geophysics. He retired in December 2010 at the age of 60 as Regional Director, CGWB.

During his 38 year professional career he has had ample opportunity to work on a variety of groundwater issues in almost all the hydrogeological terrains of India including hard rock, coastal tracts, limestone, basalts, alluvium, desert, islands and hilly tracts. In view of the scarcity of groundwater in hard rock he took up the challenging geophysical investigations of delineating fractured zones in hard rocks which cover two thirds of the country. There are several papers and reports to his credit.

He was trained in groundwater management through an Indo-British Fellowship from the UK. He has attended World Water Week, Sweden and visited the Hydro Geophysics Group (HGG), Aarhus University, Denmark for presentations on groundwater geophysics.

At Allahabad University, Central University, Patna and the Indian School of Mines, Dhanbad, he taught hydrogeology and groundwater geophysics.

After retirement he worked as a consultant to The World Bank, New Delhi and as an expert to CSIR-NGRI along with experts from the U.S. Geological Survey (USGS) and HGG, Aarhus University, Denmark for aquifer mapping in pilot projects through heliborne geophysical surveys and as advisor to WAPCOS Ltd. Govt. of India for aquifer mapping of the National Capital Region through surface and borehole geophysical surveys.

The book 'Groundwater Geophysics in Hard Rock' is based on his vast experience in delineating the fractured zones in hard rocks, subsurface characterization and locating high yielding well sites.

Chapter 1

Groundwater issues in hard rock & geophysics

1.1 INTRODUCTION

Igneous and metamorphic rocks are termed as hard rock or fractured rock. These rocks are exposed over large areas in several parts of the world, occupying about 20% – approximately 30 million sq. km – of the land area. The occurrence of groundwater in hard rock is, in general, heterogeneous and complex. Compared to sedimentary areas, hard rock possesses relatively fewer yielding, shallow discontinuous aquifers in saturated weathered and semi-weathered zones and the deeper fractured zones in underlying compact formation. Development of these aquifers of limited dimension varies from area to area, depending on their availability, water quality, storage, yielding capacity, specific need and contribution to meet the demands. Nevertheless, groundwater from these aquifers forms an indispensable component of water supply systems for drinking, irrigation and industry.

The hard rock areas under tropical and semi-arid to arid climates, viz. in Africa, Australia, India and South America are, in general, water scarce and sometimes a drought-like situation is faced in places. The inadequate availability of surface water in some of the areas has led to near-total dependency on groundwater and in places it is the only resource, playing a vital role in increasing agricultural productivity, socio-economic growth, eradication of water-borne diseases and alleviation of poverty. Nevertheless, sustenance of the resource warrants a systematic multi-disciplinary approach to its development and management.

1.2 TRENDS IN GROUNDWATER UTILIZATION

In hard rock areas of a tropical country like India, which is two-thirds hard rock, for a long time groundwater has been in use as a dependable source for drinking and irrigation. The age-old practice of groundwater abstraction is through shallow, large diameter hand-dug wells tapping weathered zone aquifer within 12–15 m depth. Increasing water demand resulted in a shift from dug wells to dug-cum-bore wells (borehole drilled within a large diameter dug well) and drilled wells with diesel, electric and hand pumps. A large number of wells tapping the weathered zone and fractured aquifers were drilled within 60–100 m depth by government agencies for community water supply and irrigation at preferred demand or consumption points. The growing demand and incentives also impelled the sinking of numerous privately owned deep wells to tap the saturated fractured zone at sites randomly selected or insisted upon either by the driller or user. The depth of the well was commanded by water requirement, money availability and local experience from previous drilling or a blind guess on depth of fractures expected to yield required water. The random drilling of wells quite often resulted in dry wells or poor yielding wells, defeating the purpose, warranting selection of alternative sites or deepening of existing wells to tap at least one water-yielding fracture. Marechal (2010) rightly points out that once drilling starts it becomes difficult to stop drilling of a dry borehole thinking that a yielding fracture will strike only after drilling another few metres and in most of the cases money is wasted. The overall expenditure on water-well drilling increases manifold due to well failures with minimal returns stressing the essentiality of scientific approach to investigate and pinpoint drilling sites for adequate yield from weathered and fractured zones. The small and marginal farmers who already experimented with such random dry-well drilling and could not manage locating a suitable site and afford sinking another yielding deep well due to additional cost, depend on rain or other sources and suffer the utmost, facing financial and food crises at times. A case study reflecting emergence of such a situation is presented by Bassi et al. (2008). The publications of the Geological Society of India (2008 a & b) present a detailed view of groundwater conditions and utilization in hard rock terrain of Peninsular India.

In parts of Africa groundwater usages are contrasting and issues are different. Giordano (2006) reports that groundwater resources are, in general, substantial in African countries and there are areas with underdeveloped as well as sustainably used groundwater and also areas with a depleting resource. Pietersen et al. (2011) report decline in groundwater level due to over-extraction in parts of South Africa and Foster et al. (2006) report it for Nairobi (Kenya) and Lusaka (Zambia).

According to Foster et al. (2006), in Sub-Saharan Africa it is mostly developed by community initiative and like India there has been a swing from hand dug wells to drilled wells with diesel and electric pumps since the 1970s which increased the groundwater use in domestic consumptions and raising livestock and thus played a key factor in extending human settlement and its growing dependency on this natural resource. Also, it helped eradicate the water borne diseases. According to Giordano, in general, groundwater for agricultural use remains underdeveloped rather than overexploited due to varied spatial distribution and other physical factors and it is used for irrigation only where there is low rainfall. In fact groundwater supports livestock economies and domestic consumptions. Foster et al. links the restricted development of groundwater to high cost of well construction, well failures due to poor well design and depth, inadequacy in water quality and quantity and unknown complex subsurface conditions. The maximum groundwater use is reported in South Africa with two-thirds population totally depending on it. The constraint for not developing groundwater to its maximum potential is the lack of reliable hydrogeological information on fractured rock aquifer (Woodford et al., 2005).

In South American countries groundwater has been in use over the centuries. It is distributed unevenly (Reboucas, 1999). According to Reboucas although the dependency on groundwater has increased over the past 3–4 decades, there is, in general, no serious water crisis. However, groundwater salinity and scarcity is felt in northeastern Brazil affected by droughts (Bocanegra and Silva Jr., 2007). Groundwater abstraction is generally through hand-dug wells and drilled wells. It is used for drinking, industry and irrigation. As such, groundwater development is focused in specific areas, e.g., private wells are plentiful in the metropolitan areas of Brazil. Wells are generally drilled up to 100 m depth to tap fractured zones of varied yield. With heterogenous climate and possible extreme climatic events, of late IGRAC (2014) indicates a challenging future sustainability of groundwater resources in South American countries.

In Australia fractured rock aquifers underlie about 40% of the continent (Ponce, 2006). Groundwater is a major source of water for drinking, industry and agriculture in a large part of the driest continent. The major problem is that of groundwater salinity. Besides, several droughts in the recent past in Australia have affected agricultural, industrial and domestic consumptions of groundwater. Accordingly, a variety of water conservation practices has been adopted including recycling and desalination and managed aquifer recharge (Bond *et al.*, 2008; Harrington and Cook, 2014).

In hard rock terrain of Sri Lanka, covering about ninety percent of its geographical area, groundwater is used for domestic, industrial and agricultural consumptions. The abstraction is mostly through large diameter (5–8 m) shallow wells. These are known as agro-wells providing readily available large water storage within the well. An agro-well construction programme was initiated in the late 1980s for supplementary irrigation (de Silva, 1998). A large number of these wells are developed in a disorganized manner causing environmental degradation and there is a need for a systematic research approach (de Silva 1998; Panabokke and Perera, 2005).

1.3 NECESSITY OF MANAGED AQUIFER RECHARGE

The increasing use of groundwater has introduced a major problem of resource sustainability in several areas and warrants management. In hard rock it is complex because of heterogeneities in quantity and quality. While salinity and contamination of groundwater is not a global issue, the resource sustainability is. The sustenance of a resource and its protection from contamination require area-specific aquifer management customized to local scales involving social aspects such as 'participatory groundwater management'. The vital management issue to be addressed in tropical countries in general and particularly in areas dominated by uncontrollable development is the sustenance of limited-storage aquifer in shallow weathered zone (the saturated thickness) for safe water supply without well-failures. For this, monitoring of groundwater level, changes in draft and water quality, recharge conditions, groundwater flow, dynamics of saturated thickness, unsaturated zone and surface water-groundwater interactions and assessment of boundary conditions and hydraulic parameters are some of the basic information required which can be generated through scientific investigations and involvement of local Water Users Associations (WUAs).

To control water level decline or recover the level of groundwater the best and most effective option is artificial recharge, also known as managed aquifer recharge (MAR), at the proper place through area- and site-specific appropriate technique to reduce surface runoff. The dried-up or dewatered aquifer zone with adequate storage capacity may also be recharged by consistently available clean or treated surface water to maximize subsurface natural storage and use it at a later date. MAR is very much in practice in almost all countries and particularly in hard rock areas of tropical continents and countries like Africa, Australia, India and South America. For this, the presence of a thick permeable weathered zone is essential as it can store water and help percolate it deeper to recharge the underlying fractured zones. The arid and semi-arid regions quite frequently face drought conditions. The recharged fractured zones may remain untapped and unaffected by vagaries of climate in places and can be tapped if locations and depths are precisely known in advance to supplement water supply during such natural disasters for managing the groundwater emergencies.

1.4 GROUNDWATER QUALITY ISSUES

In some areas a major concern is high geogenic fluoride concentration in groundwater in fractured zones at depth, particularly in granitic and gneissic terrain and in some volcanic terrains (British Geological Survey, 2001; Brunt et al., 2004; Alemayehu, 2006). High fluoride concentrations are generally encountered in arid and semi-arid areas. An adequately yielding fractured zone with high fluoride concentration is quite common. In such cases, water supply has to be mostly from an overlying thin weathered zone and alluvial capping supplemented by defluoridation of water from the fractured zones. Besides, the shallow weathered zone aquifers are exposed to possible anthropogenic contamination and saline water intrusion. While fluoride contamination requires treatment and dilution, the anthropogenic contamination and salinity require monitoring and prevention from growing up and spreading. Dilution of fluoride concentration can be an important target for MAR as suggested by Muralidharan et al. (2002). In this context it is imperative to know the groundwater flow path required to understand the movement of artificially recharged water in deeper fractured zones and also it is necessary to demarcate areas holding deeper potential fractured zones with fresh as well as contaminated groundwater through scientific investigations.

1.5 PROBLEMS IN GROUNDWATER DEVELOPMENT

As far as groundwater development in hard rock, particularly in the crystalline basement, is concerned, the terrain is hydrogeologically complex at extremely varying scale. With varied weathered zone thickness, enigmatic fracture network, low hydraulic conductivity and limited storage, a widely varying well-productivity is observed within a short distance. The fractures behave like isolated thin conduits or pipelines at depth. The interconnection between these fractures, their openness and saturation are significant for sustenance of groundwater flow and yield. The regional fractured zones are likely to encounter more interconnected fractures than localized ones. The intensity and density of fracturing vary and generally reduce with depth. With so many variables causing multi-scale heterogeneities, the geometry of fracture network forming aquifer becomes complex and carries too many uncertainties. The variations are so uncertain that wells sunk a few metres apart may have completely different yield characteristics, depending mainly on the number of productive fractures encountered. This uncertainty is the major impediment in groundwater development of weathered and fractured zone aquifers and is of very considerable concern (Chandra, 2009). In most places success in hard rock water well drilling still remains as a matter of chance. The most appropriate and ideal condition would be to locate a site with considerable thickness of saturated weathered zone and saprolite underlain by close-spaced saturated open fractures in the bedrock.

A major issue is to ascertain the presence of saturated fractured zones within a reasonably explorable depth varying from 60 to 200 m. For this, a detailed subsurface investigation at site-scale is essential to converge at precise delineation of target fractured zone and ascertain its geometry for locating well sites. According to Butler (2005) an error of 0.5 to 1.0 m in locating a fracture can result in a dry borehole. It is a tough task and obviously challenging. It requires a systematic approach integrating different methods and precision techniques, narrowing down from regional to site-scale. Besides exploring groundwater, such systematic investigations at large-scale, to some extent, also help characterize fractured zone aquifers and generate watershed hydrogeological models required for aquifer management.

The hand pumps and shallow tube wells mostly cater to the need as point sources for drinking water supply as well as small scale irrigation by tapping the near surface weathered zone of a thickness of up to about 30 to 40 m in arid and semi-arid tropical areas. The discharge from this zone is generally limited to about 1 litre per second (lps). The development of groundwater in hard rock is possible through better yielding deep wells tapping saturated fractured zones. Discharge of these deep wells may reach as high as 30 lps in places but they are very few. A few high yielding (more than 3 to 5 lps) wells may replace a number of low discharge or exhausted shallow wells if yield from fractured zones can be sustainably managed. For tapping fractured zones, wells can be drilled to a maximum depth of 200 m. This is the economic depth of drilling as observed in many areas and also indicated by Davis and Turk (1964). However, area-wise drilling-depth can also be optimized but that requires precise drilling data of existing deep boreholes in the area which is seldom available.

1.6   NEED FOR SYSTEMATIC INVESTIGATION

For groundwater exploration systematic characterization of fractured rock is generally not attempted as it is time-consuming and quite an expensive process. The exploration usually ends with successful or failed dry wells yielding point-information. Though the borehole findings help getting a generalized understanding of the area, they hardly provide any scope for extrapolation or interpolation. A thick weathered zone or a couple of fractures encountered in a borehole ascertain their presence at that borehole and may not help define lateral dimension or fracture network in the area but can obviously help conceptualize the local hydrogeological condition. Unlike sedimentaries it is difficult to define hydraulic character that represents a large area and is applicable in locating another well or estimating the yield. This brings in a constraint in large scale development and management of groundwater in hard rocks.

Attempts are to be made for systematic investigations to derive precise information on orientation, extension, width, density, possible interconnectivity of fractures and quality of saturating water to a large extent with validating support from drilling results and borehole tests. These help understand possible recharge and contamination movement paths and generate predictive models at different scales. These also help monitor and manage under different future scenarios. Systematic approaches through theoretical and laboratory studies, field investigations and modeling are in practice to characterize the fractured zones. Such studies may be necessary while carrying out investigations for groundwater exploration. MacDonald *et al.* (2008) based on their studies in Sub-Saharan Africa suggest for detailed R&D studies on age, recharge and sustainability of groundwater in basement aquifers, vulnerability of the weathered zone aquifer to pollution, delineation of deep fractures where weathered zone is thin or absent and on different types of fractures encountered in volcanic rocks. The paper by Krasny and Sharp (2007a) and book on 'Groundwater in fractured rocks' edited by Krasny and Sharp (2007b) may be referenced to comprehend the advancement in hard rock hydrogeology.

1.7 SCOPE AND ESSENTIALITY OF GEOPHYSICAL INPUT

In view of the unpredictable groundwater conditions at depth and occasional well-failures, characterization and quantification of subsurface conditions using geophysical methods has become an essential approach. Application of geophysics reduces uncertainties in borehole drilling and improves the understanding of hard rock hydrogeology at different scales. Systematic characterization of saturated fractured zone is a vital input for aquifer management and remediation (Chandra 2006). The geophysical approach to fracture characterization, in fact, got an impetus through programmes in several countries for safe storage and disposal of hazardous nuclear waste under a natural barrier of fracture-free deep crystalline environment (Soonawala, 1984). Secondly, it drew attention because of the discoveries of hydrocarbon-bearing fractured reservoirs in hard rocks in several countries (Petford and McCaffrey, 2003; Sircar, 2004; Morariu, 2012). Geophysical characterization of fracture has also been attempted in tunnel constructions and underground hydrocarbon storage. All these studies have enriched the knowledge about fracture network and their hydraulic characteristics and possible geophysical approach to characterize them rendering scope for their applications in groundwater exploration.

Geophysical methods are used to measure spatial and/or temporal variations in physical properties of the Earth and generate quantitative subsurface models which are interpreted subjectively. The application of geophysical methods and techniques in groundwater, the domain of 'groundwater geophysics' or 'hydrogeophysics' is to provide precise and comprehensive insight of the subsurface hydrogeological conditions with the least possible ambiguity to minimize the failures. The applications in groundwater are relatively new as compared to mineral and petroleum exploration. It is effective and successful when implemented selectively and thoughtfully after conceptualizing the specific groundwater problem. MacDonald *et al.* (2008) indicate the effectiveness of geophysical methods in locating water wells in hard rock of African countries. Surface geophysical investigations are non-invasive, reasonably economical and may be used from beginning of the programme of drilling site selection or site characterization. It is difficult to predict well discharge from geophysical measurement, but it may directly reduce the cost of locating a successful or the best water well site in a given hydrogeological set-up. Also, it may help optimize the depth of drilling for maximum chance of encountering a productive fracture in that hydrogeological environment. Definitely, the information obtained from surface geophysical measurement cannot replace the *in situ* point-information obtained from boreholes but it samples a larger volume for additional spatial insight and reduces the cost and time involved in drilling a large number of boreholes to obtain such close-grid information. Since the benefits arising from application of geophysics in terms of money and time are relative, as such it is difficult to quantify.

Prior to selecting and employing any geophysical method, its primary purpose and preferential hydrogeological conditions vis-a-vis its capability and limitation to deliver the desired information should be studied. It reduces occasional technical failure which may happen due to improper or over demanding application and interpretation. It makes the approach appropriate and economic. Also, to avoid over-expectation, it is necessary that the beneficiaries are informed in advance about the possible output against the objective and the limitations. In general, the effective and economic geophysical methods for surface surveys are electrical and electromagnetic. They have been used extensively. The choice of most appropriate geophysical method and technique or combination-technique depends on objective, hydrogeological conditions, scale of survey and depth up to which information and the resolution is required. At times it depends on the availability of equipment, expertise, logistics and the financial resources. Overall, the choice should be such that it gives an assertive response for the target with minimum ambiguity. The applications become expensive with increased detailing and precision. However, the expenditure on the geophysical component of the exploration programme in an area becomes insignificant compared to selection of a good number of successful sites. At times and places, application of geophysics becomes exigent as it has to ascertain availability of groundwater at users' demand point or in and around habitations.

1.8 GEOPHYSICAL DELIVERABLES

The geophysical surveys can deliver information on thickening of weathered zones, saturated weathered zone thickness and extent, bed rock topography, anomalous fractured zone locations, their orientation and lateral continuity, dykes and quartz reefs, vadose zone, suitability of a site for artificial recharge, quality of groundwater and contamination in terms of electrical conductivity variations, suitability of an area for waste dumping and landfills, dynamics of groundwater movement, dried-up aquifers, aquifer boundaries and approximate hydraulic characteristics. Monitoring of the vadose zone and the contaminated zone can also be done through geophysical studies.

1.9 PREREQUISITE TECHNICAL FIELD-GUIDANCE

A standard technical guidance needs to be imparted to initiate groundwater investigations in hard rocks in a multi-disciplinary environment. The investigations are to start with the study of geological, topographic and geomorphological maps of the area. The identification of favourable hydro-geomorphological features from satellite imageries has great significance. The lineaments are mostly associated with structural features and also with deep weathering and fracturing. They can be considered as surface manifestations of fractured zones. Hydrogeological and lineament maps are conventionally used to infer the possible trend of a major fractured zone. This is followed by an inventory of existing wells in the area for water level and quality, inspection of well sections and a dialogue with local people about their experience of groundwater availability in the area. Based on these studies and investigations, geophysical methods are selected for reconnaissance. The results of reconnaissance geophysical survey act as a guide to detailed geophysical surveys for site-scale investigations. The geophysical approach varies according to the objective and geology and therefore adequate knowledge of local geology and hydrogeological conditions is indispensable for the geophysicists. For example, the approach in granite is different from that in volcanic rock holding a number of sub-horizontal vesicular and fractured zone aquifers. The ways investigations are to be conducted are made clear with reference to the strike of geological formation and structure, scale, field procedure, data processing, methods of interpreting the data and the essentiality of support from existing borehole information to standardize the geophysical parameters. Also it is to be defined, in what form the results are to be presented or used. Presentation of results in utilizable form is very important as the users may not be well versed in the technicalities.

1.10 INTERDISCIPLINARY CONVERGENCE

The geophysical approach to groundwater exploration in hard rock has undergone a significant transformation for a much better understanding of the subsurface. Besides combining the surface geophysical methods, the air-borne geophysical methods are also integrated. Attempts are made for a holistic integration incorporating hydrogeology, geomorphology, lithology, satellite imageries, geophysics, geochemistry and borehole data to arrive at the most realistic high resolution subsurface picture possible. Also, attempts are made to visualize all the interpreted results on a single platform through GIS layers. To match the demand of fast coverage and output, it is expected that the future of groundwater geophysics will be governed by the convergence of helicopter-borne high resolution magnetic and electromagnetic surveys with on-ground validation through borehole information and support from electrical resistivity tomography and high resolution shallow reflection seismic survey to constrain the interpretations.

1.11 COST-EFFECTIVENESS VS TECHNOLOGICAL DEVELOPMENT

The cost factor is a major constraint. The primary non-recurring investment in the state-of-the-art equipment is quite high compared to data acquisition and interpretation and may be unaffordable. Therefore, traditional methods and conventional cheaper instruments are preferred and mostly used. Groundwater geophysics being still a continuously evolving science, adopting different techniques and approaches to meet the objective, it is desired that without searching for the conventional cheaper alternatives, opportunities are provided for persisting and rational applications of appropriate high resolution geophysical technologies, faster acquisition of a large volume of surface survey data at higher ratio of depth to surface spreads and their fusion at a proper scale. It will help keep pace with developments and researches to reduce inherent weakness of non-uniqueness and enhance the resolution in fractured zone differentiation and aquifer characterization – the most desired output. It will help better translation of geophysical responses into hydrogeological signatures. Also, geophysics will gain a significantly increasing responsibility in monitoring.

Though the applications of modern acquisition techniques and associated instrumentations are increasing, still groundwater geophysics is in developing stage and potentials of the techniques are yet to be fully exploited for hard rock characterization. Compared to hydrocarbon and mineral, groundwater-related specific development has been rather slow and groundwater geophysics as such has yet to develop as an independent discipline. In hard rock areas its applicability will increase by bringing in more precision in fracture characterization and monitoring, communicating to masses and enhancing awareness. With the availability of adequate financial resources, infrastructure, time and space, the integrated geophysical approach through modern techniques can provide definite solutions to most of the hydrogeological issues and problems in hard rock.

REFERENCES

Alemayehu, T. (2006) *Groundwater occurrence in Ethiopia*. UNESCO Publication. Available from: www.eah.org.et/docs/Ethiopian%20groundwater-Tamiru.pdf.

Bassi, N., Vijayshankar, P. S. &, Kumar, M.D. (2008) Wells and ill-fare: impacts of well failures on cultivators in hard rock areas of Madhya Pradesh. In: Kumar, M.D. (ed.) *Managing water in the face of growing scarcity, inequity and declining returns: exploring fresh approaches. Proceedings of the 7th Annual Partners Meet, IWMI TATA Water Policy Research Program, ICRISAT, Patancheru, Hyderabad, India, 2–4 April 2008*. Vol. 1. Hyderabad, India: International Water Management Institute (IWMI), South Asia Sub Regional Office. pp. 318–330.

Bocanegra, E. & Silva Jr., G.C. (2007) Groundwater exploitation of fractured rocks in South America. In: Krasny, J. & Sharp Jr., J.M. (eds.), *Groundwater in Fractured Rocks: Selected Papers from the Groundwater in Fractured Rocks International Conference, 15–19 September 2003, Prague, IAH Selected Paper Series, Volume 9*. The Netherlands, Taylor & Francis/Balkema. pp. 45–56.

Bond, N.R., Lake, P.S. & Arthington, A.H. (2008) The impacts of drought on freshwater ecosystems: an Australian perspective. *Hydrobiologia*, 600 (1), pp. 3–16.

British Geological Survey (2001) *Groundwater Quality: Ethiopia*. British Geological Survey, Natural Environment Research Council. Available from: www.wateraid.org/~/media/Publications/groundwater-quality-information-ethiopia.pdf.

Brunt, R., Vasak, L. & Griffioen, J. (2004) *Fluoride in groundwater: probability of occurrences of excessive concentration on global scale*. IGRAC, UNESCO Report No. SP 2004-2.

Butler, D.K. (2005) What is near-surface geophysics? In: Butler, D.K. (ed.) *Near-Surface Geophysics, Chapter 1, Series: Investigations in Geophysics: No. 33*, Tulsa, Oklahoma, U.S.A., Society of Exploration Geophysicists. pp. 1–6.

Chandra, P.C. (2006) Application of geophysical techniques in groundwater exploration & development (Keynote Paper). In: Rao B.V., Das, G.J., Sarala, C. & Girdhar, M.V.S.S (eds.) *Proceedings of 2nd Int. Conf. on Hydrology and Watershed Management-improving water productivity in the agriculture, 5–8 December 2006, Hyderabad, India*. Hyderabad, B.S. Publications. pp. 93–97.

Chandra, P.C. (2009) Geophysics reducing uncertainties-enhancing efficiency in groundwater development & management. Presented at: *IIIrd World Aqua Congress, 2–4 December 2009, New Delhi, India*. Aqua Foundation.

Davis, S. N. & Turk, L.J. (1964) Optimum depth of wells in crystalline rocks. *Ground Water,* 2 (2), 6–11.

De Silva, C.S. (1998) Development of guidelines for efficient use of agro-wells. In: Samad, M., Wijesekera, N.T.S. & Birch, A. (eds.) *Proceedings of the National Conference on 'Status and Future Direction of Water Research in Sri Lanka' 4–6 November 1998, BIMCH Colombo, Sri Lanka,* Publication of International Water Management Institute.

Foster, S., Tuinhof, A. & Garduno, H. (2006) *Groundwater development in Sub-Saharan Africa: a strategic overview of key issues and major needs.* GW-MATE Case Profile Collection No. 15, Sustainable Groundwater Management-Lesson from Practice report, The World Bank.

Geological Society of India (2008 a) Drinking water and food security in hard rock areas of India. Das, S. (ed.) *Golden Jubilee Volume.* Bangalore, Geological Society of India.

Geological Society of India (2008 b) Changing geohydrological Scenario: hardrock terrain of Peninsular India. Das, S. (ed.), *Golden Jubilee Volume.* Bangalore, Geological Society of India.

Giordano, M. (2006) Agricultural groundwater use and rural livelihoods in sub-Saharan Africa: a first-cut assessment. *Hydrogeology Journal,* 14 (3), pp. 310–318.

Harrington, N. & Cook, P. (2014) *Groundwater in Australia.* Adelaide, South Australia, National Centre for Groundwater Research and Training, Australia.

IGRAC (2014) *Groundwater monitoring in Latin America.* Summary report of information shared during the Regional Workshop on Groundwater Monitoring. International Groundwater Resources Assessment Centre (IGRAC), 11–12 December 2013, UNESCO facilities, Montevideo, Uruguay.

Krasny, J. & Sharp Jr., J.M. (2007 a) Hydrogeology of fractured rocks from particular fractures to regional approaches: State-of-the-art and future challenges. In: Krasny, J. & Sharp Jr., J.M. (eds.), *Groundwater in Fractured Rocks: Selected Papers from the Groundwater in Fractured Rocks International Conference, 15–19 September 2003, Prague, IAH Selected Paper Series, Volume 9.* The Netherlands, Taylor & Francis/Balkema. pp. 1–30.

Krasny, J. & Sharp Jr., J.M. (eds.) (2007 b) *Groundwater in Fractured Rocks: Selected Papers from the Groundwater in Fractured Rocks International Conference, 15–19 September 2003, Prague, IAH Selected Paper Series, Volume 9.* The Netherlands, Taylor & Francis/Balkema.

Marechal, J.-C. (2010) The sunk cost fallacy of deep drilling: Editor's message. *Hydrogeology Journal,* 18 (2), 287–289.

MacDonald, A.M., Davies, J. & Calow, R.C. (2008) African hydrogeology and rural water supply. In: Adelana, S.A.M. & MacDonald, A.M. (eds.) *Applied Groundwater Studies in Africa, Chapter 9, IAH Selected Papers on Hydrogeology, Vol. 13,* DFID, National Environment Research Council, British Geological Survey, The Netherlands, CRC Press/Balkema. pp. 127–149.

Morariu, D.C. (2012) Contribution to hydrocarbon occurrence in basement rocks. *Neftegasovaa geologia, Teoria i practika (RUS).* ISSN 2070-5379, http://www.ngtp.ru/rub/9/51_2012.pdf.

Muralidharan D., Nair, A.P. & Sathyanarayana, U. (2002) Fluoride in shallow aquifers in Rajgarh Tehsil of Churu District, Rajasthan – an arid environment. *Current Science,* 83 (6), pp. 699–702.

Panabokke, C.R. & Perera, A.P.G.R.L. (2005) *Groundwater resources of Sri Lanka.* Water Resources Board, Sri Lanka.

Petford, N. & McCaffrey, K. (2003) Hydrocarbon in crystalline rocks: an introduction. In: Petford, N. & McCaffrey, K.J.W. (eds.) *Hydrocarbons in Crystalline Rocks. Geological Society Special Publication 214.* Bath, U.K., The Geological Society Publishing House. pp. 1–6 (https://books.google.co.in).

Pietersen, K. Beekman, H. E. & Holland, M. (2011) *South African groundwater governance case study.* Report prepared for the World Bank in partnership with the South African Department of Water Affairs and the Water Research Commission, WRC Report No. KV 273/11, ISBN 978-1-4312-0122-8.

Ponce, V.M. (2006) Groundwater utilization and sustainability. Available at: http://goundwater. sdsu.edu

Reboucas, A. da C. (1999) Groundwater resources in South America. *Episodes*, 22 (3), pp. 232–237.

Sircar, A. (2004) Hydrocarbon production from fractured basement formation. *Current Science*, 87 (2), pp. 147–151.

Soonawala, N.M. (1984) An overview of the geophysics activities within the Canadian Nuclear Fuel Waste Management Program. *Special Issue: Nuclear Fuel Waste Management, Geoexploration*, 22 (3 & 4), 149–168.

Woodford, A. Rosewarne, P. & Girman, J. (2005) *How much groundwater does South Africa have?* SRK Consulting, Cape Town & Department of Water Affairs & Forestry, Pretoria, South Africa.

Chapter 2

Introduction to the hydrogeology of hard rock

2.1 INTRODUCTION

The term 'hard rock' is used, in general, for all geological formations which are compact and massive and do not possess any primary porosity. Hard rock is also known as fractured rock. According to Larsson (1984) all kinds of igneous and metamorphic rocks are considered as hard rocks, excluding volcanic and carbonate rocks which may have primary porosity. The generalized definition of hard rock stressing its hydrogeological character as given by Gustafsson (1993) is that hard rock *"includes all rocks without sufficient primary porosity and (hydraulic) conductivity for feasible groundwater extraction"*. Accordingly, Archaean and Precambrian sedimentaries, carbonate rocks and younger volcanics possessing similar hydrogeological character can also be considered as hard rock. Hard rocks cover large areas in several parts of the world. There are shield areas like Canadian, Guianan and Amazonian, Baltic, African, Indian, Australian, Angaran (Siberian) and Antarctic where hard rocks are widely exposed. The other significant exposures of hard rock are Armorican Massif, Spanish Meseta, Massif Central in France and Bohemian Massif. The total area of shield-rock outcrop is about 30 million sq. km covering around 20 percent of the land surface (Gustafson and Krasny, 1993; Krasny and Sharp, 2007). In this book the hard rocks, viz., granite, gneiss, schist, phyllite, quartzite, basalt etc. excluding carbonate rock are discussed. However, in the following chapters while explaining the geophysical methods some carbonate rock case studies are given. The methods described can be used appropriately in carbonate rocks also.

2.2 WEATHERED ZONE AND FRACTURES

In hard rock areas fresh and compact formations are, in general, capped by a layer of porous, weathered material. This is attributed to *in situ* structural and mineralogical disintegration and alteration of rocks by combined physical, chemical and biological processes at or near the surface (Olvmo, 2010, Hamblin and Christiansen, 2009), over a long span of geological time. The weathered material may be overlain in places by a soil cover and/or alluvial deposits. However, the hard rock area which has more than 20–25 m thick alluvial cover is not considered as hard rock for water well drilling purpose (Saha, personal communication). The character of the weathered zone depends on complex interactions of factors like mineralogical composition and grain size of the parent rock that has undergone weathering, type and degree of weathering, tectonic controls, palaeo-climatic history and the geomorphological set up (Gustafson and Krasny, 1993). The fractures, fissures and joints cause exfoliation and climatic conditions help activate chemical processes of weathering. The degree of weathering decreases with depth. With intense fracturing development of weathered zone becomes prominent and deep weathering gets initiated by the structural deformities. In fact, physical processes facilitate the chemical ones and vice versa. However, the mechanical processes predominate in shallow parts and the chemical processes prevail in deeper parts just above the unweathered rock (Braga et al., 2002).

Granites, primarily consisting of felsic minerals, like quartz, orthoclase feldspar, and mica (muscovite) with minor amount of other minerals, produce weathered material comprising clays decomposed from feldspars and mica and unaltered residual quartz-sand grains which are resistant to weathering. In general, the thickness of the weathered zone is more in coarse grained granite compared to metamorphics like quartz mica schists and thin-bedded quartzites (Larsson, 1984). Basalts mostly composed of mafic minerals like plagioclase feldspar, olivine and pyroxene produce clay and iron oxides as weathered material (Hamblin and Christiansen, 2009). The mafic minerals generally weather much faster compared to felsic minerals (McQueen, 2009) and therefore weathering in basalt could be deep and faster than granite. The term saprolite is used for the bottom part of the weathered zone where chemical weathering prevails and which retains the fresh bedrock structure and texture (Dosseto et al., 2012, Dixon et al., 2009, Braga et al., 2002). Saprolite is the transition zone of varying thickness between weathered zone and fresh rock. Wright (1992) further divides the saprolite into upper saprolite with a higher proportion of secondary clay and lower saprolite with a greater abundance of primary mineral with early forms of secondary clay. The variation from weathered rock to fresh rock can be transitional (Jones, 1985) or sharp with a weathering front (Twidale and Campbell, 1993, Houston and Lewis, 1988) occupying a minimum of a few centimetres in transition. Hall (1986) reports its development as thick transition zone in humid temperate climate of Scotland and sharp weathered front in warmer environment. In general, weathering occurred across the entire palaeo-surface and gradually extended to depths from a few metres to a few tens of metres through the favourable pre-existing fractured zones (Olesen et al., 2007), joints and cracks. The thickness of weathering varies depending on the irregularity of weathering front and erosion. It was further pointed out by Olesen et al. (2007) that the alteration process initiates at the top and penetrates downward. The intermediate zone between uppermost highly weathered part and deeper saprolite holds horizontal fractures (Lachassagne et al., 2001).

Climate and groundwater dynamics control the process as well as rate of weathering (Olvmo, 2010; Robineau et al., 2007). This predominates near the surface and hence weathering starts from the surface downwards with most weathered material occurring near the surface. The humid tropical climate with an abundance of water and favourable temperature render maximum to intense and thick weathering of rocks and also formation of laterite capping. Generally, in tropical sub-humid and humid climates the weathered zone may extend beyond 30 to 40 m depth from the surface, if it is not eroded away. Where erosion is prominent the weathered zone is thin and where erosion is less the weathered zone is thick. This thick weathered zone is also found in parts of North and South America and Europe which are presently under a temperate climate (Larsson, 1984; Hall, 1986). The presence of a thick weathered zone indicates a favourable palaeo-climate for its formation. Jones (1985) reports the effect of weathering even beyond 90 m depth. Rocks can be weathered to depths of hundreds of metres in tectonically disturbed areas by water through hydrolysis and dissolution. According to Petrov (1991) the thickness of palaeo-weathering can reach up to 150 m, if it is not eroded away. The entire column, including transported and deposited material, also called regolith (Scott and Pain, 2009) could be as thick as 100 to 200 m found in Brazil, West Africa and Australia (Twidale and Campbell, 1993).

The weathering may or may not have a correlation with topography and drainage. Generally, the erosion of weathered material leads to topographic depressions. Though the generalized vertical sequence is of weathered zone, fractured zone and compact and massive formation with very little fracturing, under specific climatic and geomorphic conditions either a weathered zone or a fractured zone may not form at all, i.e., one of them may be missing. The succession of layers from top of weathered zone to fresh rock at depths is known as the weathering profile. The weathered profile will have horizontal as well as vertical heterogeneity. Larsson (1984) has however, given a generalized depth-wise division of the weathered profile into 4 zones, viz., a, b, c and d which are presented through a schematic diagram (Fig. 2.1) by Acworth (1987). The interfaces between these zones are, in general, transitional indicating progressive degradation by weathering from compact rock at the bottom of the weathered profile to the top soil layer.

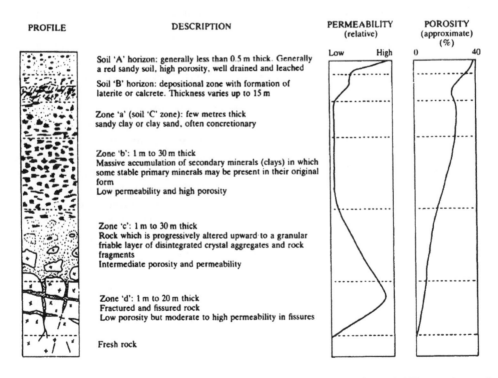

Figure 2.1 The weathering profile showing depth-wise division of zones a, b, c and d (Source: Acworth, 1987).

Figure 2.2 The jointed and fractured granite, Jharkhand, India (Courtesy: Dr D. Saha, CGWB, India).

In parts of Jharkhand, India thickness of weathered zone ranges up to about 25 m. On an average it varies from 12 to 16 m (Chandra *et al.*, 1994). Thickness of weathered zone in metasediments in this area varies widely. It is less than 5 m in quartzites and around 17 m in quartz-mica-feldspar schist/gneiss. In metavolcanics of this area weathered zone is around 12–14 m thick. The average depth of occurrence of saprolite is 16 m in younger granites and 12 m in older granites. Its average thickness is 4 m. Drilling data reveals thicker saprolite (5 to 8 m) in younger granites which are more acidic in nature with abundant free silica than the older granites (Dev Burman and Das, 1990). Uhl *et al.* (1979) report weathered zone thickness varying from 1 to 35 m in parts of central India and about 27 m in southern India. Woodford *et al.* (2005) report an average 33 m thick weathered zone in South Africa.

The compact impervious rock, underlying the weathered zone may possess a well-preserved network of joints and fracture openings. This is known as secondary porosity or fracture porosity. Figure 2.2 presents a view of joints and fractures in granite from Jharkhand India. These result due to tectonic stresses at different geologic times, physical and geochemical properties of the rock, percolating groundwater and palaeo-climatic conditions. The intensity of fracturing in lower grade metamorphic rocks is more than higher grade metamorphics, also in fine grained rocks fracturing is dense and of limited length, while in coarse grained rocks they are wide-spaced and long (Larsson, 1984). The fractures, linear in nature, mostly originate with maximum intensity at the faults and extend laterally as well as vertically to large distances. Fractures at depth may occur as single planer features with openings in the order of millimetres to centimetres or as a group of discrete close-spaced planer features within a metre or so. The latter is known as fractured zone. They are mostly confined within a few hundred metres from the earth surface and may be interconnected maintaining hydraulic continuity over a large area. On the surface the fractured zones could be quite wide as well as long, ranging in scale from micro-level to regional over kilometres. Faults and fractures being the zones of weakness, they are prone to deep weathering and allow deeper percolation of water. Therefore, fault and vertical to sub-vertical fractured zones associated with intense and deep weathering are also associated with topographic depressions due to subsequent erosion which get occupied by the drainage systems of the area. Fractures continue deep underneath the weathered zone and generally their density, intensity and dimension diminish with depth depending on lithology and tectonic history of the area (Clark, 1985).

2.3   GROUNDWATER OCCURRENCES

Groundwater is stored in the superficial weathered zone capping which has sufficient porosity and behaves as granular aquifer. The deeper groundwater flow is facilitated by the secondary porosities developed through the process of fracturing and jointing in the compact formation underlying the weathered zone. The tectonic deformations like faults, shear zones etc. may improve the overall storage. The flow-path is developed and increased by fracture connectivity. The other structures which can facilitate the movement of groundwater are litho-contacts and foliation. The basic dykes and quartz reefs either act as barrier or conduit and groundwater repository depending on their orientation with respect to groundwater flow direction and their thickness, weathering and fracturing.

2.3.1   Weathered zone aquifers in granitic terrain

The weathered zone, whose porosity can go up to 50% (Davis and Turk, 1964), forms the near-surface unconfined aquifer. The recharging water fills the pores in the weathered zone and percolates down to the fracture network. Since saturated thickness of the weathered zone forms near surface aquifer, the greater the thickness and lateral extents the higher is the storage capacity of this aquifer and the better is the well yield. The hydraulic characteristics of the weathered zone vary laterally as well as with depth depending on saturated thickness and lithological character. An irregular change in the hydraulic properties of the weathered zone on a local scale is common (Krasny, 1997). The upper part may have high porosity and water retention capacity while the lower part may have high permeability. The smaller saturated thickness or reduction in saturated thickness, uneven weathering front and outcropping compact rock surroundings in places limit the weathered zone aquifer locally. However, the near surface, unconfined, discontinuous weathered zone aquifer can be considered regionally extensive because of its presence almost everywhere in hard rock terrain. As a convenient resource for local water supply, it is developed by shallow hand pumps, bore wells and large-diameter dug wells (Fig. 2.3). The groundwater level in the weathered zone unconfined aquifer generally follows the topography and if regionally extensive, it conforms to the surface drainage. Being close to the surface, the weathered zone aquifer is affected maximally by climatic conditions. The water level declines in summer months and the saturated thickness is reduced. The well yield depends mainly on available saturated thickness which is sustained by annual recharge from precipitation and other sources. Groundwater in the weathered zone is mostly supported by local recharge. Though the weathered zone aquifer may have good hydraulic conductivity, its storage capacity and yield vary because of limiting lateral extent and the transmissivity varying with saturated thickness that changes topographically as well as seasonally. According to Larsson (1984), generally, a weathered zone less than 10 m thick does

Figure 2.3 A large diameter dug well in weathered zone and saprolite under construction, Jharkhand, India (Courtesy: Dr D. Saha, CGWB, India).

not sustain an aquifer effectively unless there is high recharge. Further, he says that only 10 to 15% of the total weathered zone thickness contains sufficient permeable material and contributes to the well yield. With reducing grain size upward, in most of the cases, active groundwater flow is at the base of the weathered zone (Jones, 1985) which though may have less porosity but with higher hydraulic conductivity. According to Acworth (1987) 10 to 15 m thick zone 'c' shown in figure 2.1, comprising granular, friable layer of disintegrated crystal aggregates is productive. In general, the yield from 10 to 15 m thick saturated weathered zone aquifer ranges from 0.25 to 0.5 litres per second (lps) in Malawi (Grey *et al.*, 1985). In hard rocks of Jharkhand, India also yield from weathered zone is of the same order. However, the cumulative yield from bore wells in weathered zone and saprolite was observed up to 2.7 lps (Dev Burman and Das, 1990; Chandra *et al.* 1994) and a similar yield was obtained from dug wells in the hard rocks of South America (Reboucas, 1999). Singhal and Gupta (1999) based on the studies by Subramanian (1992) indicate a wide range of dug well yield from 0.1 to 9 lps varying with rock type.

2.3.2 Fractured zone aquifers in granitic terrain

The igneous and metamorphic rocks which are basically heterogeneous and anisotropic are made more so by the fractures but ironically the fractures inducing heterogeneity and anisotropy only are responsible for high groundwater yield. The hydrogeology of the saturated fractured zone aquifer is quite complex. The fracture behaviour is controlled by lithological characteristics, antiquity of rock mass and structural features (Dev Burman and Das, 1990). The geometry (i.e., length and orientation) of fractures, their lateral and depth-wise distribution (i.e., frequency or density of occurrence), openness, volume of secondary material filling them and their interconnection at local and regional scale control the storage and movement of groundwater generally received through the overlying weathered zone. While the vertical fractures and joints immediately underlying the weathered zone generally transmit water to a deeper fracture network, the horizontal and vertical fractures and joints at depth control deeper movement and direction of the groundwater flow. The fractured zones in compact rock associated with micro fracturing which may have different orientations create or enhance the local permeability and behave as conduits as well as groundwater storage and form potential aquifers. Therefore, ascertaining their presence and character is very important for groundwater development and also for understanding the complexities in groundwater flow. The experience from groundwater exploration within 200 to 300 m depth in hard rock areas reveals that fractures gradually close with depth, diminishing in intensity, frequency and dimension and in total the probability of intersecting fractures abruptly reduces with depth. Nevertheless, an individual fracture encountered at depth can dramatically increase the yield and only with this hope drilling is continued deeper. By volume, the amount of water in the fracture will be hardly a few percent of total rock volume but the flow through it could be higher. That is, to sustain the yield, fractures must have regional connectivity with other fractures. The yielding fractures appearing locally isolated could have sufficient regional hydraulic connection with other interconnected fractures getting recharged through the weathered zone. They could be highly productive and contribute maximum yield. The probability of such occurrences increases in areas with a thick weathered zone which could be associated with tectonics. However, Pietersen *et al.* (2011) based on observations in South Africa do not associate higher yield from fractured zone with the thickness of overlying weathered material. They further indicate that fractured aquifers in parts of South Africa are highly productive with large storage potentials.

The hydrogeological variations even within the fractured zone are so uncertain that wells sunk a few meters apart may have completely different yield characteristics, depending mainly on the number of productive fractures encountered. The findings of one well cannot be extrapolated to another. Also, it is not always necessary that large tectonic zones would create high hydraulic conductivity as mylonite associated with shear zone has very low hydraulic conductivity (Gustafson & Krasny, 1993). As far as hydraulic conductivity of fractures is concerned, in general, the deeper the fractures the lesser the hydraulic conductivity. The more regional the fracture and the higher the density of the fracture with better opening, the better is the hydraulic conductivity. As such, the areas or zones with long fractures, larger density of fracturing and greater fracture aperture have higher hydraulic conductivity (Krasny and Sharp, 2007). Because of the variations in fracturing and interconnections hydraulic conductivity is highly anisotropic (Gustafson & Krasny, 1993). Cook (2003) presents a detailed description of hydraulic conductivity of fractured rocks. In general hydraulic conductivity is quite low and ranges from 10^{-11} to 10^{-2} m/s. The hydraulic conductivity of granites and metasediments of Jharkhand, India ranges from about 1×10^{-7} to 0.7×10^{-6} m/s and transmissivity ranges from 0.23×10^{-4} to 3.24×10^{-4} m²/s (CGWB, 2012a).

Except the individual fractured zone, intuitively, there should be in general a large decrease in hydraulic conductivity of the rock mass with depth. This is clearly evidenced by the study carried out at Fjallveden in Sweden by Ahlbom *et al.* (1991) up to a depth of 700 m through deep boreholes in sedimentary gneiss (banded gneiss composed of alternating medium grained quartzo-feldspatic layers and fine grained biotite rich layer) and granite gneiss. Generally, the fractured zones within the top 200 m depth zone have higher hydraulic conductivity than those beyond this depth. It means that beyond a certain depth the water well drilling will be, in general, uneconomical, but this depth will vary from region to region. For example, in Karamoja, Uganda, it is 90 m and in Nigeria it is 50 m only (Clark, 1985). In northeastern Precambrian basement of South America Reboucas (1999) suggests 60 m depth for drilling as fractured zones are mostly within 30 m depth with maximum discharge around 3 lps, whereas in southern Precambrian basement the yield may go up to 10 lps.

In hard rocks of Jharkhand, India, Dev Burman and Das (1990) report the occurrence of saturated fractured zones in the depth range of 12.66 to 230 m and unsaturated ones up to 285 m depth which are either non-productive or completely dry. The former bears the mark of water passage and oxidation. Depth-wise saturated fractured zones are qualitatively grouped as shallow and deep. Conventionally, the shallow ones occurring within 50 to 60 m depth are tapped by hand pumps. It can also be grouped as shallow, medium depth and deep. In Jharkhand, India, saturated fractured zones occurring up to 110 m depth in younger granites are prominent, while in older granites they occur in the depth range of 100 to 140 m and those that are poorly interconnected yield for a short while only to become dewatered. The reported maximum discharges from individual zone in younger and older granites are 3 and 6 lps respectively.

2.3.2.1   Deeper fractured aquifers

The deeper compact and massive rocks possess a limited number of fractured zones. Whatever fracture and fault zones are present they act as isolated deep conduits for groundwater flow, at places forming an interconnected network for regional and continental deep groundwater movement (Krasny and Sharp, 2007). The core-drilled boreholes to a depth of around 700 m in Sweden encountered fractures at depths beyond 600 m and reveal a decrease in fracture frequency beyond 200 m depth (Ahlbom *et al.*, 1991). In Sardinia highly fractured zones were encountered up to 1000 m depth in a borehole and a fracture with high hydrostatic head (18 m amsl) and high temperature groundwater was found at a depth of 735 m bmsl (Barrocu, 2007). Such information on deep fractured zones which can contribute substantially to well yield is sporadic. The Central Ground Water Board (2012b) drilled a good number of deep water wells to a maximum depth of around 500 m in granites, gneisses and schists of Karnataka state in the southern part of India. The maximum cumulative yield of a well was reported as 15 lps. Yielding fractured zones were encountered as deep as 389 m. The weathered zone in this area is almost dried up with the water table as deep as 68 m in places and therefore fractures immediately underlying the weathered zone within 75 m depth are either dry or have a very poor yield (CGWB, 2009).

2.3.3   Aquifers in metasediments

The aquifer in weathered and fractured zone of metasediment, viz., schist, quartzite and phyllite has varied potential. In addition to weathered zone, joints and fractures, weak planes of schistosity, foliation and contacts form groundwater repositories. The weathered material of phyllite, schist and slate is predominantly clay and therefore does not yield much. Phyllite is generally considered weak foliated rock. In this, the open fractures without clay filling can form good aquifers. According to Dev Burman and Das (1990), in metasediments the depth persistence of deeper fractures is controlled

Figure 2.4 The deformed quartz-schist, Bihar, India with little possibility of groundwater in it (Courtesy: Dr D. Saha, CGWB, India).

more by structures like folds. Further, they indicate that fractures are better preserved in harder formation than in low grade softer schist. Generally, in low grade schist, due to intense weathering clays impede the passage of water which may result in deeper dry fractured zones. Figure 2.4 presents a view of deformed quartz-schist from Bihar, India with very little scope of getting groundwater in it. However, the high yielding zones in quartzite and phyllite are also reported in places. The cumulative yield from such zones occurring between 27.5 and 147 m depth at a site in quartzite-phyllite of Jharkhand, India was quite high, in the range of 7 to 25 lps.

2.3.4 Aquifers in quartz reefs and dykes

Basic dykes and quartz reefs are very common in hard rock areas and occur as discordant bodies. They may act either as barriers or conduits for groundwater flow and even a group of such features can form close and discrete flow systems within a catchment area. Figure 2.5 presents a view of a highly fractured quartz reef from Jharkhand India. The contact of quartz reef with the country rock being more susceptible to weathering and fracturing can form a potential groundwater zone. Chandra *et al.* (2008) and Dewandel *et al.* (2009) investigate quartz reefs in India and find abrupt deepening of weathering front at the contact of quartz reef with granitic country rock, confirmed by high discharge (5 lps) wells in the contact zone. Verma *et al.* (1980) report that the central portion of the quartz reefs, if massive and compact, acts as a barrier to groundwater flow and the brecciated zone on either side forms a high yielding zone.

Many areas in hard rock terrain are intruded by numerous criss-crossing basic dykes along macro-fractures. There are areas with dyke swarms complicating the groundwater flow. Though these dykes compartmentalize the groundwater flow system, the relation with local groundwater flow depends on their character. Figure 2.6 shows one such area from crystalline part of Odisha, India. Dev Burman and Das (1990) indicate that numerous dykes or swarm of dykes occur in older granites and

Figure 2.5 The highly fractured quartz reef, Jharkhand, India (Courtesy: Dr D Saha, CGWB, India).

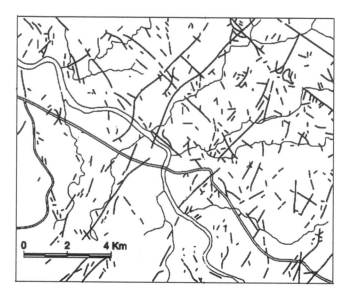

Figure 2.6 The dyke infested hard rock area, Odisha, India (Chandra *et al.*, 1994).

are absent in younger granites. These dykes may hold primary fractures. While massive and unweathered dykes form barriers to groundwater flow, the weathered and fractured ones may form groundwater repositories. Dewandel *et al.* (2011) however indicate a local lowering of hydraulic conductivity by weathering of dolerite dyke. Larsson (1984) points out that fractures parallel to dykes could be water bearing if not filled up by dyke material. Dev Burman and Das (1990) also support this and reveal that besides near-surface fractured part of the dykes, the saturated deep fractures occurring along the margin or contact of dyke with the granitic country rock are also potential groundwater zones. The contacts with country rock allow deeper infiltration to form fractured aquifer systems. Buckley and Zeil (1984) also report better yield from wells drilled close to dolerite intrusions which create parasitic fractures. Dykes of thickness less than 10 m if affected by fracturing and jointing may not act as barriers, while thicker dykes may act as barriers (Bromley et al., 1994). The experimental drilling of a number of boreholes across a dolerite dyke in dyke swarm area of Northampton, Australia indicates that though dykes may act as a barrier there is presence of yielding fractured zones also, within 20 to 40 m depth in proximity of the dyke (Ventriss et al., 1982). These fractured zones get recharged from the weathered zone. A well drilled very close to a dyke in Jharkhand, India encountered a dry weathered zone and saprolite up to 13 m depth and saturated fractures at depths of 39, 59, 83, 105, 131 and 140 m with individual discharges of 0.41, 0.25, 0.34, 1.0, 2.1 and 5.4 lps respectively during drilling confirming that dykes also can act as aquifer and sustain high yielding shallow bore wells. Interestingly, dry fractures were also encountered in this borehole at 26 m, 70 m and 160 m depth (Dev Burman and Das, 1990). High yielding fractures are also encountered in areas where quartz reefs and dykes crisscross. Boreholes drilled near the intersection of a NW-SE trending dyke and a NE-SW trending quartz reef in the Bundelkhand granites of Uttar Pradesh, India encountered 4–5 fractured zones in the depth range of 29 to 123 m. The cumulative discharge from these fractured zones varied between 6 and 8 lps during drilling (Chandra et al., 1992).

2.3.5   Volcanic rock aquifers

Basalt is a common volcanic rock. There are several areas throughout the world occupied by flood basalts. The two classic examples are the extensive Deccan Trap Basalts (DTB) in west central India occupying about 500,000 sq km and the Columbia River Basalt Group (CRBG) occupying about 164,000 sq km in northwestern USA. The DTB were formed 67 to 60 My ago towards the end of the Cretaceous period while the CRBG was formed only 17 to 6 My ago during the late Miocene to early Pliocene period. The volcanic rocks were formed by solidification of molten magma on the existing palaeo-surface. The repeated eruptions at varying time intervals, flooding of molten lava over the existing landscape and stacking of lava flows in a time sequence one over the other created a succession of lava flow stratification. The DTB and CRBG comprise a number of such horizontal to sub-horizontal lava flows. The thickness of DTB increases towards the west and is at its maximum about 2000 m near the western coast of India and that of CRBG is about 5000 m in the central part of the Columbia basin in the Pasco Basin area (Reidel et al., 2002). In the southern part of DTB a total of 31 flows are identified (Jayaprakash, 2007). Singhal (1997) reports 32 flows. CGWB (1981, 1982) mapped 27 flows in 538 m elevation difference in the southeastern part and 19 flows in 330 m elevation difference in the northeastern part of DTB. Boreholes drilled to a depth of around 200 m by CGWB (1982) in the southeastern fringe area of DTB encountered only 5 flow units. All this indicates restricted extents of flows, though some can be traced for more than 100 km. There could be many more flows with the maximum number occurring in the western part of DTB (Mahoney, 1988). The individual DTB lava flow varies in thickness from 1 to 35 m (Singhal, 1997). According to Kale and Kulkarni (1992) an individual flow unit of DTB can be up to 100 m thick and may pinch out or thicken locally as well as laterally. CGWB (1982) indicates variation in flow thickness from 10 to 42 m. The variations in thickness are due to

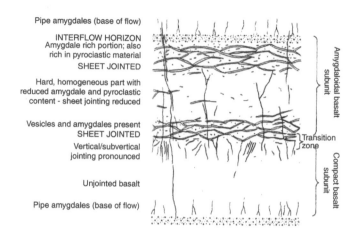

Pipe amygdales (base of flow)

INTERFLOW HORIZON
Amygdale rich portion; also
rich in pyroclastic material
SHEET JOINTED

Hard, homogeneous part with
reduced amygdale and pyroclastic
content - sheet jointing reduced

Vesicles and amygdales present
SHEET JOINTED
Vertical/subvertical
jointing pronounced

Unjointed basalt

Pipe amygdales (base of flow)

Amygdaloidal basalt subunit

Transition zone

Compact basalt subunit

Figure 2.7 Schematic vertical cross-section of a flow unit of Deccan Trap basalt (Source: Kale and Kulkarni, 1992).

undulating palaeo-topography on which the flow took place and actions of palaeo-weathering and erosion. The maximum flow thickness reported is 160 m (Mahoney, 1988). The flow units may rest one over the other or be separated by interflow horizon of tuffaceous, scoriaceous or pyroclastic (Kale and Kulkarni, 1992) material which got accumulated during the hiatus between two successive flows (Adyalkar *et al.*, 1975). It is locally known as 'red bole' or 'green bole'. Singhal (1997) and Singhal and Gupta (1999) report that bole rich in clay could be ancient soil – a product of atmospheric weathering or it could be a product of hydrothermal alteration of amygdaloidal basalt or pyroclastic material. However, based on chemical composition Ghosh *et al.* (2006) conclude that red bole is formed by weathering of basalt during a hiatus and green bole is a mixture of weathered basalt and remnants of volcanic ashes. The sheet joints in the bole bed make it friable (Kale and Kulkarni, 1992). The thickness of the inter-flow red bole bed varies from a few centimetres to a metre and this being regionally persistent forms the marker bed to demarcate the flows. One interflow to another interflow is taken as one flow unit. Generally, the flow unit is massive at its bottom and vesicular towards the top. However, a flow unit may comprise three sub-units of varying thicknesses. These sub-units are pipe-amygdaloidal base, compact middle part and sheet-joint amygdaloidal top with transitional boundaries in between (Kale and Kulkarni, 1992, Fig. 2.7). Some of the flows may be without compact middle part making the entire flow as amygdaloidal while the other without top and bottom amyg-daloidal parts making it entirely compact. The amygdaloidal parts hold horizontal to sub-horizontal sheet joints while the compact part is devoid of sheet joints, instead it may have vertical to sub-vertical joints and fractures. The basalts are weathered at the top. DTB are traversed by dykes and regional as well as local tectonic fractured zones. The dykes form linear ridges of moderate relief (Gupta *et al.*, 2012). According to Tolen *et al.* (2009) in the CRBG there are about 300 flood basalt flows and the flow units comprise three parts, the permeable flow top, followed by a dense, relatively impermeable part of the flow and then flow bottom. The flow top can be vesicular or brecciated. The dense impermeable flow part called 'flow interior' is non-vesicular and holds cooling joints. The flow bottom could be a thin zone of glassy to very fine-grained basalt with very small amount of vesicles or a 'pillow lava complex' depending on the palaeo-environmental conditions. The top of underlying and bottom of overlying flow units in combination make the interflow zone. The thickness of flow unit varies from 20 to 100 m (Reidel *et al.*, 2002). For further details on the CRBG readers should refer to the publications of Tolen *et al.* (2009) and Reidel *et al.* (2002).

Groundwater conditions in volcanic rocks, predominantly basaltic, differ from those in granitic terrain. As observed above, basalt flows produce stratification and therefore hydrogeological behaviour of volcanic rocks ranges from a typical crystalline fractured rock to heterogeneous, anisotropic near-porous rock (Custodio, 2007). The basalts have little primary porosity developed in the vesicular part and can otherwise be considered impervious. Added to the non-filled interconnected vesicles, the weathering, fracturing, open cooling and sheet joints in basalts make them porous. According to Saha and Agrawal (2006) the vesicular part constitutes about one-fourth to one-third of a flow unit and the sheet joints increase its transmissive capacity. As such, transmissivity of weathered basalt is more than vesicular and fractured-jointed basalts (Deolankar, 1980, Surinaidu *et al.*, 2012). The maximum yield of a dug well in DTB is reported around 8 lps. In parts of DTB the dug well yield ($\approx$1 lps) is more than shallow bore well yield ($\approx$0.8 lps) (Dhonde, 2009).

On the basis of structural and hydrological properties DTB are grouped in five classes viz., weathered, vesicular, amygdaloidal, fractured and jointed and compact basalt (Deolankar, 1980). The weathered, vesicular, fractured and jointed basalts form the aquifers. The interflow zones without clay predominance also act as aquifer. The amygdaloidal basalts do not usually form a good aquifer (Kulkarni and Deolankar, 1995). The accumulation of groundwater with dominant horizontal flow through sheet joints and the vertical to sub-vertical fractures and joints in the compact basalts acting as conduits for vertical flow to the deeper sheet joints form the basalt-aquifer system. Joints and fractures are not found to pass into adjacent flow units (CGWB, 1982). Though the vertical joints produce good hydraulic conductivity, the horizontal hydraulic conductivity is generally higher than the vertical hydraulic conductivity (Cook, 2003). According to Foster *et al.* (2007) availability of water in shallow aquifers in DTB is controlled by topography; the lower ground holding deep weathering forms a continuous perennial aquifer. Besides top weathered zone aquifer, hydrogeology of trap basalts reveals a multiple semi-confined to confined aquifer system with massive part repeatedly separating the aquifers in vesicular, fractured and sheet-jointed parts. In the CRBG the aquifers are formed by sedimentary interbeds, interflow zones and the flow tops, predominant among which are the laterally extensive, 10–20 m thick interflow zones with higher permeability (Tolen *et al.*, 2009). Like DTB this gives rise to a stacked multiple semi-confined to confined aquifer system.

2.4 GROUNDWATER DEVELOPMENT

In areas with a sufficiently thick saturated weathered zone and shallow water table, large diameter dug wells, dug-cum bore wells and shallow bore wells can be constructed within this zone. Developing a shallow weathered zone aquifer is easy, economical and without any financial risk. Figure 2.3 shows a large diameter dug well in a weathered zone and saprolite under construction. These wells can cater to the needs of local domestic water supply as well as irrigation. Generally, the weathered zone aquifers are localized and may not support water supply over a large area. Therefore, for a better yield deeper fractured zones along with saturated weathered zone are tapped. As already mentioned, yield from fractures sustains if the fractures tapped are interconnected with other fractures, not filled with material of low permeability and connected to a weathered zone and saprolite somewhere. The weathered zone acts as a recharging body for most of the underlying fractures. It is necessary that location of deep well drilling site is decided by hydrogeologist and geophysicist together and that the hydrogeologist supervises the drilling and designs the well based on his observations during drilling. When locating a well site it is necessary to consider the purpose, the beneficiaries and potential hazards like contamination, submergence etc.

The well yield depends on many factors, like rock composition, weathering, proximity to lineament, joint and fracture presence, depth drilled, topography etc. and therefore the range of yield varies as a complex multiplicity of these effects. However, in general, the yield of shallow wells from weathered zone is around 1 lps (Clark, 1985) and mostly caters to local needs. Generally, wells with a discharge lower than 1 lps in granitic terrain are classified as low yielding or failure wells. Wells with 1 to 5 lps discharge can be considered as moderate-yielding and more than 5 lps as high-yielding. Naik et al. (2010) define the minimum limit of 3 lps for high yield in the eastern part of India. However, this definition of success or failure and grading is arbitrary. Quantitatively, it varies from place to place with time and nature and type of demand oriented socially, geologically and logistically. For example, while a well with yield of 0.6 lps in granites at a particular time for a specific purpose is successful, it could be termed a failure in the same litho-unit at another time for another purpose. In Kandy district, Sri Lanka, a well yield of 0.07 lps is considered low and more than 1.7 lps is considered high for a drilling up to 90m depth (Jayawardena, 2004). In Uganda a well yield less than 0.13 lps is considered a failure. In New Hampshire granitic terrain 2.5 lps well yield is considered high (Drew et al., 2001, Moore et al., 2002). Uhl et al. (1979) report a maximum yield of 12.6 lps from the Satpura hills in central India and 4.7 to 9.45 lps from southern India.

In areas where the water table is deep and the remaining saturated weathered zone thickness is only a few metres or where the weathered zone has gone totally dry, wells are drilled deep to tap the underlying saturated fracture zones. That is, depth of drilling (DoD) is more where the yield from the top weathered zone is poor. Beside this, it is demand-based and other than groundwater condition, it is also controlled by the money available for it. As such, DoD can be defined as minimum or economic that gets groundwater yield of adequate quantity and acceptable quality. In hard rock terrain of India wells as deep as 300 m and in places up to 500m are drilled to tap yielding fractures but they are only a few in number. The deeper drilling being costly, unless it is essential to tap a deep fracture (within the capacity of drilling machine deployed), generally, the boreholes for drinking water supply are drilled to a depth of 60 to 100 m. A study conducted by Davis and Turk in the USA as early as 1963 indicates that economic depth of drilling for a high yielding well could be maximal up to 600 feet (can be approximated as 200 m). While Drew et al. (2001) based on statistical analysis show that mean borehole depth in New Hampshire ranges from about 90 to 120 m and well yield increases up to approximately 130 m depth. The maximum yield of 31.5 lps is reported from this area (Moore *et al.*, 2002). Based on drilling details of around 1000 wells Ronka (1983) concludes that in Finland where the weathered zone is either absent or thin or the bedrock is capped by a thin layer of glacial till, the depth of water well drilling is on average 60 m. However, Makela (2012) based on his studies in central Finland reports a median drilled well depth and yield as 73 m and 0.19 lps respectively. The soil type and overburden thickness at the well site do not have much influence on well yield. According to Houston and Lewis (1988) significant yield may not be obtained if the thickness of the weathered zone is less than 10 m. The maximum yield is obtained from thick weathered zone which could be up to 40 m with considerable saturated thickness. The drilling can be continued deeper up to 70 or 80 m if considerable fractures are encountered in the top 10 m of the bed rock. Summers (1972) suggests that if water bearing shallow fractures are not encountered in a borehole it is unlikely that deeper fractures encountered will be productive. Contrary to it in parts of Jharkhand, India, deeper yielding fractures were encountered at several sites in absence of shallow fractures (Fig. 2.8). Read (1982) suggests an estimation of optimum drilling depth from existing data and presents a method of calculating depth where drilling is to be stopped. It is 45 m for the crystalline

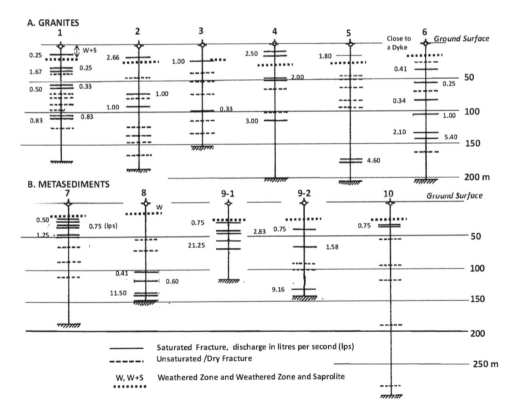

Figure 2.8 Saturated and dry fractures encountered in drilling in hard rocks at various locations in eastern part of India (modified after Dev Burman and Das, 1990).

rocks and 75 m for basalts of Satpura hills, India. At places topography also plays a role in well yield. Study of topography of a large number of bed rock wells in parts of North Carolina by Moore *et al.* (2002) and in Norway and Finland by Henriksen (1995) and Makela (2012) reveals that wells in valleys have much larger yield than those located on hills and ridges. That is, wells located in the discharge zone have higher yield than in the recharge zone. However, Yin and Brook (1992) do not support the link between topography and well yield. Makela (2012) in fact groups as large as 60 well factors like construction, geology and tectonics, topography, lineament and catchment etc. affecting the well yield.

The exploration carried out by Dev Burman and Das (1990) in Jharkhand state of India indicates 0.2 to 2.7 lps yield from weathered and saprolite zone up to a depth of 25 m. The fractures are encountered to a maximum depth of 285 m. The general depth range of occurrences of yielding fractures is reported as 20 to 150 m with yield of individual fracture ranging from 0.25 to 11.5 lps (Fig. 2.8). It is also interesting to note that yield of saprolite and shallow fractures increased and that of deep fractures (>100 m depth) decreased during pumping as compared to that obtained during drilling. Dev Burman and Das (1990) opine that since pumping test represents average macroscopic characteristics of a large volume of rock surrounding the well, increase in yield of saprolite and shallow fractures during pumping test indicates their larger and wider connectivity as compared to deeper fractures. The drawdown for saprolite was less as compared to that for fractures and also recovery was faster.

Generally, in fractured rock a continuous increase in yield with depth is not observed. Mostly there will be constant yield through a certain depth range drilled with sudden increase in yield as soon as another contributing saturated fracture is pierced through. It continues and further increase or decrease in yield with depth depends on relative hydrostatic head of the deeper fracture. Dev Burman and Das (1990) report several dry and non-yielding fractures from eastern India. Encountering a dry open fracture will reduce the yield as it will draw water from the other yielding fractures through the borehole that connects them. Usually the number of fractures per unit depth of drilling decreases with depth and yield may not increase. So the depth of drilling in hard rock could be, in general, 60 to 100 m for shallow wells and can go up to 200m for deep wells. It can be continued beyond 200m depth if there is certainty of encountering isolated deeper yielding fractures evidenced by hydrogeological and geophysical investigations. In volcanic rocks like DTB boreholes are generally drilled up to about 200 m depth. The reported maximum depth of drilling in DTB is 352 m (Parchure, 2010). The borewell discharge varies over a wide range from less than 1 lps to 49 lps. The wells in the basalt flows covering Parana sedimentary basin, Brazil tapping inter-flow zones yield in the range of 1.5 to 5.5 lps and depth of boreholes drilled is about 100 m (Reboucas, 1999). In high yielding wells drilling gets stopped when back pressure of formation water starts affecting the hammer drilling efficiency.

2.5 GROUNDWATER QUALITY

The quality of groundwater is assessed to determine its suitability for domestic consumption, irrigation and industry. The chemical quality of groundwater is dominated by its interaction with rock and hence by the composition of mineral constituents and the dissolution. Frengstad and Banks (2007) based on their study on crystalline aquifers of Norway reveal the dependency of chemical composition of groundwater in crystalline silicate rock aquifers on five factors, viz., partial pressure of CO_2 of initial recharge water, openness of aquifer geochemical system to CO_2, CO_2 consumed by plagioclase hydrolysis, continuation of feldspar hydrolysis and availability and composition of hydrolysable silicate phases. Bath (2007) suggests conjunctive use of geochemical and isotopic indicators for understanding the geochemical evolution of groundwater in fractured rock and its palaeo-hydrogeological modeling.

The study of spatial and temporal variations in quality of water in aquifers, helps understand the natural groundwater flow processes and geogenic and anthropogenic contaminations, defines the protection criteria, helps select areas and depth zones for development and supports management strategies. The quality of groundwater in terms of chemical constituent concentrations is analyzed through samples collected methodically from weathered zone hand-dug wells in use and also from fractured zones encountered in a borehole. The groundwater samples are analyzed for inorganic, trace elements, organic and radioactive (radon content) constituents. Biological constituents are also analyzed. The results are compared with the standards of drinking water for health risk assessment and also checked for irrigation purposes. For concentrations higher than the permissible limits remedial measures are adopted and vulnerability assessment is made. Different diagrams and statistical methods are applied to understand the dominating factors.

The quality of water in weathered zone may vary from that in fractured zones. Also the quality of water can vary from one fracture to another. Jayasena (1993) observes that due to flushing and movement of groundwater through fracture network, in general, locations with high fracture density hold better quality groundwater and local structural features and intrusive bodies control the quality of groundwater. The quality of water in shallow weathered zone aquifer varies with lithology. It may also vary considerably with space and time depending on near surface industrial or sewage contaminations, pesticides and the local seasonal recharge. The variations in anthropogenic contaminant concentrations may be more than geogenic contaminant concentrations. Groundwater salinity in hard rock is generally low except in coastal tracts and in some arid areas. Morgan and Jankowski (2007) relate the possible origin of sodium-magnesium chloride rich groundwater in deep fractures to long residence time and extensive water-rock interaction. Higher fluoride concentration in groundwater in hard rock is a common quality issue all over the world. Brunt *et al.* (2004) detail the factors like geology, contact time and climate controlling the fluoride concentrations and the areas of high probability. According to them the maximum probability is in hyper-arid and arid zones.

REFERENCES

Acworth, R.I. (1987) The development of crystalline basement aquifers in a tropical environment. *Quarterly Journal of Engineering Geology and Hydrogeology*, 20, 265–272.
Adyalkar, P.G., Rane, V.V., Mani, V.V.S. & Dias, J.P. (1975) Evaluation of the aquifer characteristics of the basaltic terrain of Maharashtra in India. *Proc. Indian Academy Sciences*, Section A, 81 (3), 108–117.

Ahlbom, K., Andersson, J.E., Nordqvist, R., Ljunggren, C., Tiren, S. & Voss, C. (1991) *Fjallveden study site, scope of activities and main results.* SKB Technical report: 91–52.

Barrocu, G. (2007) Hydrogeology of granite rocks in Sardinia. In: Krasny, J. & Sharp J.M. (eds.), *Groundwater in Fractured Rocks: Selected Papers from the Groundwater in Fractured Rocks International Conference, 15–19 September, 2003, Prague, IAH Selected Paper Series, Volume 9.* The Netherlands, Taylor & Francis/Balkema. pp. 33–44.

Bath, A. (2007) Interpreting the evolution and stability of groundwaters in fractured rocks. In: Krasny, J. & Sharp Jr., J.M. (eds.), *Groundwater in Fractured Rocks: Selected Papers from the Groundwater in Fractured Rocks International Conference, 15–19 September, 2003, Prague, IAH Selected Paper Series, Volume 9.* The Netherlands, Taylor & Francis/Balkema. pp. 261–274.

Braga, M.A.S., Paquet, H. & Begonha, A. (2002) Weathering of granites in a temperate climate (NW Portugal): granitic saprolites and arenization. *Catena,* 49 (1 & 2), 41–56.

Brunt, R., Vasak, L. & Griffioen, J. (2004) *Fluoride in groundwater: probability of occurrences of excessive concentration on global scale.* IGRAC, UNESCO Report No. SP 2004-2.

Bromley, J., Mannstrom, B., Nisca, D. & Jamtlid, A. (1994) Airborne geophysics: Application to a ground-water study in Botswana. *Ground Water,* 32 (1), 79–90.

Buckley, D.K. & Zeil, P. (1984) The character of fractured rock aquifers in eastern Botswana. In: Walling, D.E., Foster, S.S.D. and Wurzel, P. (eds.) *Challenges in African Hydrology and Water Resources: Proceedings of the Harare Symposium, July, 1984,* IAHS Publication No. 144, 25–36.

Central Ground Water Board (1981) *Groundwater studies in the Upper Betwa River Basin, Madhya Pradesh and Uttar Pradesh – project findings and recommendations.* Indo-British Betwa Groundwater Project, CGWB, Min. of Irrigation, Govt. of India Project Report.

Central Ground Water Board (1982) *Groundwater studies in the Sina and Man river basins, south Maharashtra-project findings and recommendations.* CGWB, Min. of Irrigation, Govt. of India, Project Report No. P-1.

Central Ground Water Board (2009) *Annual report (2008-09).* CGWB, Min. of Water Resources Govt. of India.

Central Ground Water Board (2012a) *Groundwater studies in Kasai-Subarnarekha river basins.* CGWB, Min. of Water Resources, Govt. of India, published report.

Central Ground Water Board (2012b) *Ground Water Information Booklet, Kolar District.* CGWB, Min. of Water Resources, Govt. of India.

Chandra, S., Kumar, D., Ahmed, S., Perrin, J., & Dewandel, B. (2008) Contribution of geophysical methods in exploration and assessment of groundwater in hard rock aquifers. Presented at: *The 3rd International Conference On Water Resources and Arid Environments and the 1st Arab Water Forum, 16–19 November, 2008,Riyadh Saudi Arabia,* Arab Water Forum.

Chandra, P.C., Reddy, P.H.P. & Singh, S.C. (1994) *Geophysical studies for groundwater exploration in Kasai and Subarnarekha River basins* (UNDP Project) Central Ground Water Board, Min. of Water Resources, Govt. of India Tech. Report

Chandra, P.C., Srivastava, M.M. & Adil, M. (1992) *Electrical resistivity surveys for the delineation of deeper groundwater zones at Jarar and Parsahar, district Banda, Uttar Pradesh.* Central Ground Water Board, Min. of Water Resources, Govt. of India, Tech. Report.

Clark, L. (1985) Groundwater abstraction from basement complex areas of Africa. *The Quarterly Journal of Engineering Geology,* 18 (1), 25–34.

Cook, P.G. (2003) *A guide to regional groundwater flow in fractured rock aquifers.* CSIRO Land and Water, Australia.

Custodio, E. (2007) Groundwater in volcanic hard rocks. In: Krasny, J. & Sharp Jr., J.M. (eds.) Groundwater in Fractured Rocks: Selected Papers from the Groundwater in Fractured Rocks International Conference, 15–19 September, 2003, Prague, IAH Selected Paper Series, Volume 9. The Netherlands, Taylor & Francis/Balkema. pp. 95–108.

Davis, S.N. & Turk, L.J. (1964) Optimum depth of wells in crystalline rocks. *Ground Water,* 2 (2), 6–11.

Deolankar, S.B. (1980) The Deccan basalts of Maharashtra, India- their potential as aquifers. *Ground Water,* 18 (5), 434–437.

Dev Burman, G.K. & Das, P.K. (1990) Groundwater exploration in hard rock terrain: an experience from eastern India. In: Shamir, U. and Jiaqi, C. (eds.) *The Hydrological Basis for Water Resources Management: Proceedings of the Beijing Symposium, October, 1990.* IAHS Publication. No. 197, pp. 19–30.

Dewandel, B., Lachassagne, P., Jaidi, F. K. & Chandra, S. (2011) A conceptual hydrodynamic model of a geological discontinuity in hard rock aquifers: example of a quartz reef in granitic terrain in South India. *Journal of Hydrology,* 405 (3–4), 474–487.

Dewandel, B., Perrin, J., Ahmed, S., Chandra, S., Kumar, D., Lachassagne, P. & Marechal, J.C. (2009) A new look into hard rock aquifers: geological model validated by geophysics and hydraulic experiments. In: *Proceedings of Joint International Convention of 8th IAHS Scientific Assembly and 37th IAH Congress Water: A vital resource under stress – How science can help, September 6–12, 2009, Hyderabad, India.*

Dhonde, U.V. (2009) *Groundwater information, Nashik district, Maharashtra.* 1614/DBR/2009, Central Ground Water Board, Min. of Water Resources, Govt. of India.

Dixon, J.L., Heimsath, A.M. & Amundson, R. (2009) The critical role of climate and saprolite weathering in landscape evolution. *Earth Surface Processes and Landforms,* 34 (11) 1507–1521.

Dosseto, A., Buss, H.L. & Suresh, P. O. (2012) Rapid regolith formation over volcanic bedrock and implications for landscape evolution. *Earth and Planetary Science Letters,* 337–338, 47–55.

Drew, L.J., Schuenemeyer, J.H., Armstrong, T.R. & Sutphin, D.M. (2001) Initial yield to depth relation for water wells drilled into crystalline bedrock-Pinardville quadrangle, New Hampshire. *Ground Water,* 39 (5), 676–684.

Foster, S., Garduno, H. & Tuinhof, A. (2007) *Confronting the groundwater management challenge in the Deccan Traps country of Maharshtra-India, Sustainable Groundwater Management, Lessons from Practice.* The GW-MATE Cross profile Collection No. 18, The World Bank Publication.

Frengstad, B. & Banks, D. (2007) Universal controls on the evolution of groundwater chemistry in shallow crystalline rock aquifers: the evidnce from empirical and theoretical studies. In: Krasny, J. & Sharp Jr., J.M. (eds.), *Groundwater in Fractured Rocks: Selected Papers from the Groundwater in Fractured Rocks International Conference, 15–19 September, 2003, Prague, IAH Selected Paper Series, Volume 9.* The Netherlands, Taylor & Francis/Balkema. pp. 275–289.

Ghosh, P., Sayeed, M.R.G., Islam, R. & Hundekari, S.M. (2006) Inter-basaltic clay (bole bed) horizons from Deccan traps of India: Implications for palaeo-weathering and palaeo-climate during Deccan volcanism. *Palaeogeography, Palaeoclimatology, Palaeontology,* 242 (1–2), 90–109.

Grey, D.R.C., Chilton, P.J., Smith-Carington, A.K. & Wright, E.P. (1985) The expanding role of the hydrogeologist in the provision of village water supplies: an African perspective. *The Quarterly Journal of Engineering Geology,* 18 (1), 13–24.

Gupta, G., Erram, V.C. & Kumar, S. (2012) Temporal geoelectric behaviour of dyke aquifers in northern Deccan volcanic province, India. *Journal of Earth System Science,* 121 (3), 723–732.

Gustafson, G. & Krasny, J. (1993) Crystalline rock aquifers: their occurrence, use and importance. In: Banks, D and Banks Sheila (eds.) *Hydrogeology of Hard Rocks: Memoires of the XXIV Congress of IAH, 28 June–2 July, As, Oslo, Norway,* IAH Press. pp. 3–20.

Gustafsson, P. (1993) SPOT satellite data for exploration of fractured aquifers in a semi-arid area in southeastern Botswana. In: Banks, D. and Banks Sheila (eds.) *Hydrogeology of Hard*

Rocks: *Memoires of the XXIV Congress of IAH, 28 June–2 July, As, Oslo, Norway,* IAH Press.pp 562–576.

Hall, A.M. (1986) Deep weathering pattern in north-east Scotland and their geomorphological significance. *Zeitschrift fur Geomorphologie, NF,* 30 (4), 407–22.

Hamblin, W.K. & Christiansen, E.H. (2009) *Weathering.* Chapter 10, Earth's Dynamic System, Prentice Hall.

Henriksen, H. (1995) Relation between topography and well yield in boreholes in crystalline rocks, Sogn og Fjordane, Norway. *Ground Water,* 33 (4), 635–643.

Houston, J.F.T. & Lewis, R.T. (1988) The Victoria province drought relief project, II: borehole yield relationships. *Ground Water,* 26 (4), 418–426.

Jayapraksh, A.V. (2007) Flow stratigraphy of Deccan Traps in Karnataka. *Journal of Geological Society of India,* 70 (2), 378–381.

Jayasena, H.A.H. (1993) Geological and structural significance in variation of groundwater quality in hard crystalline rocks of Sri Lanka. In: Banks, D and Banks Sheila (eds.) *Hydrogeology of Hard Rocks: Memoires of the XXIV Congress of IAH, 28 June–2 July, As, Oslo, Norway,* IAH Press. pp. 450–471.

Jayawardena, U. de S. (2004) Sources of groundwater in crystalline hard rocks of Kandy area, Sri Lanka. *Asian Jour. of Water, Env. and Pollution,* 1 (1&2), 119–122.

Jones, M.J. (1985) The weathered zone aquifers of the basement complex areas of Africa. *The Quarterly Journal of Engineering Geology,* 18 (1), 35–46.

Kale, V.S. & Kulkarni, H. (1992) IRS-1A and landsat data in mapping Deccan Trap flows around Pune, India: implications on hydrogeological modelling. In: Fritz, L.W. & Lucas, J.R. (eds.) *Interpretation of Photographic and Remote Sensing Data, Technical Commission VII, ISPRS Archives-Vol. XXIX, part B7, Proceedings of the XVIIth ISPRS Congress, August 2–14, 1992, Washington, DC.* pp. 429–435.

Krasny, J. (1997) Transmissivity and permeability distribution in hard rock environment: a regional approach. In: Pointet T. (ed.) *Hard Rock Hydrosystems, Proceedings 5th IAHS Int. Symp. April–May 1997, IAHS Publ. No. 241, Morocco, Rabat,* Oxfordshire, UK, Institute of Hydrology, IAHS Press, pp. 81–90.

Krasny, J. & Sharp Jr., J.M. (2007) Hydrogeology of fractured rocks from particular fractures to regional approaches: State-of-the-art and future challenges. In: Krasny, J. & Sharp Jr., J.M. (eds.), *Groundwater in Fractured Rocks: Selected Papers from the Groundwater in Fractured Rocks International Conference, 15–19 September, 2003, Prague, IAH Selected Paper Series, Volume 9.* The Netherlands, Taylor & Francis/Balkema. pp. 1–30.

Kulkarni, H. & Deolankar, S.B. (1995) Hydrogeological mapping in the Deccan Basalt – an appraisal. *Jour. Geol. Soc. India,* 46 (4), 345–352.

Lachassagne, P., Wyns, R., Berard, P., Bruel, T., Chery, L., Coutand, T., Desprats, J.-F. & Strat, P.L. (2001) Exploitation of high-yields in hard rock aquifers: downscaling methodology combining GIS and multicriteria analysis to delineate field prospecting zones. *Ground Water,* 39 (4), 568–581.

Larsson, I. (1984) *Groundwater in hardrocks.* Project 8.6 of the IHP, UNESCO, Studies & Reports in Hydrology 33. ISBN 92-3-101980-5. France, UNESCO Publication.

Mahoney, J.J. (1988) Deccan traps. In: Macdougall, J.D. (ed.) *Continental Flood Basalts* [Online] The Netherlands, Kluwer Academic Publishers. pp. 151–194.

Makela, J. (2012) Drilled well yield and hydraulic properties in the Precambrian crystalline bedrock of Central Finland. *Thesis, Dept. of Geography and Geology, Faculty of Mathematics and Natural Sciences, University of Turku, Finland.*

McQueen, K.G. (2009) Regolith geochemistry. In: Scott, K.M. & Pain, C.F. (eds.) *Regolith Science.* Australia & New Zealand, CSIRO Publishing & The Netherlands, Springer, pp. 73–104.

Moore, R.B., Schwarz, G.E., Clark Jr., S.F., Walsh, G.J. & Degan, J.R. (2002) *Factors related to well yield in the fractured-bedrock aquifer of New Hampshire.* U.S. Geological Survey Professional Paper 1660.

Morgan, K. & Jankowski, J. (2007) Origin of salinity in a fractured bedrock aquifer in the Central West Region of NSW, Australia. In: Krasny, J. & Sharp Jr., J.M. (eds.), *Groundwater in Fractured Rocks: Selected Papers from the Groundwater in Fractured Rocks International Conference, 15–19 September, 2003, Prague, IAH Selected Paper Series, Volume 9.* The Netherlands, Taylor & Francis/Balkema. pp. 329–341.

Naik, P.K., Naik, K. C., Choudhury, A. & Chakraborty, D. (2010) Solving drinking water crisis in hard rock terrain of Peninsular India. *Current Science,* 99 (10), 1309–1310.

Olesen, O., Dehls, J.F., Ebbing, J., Henriksen, H., Kihle, O. & Lundin, E. (2007) Aeromagnetic mapping of deep-weathered fracture zones in the Oslo Region – a new tool for improved planning of tunnel. *Norwegian Journal of Geology,* 87 (1–2), 253–267.

Olvmo, M. (2010) Review of denudation processes and quantification of weathering and erosion rates at a 0.1 to 1 Ma time scale. *SKB Tech Report TR-09-18.*

Parchure, P.K. (2010) Groundwater information Ahmadnagar district, Maharashtra. 1645/DBR/2010, Central Ground Water Board, Min. of Water Resources, Govt. of India.

Petrov, V.P. (1991) The nature of thick zones of paleoweathering. *International Geology Review,* 33 (1), 49–61.

Pietersen, K., Beekman, H. E. & Holland, M. (2011) *South African groundwater governance case study.* Report prepared for the World Bank in partnership with the South African Department of Water Affairs and the Water Research Commission, WRC Report No. KV 273/11, ISBN 978-1-4312-0122-8.

Read, R.E. (1982) Estimation of optimum drilling depth in fractured rock. In: AWRC Groundwater Committee: *Conference Series, Australian Water Resources Council, No. 5: Papers of the Groundwater in Fractured Rock Conference, 31 August–3 September, 1982, Canberra,* Australian Govt. Pub. Service

Reboucas, A. da C. (1999) Groundwater resources in South America. *Episodes,* 22 (3), pp. 232–237.

Reidel, S.P., Johnson, V.G. & Spane, F.A. (2002) *Natural gas storage in basalt aquifers of the Columbia Basin, Pacific Northwest USA: a guide to site characterization.* Pacific Northwest National Laboratory report PNNL-13962, U.S. Dept. of Energy.

Robineau, B., Join, J.L., Beauvais, A., Parisot, J.-C. & Savin, C. (2007) Geoelectrical imaging of thick regolith developed on ultramafic rocks: groundwater influence. *Australian Journal of Earth Sciences,* 54 (5), 773–781.

Ronka, E. (1983) *Drilled wells and groundwater in the Precambrian crystalline bedrock of Finland.* Publications of The Water Research Institute, National Board of Waters, Finland, Report No. 52.

Saha, D. & Agrawal, A.K. (2006) Determination of specific yield using water balance approach – case study of Toria Odha watershed in the Deccan Trap province, Maharashtra State, India. *Hydrogeology Journal,* 14 (4), 625–635.

Saha, D. Member, Central Ground Water Board, Government of India, Faridabad, India (Personal Communication, June, 2015).

Scott, K.M. & Pain, C.F. (2009) Introduction. In: Scott, K.M. & Pain, C.F. (eds.) *Regolith Science.* Australia & New Zealand, CSIRO Publishing & Springer, The Netherlands. pp. 1–6.

Singhal, B.B.S. (1997) Hydrogeological characteristics of Deccan Trap formations of India. In: Pointet T. (ed.) *Hard Rock Hydrosystems, Proceedings 5th IAHS Int. Symp. April–May 1997, IAHS Publ. No. 241, Morocco, Rabat.* Oxfordshire, UK, Institute of Hydrology, IAHS Press, pp. 75–80.

Singhal, B.B.S. & Gupta, R.P. (1999) *Applied hydrogeology of fractured rocks*. The Netherlands, Kluwer Academic Publishers.

Subramanian, P. (1992) Hydrogeology and its variations in the granites and associated rock formations in India. In: *Proceedings of the Workshop on Artificial Recharge of Groundwater in Granitic Terrain, 19 October 1992, Bangalore*. pp. 1–13.

Summers, W.K. (1972) Specific capacities of wells in crystalline rocks. *Ground Water*, 10 (6), 37–47.

Surinaidu, L., Bacon, C.G.D. & Pavelic, P. (2013) Agricultural groundwater management in the Upper Bhima basin, India: current status and future scenarios. *Hydrology & Earth System Sciences*, 17 (2), 507–517.

Tolen, T., Lindsey, K. & Porcello, J. (2009) *A summary of Columbia River Basalt Group geology and its influence on the hydrogeology of the Columbia River Basalt Aquifer System: Columbia Basin Ground Water Management Area of Adams, Franklin, Grant, and Lincoln counties*. The CB GWMA of Adams, Franklin, Grant, and Lincoln counties Report.

Twidale, C.R. & Campbell, E.M. (1993) Australian Landforms. *Rosenberg Publishing, ISBN 1 877058 32 7*, pp. 72–73.

Uhl Jr., V.W., Nagabhushanam, K. & Johansson, J.O. (1979) Hydrogeology of crystalline rocks: case studies of two areas in India. *Nordic Hydrology*, 10 (5), 287–308.

Ventriss, H.B., Collett, D.B. & Boyd, D.W. (1982) Relationship between groundwater occurrence and a dolerite dyke in the Northampton area of Western Australia. In: AWRC Groundwater Committee: *Conference Series, Australian Water Resources Council, No. 5: Papers of the Groundwater in Fractured Rock Conference, 31 August–3 September 1982, Canberra* Australian Govt. Pub. Service

Verma, R.K., Rao, M.K. & Rao, C.V. (1980) Resistivity investigations for groundwater in metamorphic areas near Dhanbad, India. *Ground Water*, 18 (1), 46–55.

Woodford, A. Rosewarne. P. & Girman, J. (2005) *How much groundwater does South Africa have?* SRK Consulting, Cape Town & Department of Water Affairs & Forestry, Pretoria, South Africa.

Wright, E.P. (1992) The hydrogeology of crystalline basement aquifers in Africa. In: Wright, E.P. & Burgess, W.G. (eds.) *Hydrogeology of Crystalline Basement Aquifers in Africa*. Special Publication No. 66. Geological Society, London, pp. 1–27.

Yin, Z.-Y. & Brook, G.A. (1992) The topographic approach to locating high-yield wells in crystalline rocks: does it work? *Ground Water*, 30 (1), 96–102.

Chapter 3

Introduction to geophysical investigations in hard rock

3.1 INTRODUCTION

Application of geophysical methods in groundwater investigations is based on identifying the anomalies caused by subsurface variations in physical properties and resolving hydrogeological conditions through subjective interpretations of the anomalies. A significant contrast in physical property and larger dimension of target (related to its depth) leading to an appreciable geophysical anomaly are required to resolve the target definition precisely. The efficacy of a geophysical method including instrumentation, data acquisition and interpretation lies in its ability to pick up anomalies due to small and deep targets with less-contrasting physical property and resolve them with least ambiguity.

Geophysical methods are being used extensively in hard rock groundwater investigations providing appropriate information for various objectives. In the realm of groundwater geophysics, application starts from aquifer mapping, site selection for constructing bore wells and monitoring groundwater dynamics through suitable time-lapsed measurements at different scales and levels. Mostly, the geophysical methods are non-destructive or non-invasive. The most significant advantage of geophysical applications is the cost effective generation of high density data on subsurface physical property variations over a large area which cannot be done by any other means. The geophysical methods and techniques, used in groundwater investigation up to about 500 m depth, mostly fall under near-surface geophysics. The concepts developed in groundwater geophysics are within the broad spectrum of applications of near-surface geophysics which includes environmental geophysics and agriculture geophysics also. Of late groundwater geophysics is becoming identified as a separate sub-discipline of exploration geophysics. In recent years there has been a remarkable advancement in theory, instrumentation for fast, high density data acquisition and coverage, storage, processing and interpretation. All these have strengthened the ability and enhanced the applications in subsurface characterization, acquiring precise subsurface hydrogeological information cost effectively. However, there is still enough scope for improvement in acquisition and accuracy and precision in interpretation. Overall, the future lies in deeper investigations with minimum surface-spread and better accuracy in predictive aspects of resolving thin targets through an integrated approach.

The geophysical methods, being quite fascinating in unravelling hidden subsurface hydrogeological complexities, have been used indiscriminately in places. Also, there are cases where even the essential minimum geophysical method has not been used due to lack of awareness. For hard rock water well drilling, the success of geophysical applications is generally measured by the beneficiaries in qualitative terms of well yield. Whatever geophysical method is used, if the yield of the well drilled at the pinpointed site is copious, the application is successful. At times, points of drilling near habitations may not be geophysically suitable but drilled because of the demand. The drilling of wells is obligated on 'local best' anomalies that may yield inadequately and in the process geophysical application comes under doubt. Applications also face dissatisfaction and criticism of technical inadequacy and inaccuracy in places. It is mainly because of over-expectations, without knowing the limitations of the method and technique applied vis a vis the objective or what is actually wanted from the geophysical method applied and what it can deliver. This led to occasional non-acceptance and there are instances where geophysical investigations were not a priority in groundwater exploration programmes. A large amount of money is spent on the drilling of alternative well sites and the geophysical investigations are introduced later to find a solution. A relevant awareness is essential for the beneficiaries. Definitely, geophysics contributes significantly and has the potential to face challenges of precise characterization and reducing well failures. As already mentioned the challenges are critical technically as well as socially, because whatever is predicted is proved or disproved immediately by drilling; everyone eagerly waiting for the water to ooze out of the borehole. Butler (2005) has rightly pointed out that "... *public interest gives more exposure and scrutiny to near-surface geophysics (which encompasses groundwater geophysics) than other areas of geophysics*".

3.2   GEOPHYSICAL METHODS AND PHYSICAL PROPERTY MEASUREMENTS

The geophysical methods generally used are electrical, electromagnetic, magnetic, gravity and seismic. The electrical resistivity and electromagnetic methods measure the variations in electrical conductivity of rock formations caused by changes in lithology, water content, porosity and water salinity; the magnetic method measures the variations in magnetic field caused by differences in magnetic susceptibility; the gravity method measures the variations in gravity field caused by disparities in formation density and the seismic method measures the variations in seismic velocity caused by the differences in mechanical properties. The other methods in use are induced polarization (IP), self potential (SP) and mise-a-la-masse (MAM). The IP method measures variations in electrical polarizability or chargeability caused by differences in clay content; the SP method measures the variations in natural electrical potential developed due to changes in lithology, water quality and water movement and the MAM method measures the potential developed due to preferential passage of electric current in a particular direction of electrical conductivity discontinuity. Also, the variations in natural radioactivity caused by varied clay and the radioactive mineral contents are measured by radiometric method.

The geophysical methods are classified as active and passive. The active methods require energization of the ground while passive methods make use of natural fields of the Earth. The ambient natural fields include gravity, magnetic and electrical.

The active methods are electrical resistivity, induced polarization, controlled or remote-source frequency and time domain electromagnetic and the seismic. Peterson and Ronka (1971) consider remote-source electromagnetic (VLF) also as passive. Working with natural field methods is easier as there is no need of a source to energize the ground and they are effective in delineating the targets which are associated with variations in those fields. Mostly, active methods are used as the source and receiver power and configuration can be manipulated according to the expected character of the target. The geophysical methods are selected and surveys are designed depending on terrain condition, target geometry, expected physical property variations and their order and limitations of individual method. Generally, combination geophysical methods and techniques and integration with other disciplines are preferred. Recently a few modern methods and data acquisition technologies have come up in near surface geophysics like electrical resistivity imaging, high resolution shallow reflection seismic, passive seismic, shear wave splitting analysis, ground penetrating radar, nuclear magnetic resonance and *in situ* borehole measurements which have larger applications in groundwater exploration, subsurface characterization and monitoring. However, these approaches being quite expensive are used in a restrictive manner. The general applications of geophysical methods for investigating groundwater related subsurface structures in hard rock are shown in Figure 3.1.

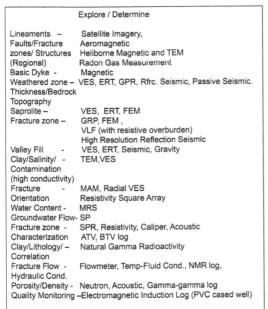

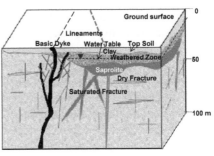

Explore / Determine	
Lineaments –	Satellite Imagery,
Faults/Fracture	Aeromagnetic
zones/ Structures	Heliborne Magnetic and TEM
(Regional)	Radon Gas Measurement
Basic Dyke -	Magnetic
Weathered zone –	VES, ERT, GPR, Rfrc. Seismic, Passive Seismic.
Thickness/Bedrock Topography	
Saprolite –	VES, ERT, FEM
Fracture zone –	GRP, FEM ,
	VLF (with resistive overburden)
	High Resolution Reflection Seismic
Valley Fill -	VES, ERT, Seismic, Gravity
Clay/Salinity/ -	TEM,VES
Contamination (high conductivity)	
Fracture -	MAM, Radial VES
Orientation	Resistivity Square Array
Water Content -	MRS
Groundwater Flow-	SP
Fracture zone -	SPR, Resistivity, Caliper, Acoustic
Characterization	ATV, BTV log
Clay/Lithology/ –	Natural Gamma Radioactivity
Correlation	
Fracture Flow -	Flowmeter, Temp-Fluid Cond., NMR log,
Hydraulic Cond.	
Porosity/Density -	Neutron, Acoustic, Gamma-gamma log
Quality Monitoring –	Electromagnetic Induction Log (PVC cased well)

VES : Vertical Electrical Sounding , ERT : Electrical Resistivity Tomography , GPR : Ground Penetrating Radar ,
FEM: Frequency Domain Electromagnetic, TEM : Time Domain Electromagnetic, VLF : Very Low Frequency,
GRP : Grad. Res. Profiling, MAM: Mise-a-la-Masse MRS: Magnetic Resonance Sounding , SP : Self Potential,
SPR: Single Point Resistance, NMR : Nuclear Magnetic Resonance, ATV : Acoustic Televiewer, BTV: Borehole Televiewer

Figure 3.1 Schematic model of hard rock area showing groundwater related features and general applications of geophysical methods.

3.3 APPLICATIONS OF GEOPHYSICAL METHODS

The applications of geophysics for groundwater in hard rocks are grouped into 3 categories, viz., the airborne, surface and borehole. Some of the applications are described in an SEG publication on Geotechnical and Environmental Geophysics edited by Ward (1990), Springer publications on Groundwater Geophysics edited by Kirsch (2009) and Hydrogeophysics edited by Rubin and Hubbard (2005).

3.3.1 Airborne geophysics

The airborne geophysical survey provides a large areal coverage with high data density and information, taking a relatively much shorter time for extensive and detailed survey operation and delivery of results. Also it gives access to remote inaccessible areas. In fact it presents a holistic view of subsurface geological structures over the entire area. It has been in practice for mineral and petroleum exploration over the past several decades and now its usefulness in groundwater exploration is also established. However, its use exclusively for groundwater exploration has been sporadic. The interpretation of high resolution (low-altitude, close-spaced flight lines) airborne data integrated with photo-lineament and geological maps can yield accurate information on small-scale hydrogeological controls like thick weathered zone, lithological contacts, faults, folds, dykes and quartz reefs, palaeo-channels and regional network of hidden prominent fractured zones. Some of the airborne surveys conducted for hydrogeological purposes are by Astier and Paterson (1987), Paterson and Bosschart (1987), Bromley et al. (1994), Paine et al. (2003), Simth et al. (2007), Cresswell and Gibson (2004), d'Ozouville et al. (2008), Viezzoli et al. (2009), Jorgensen et al. (2012) and Ahmed (2014). These studies include magnetic and frequency (FEM) and time domain (TEM) electromagnetic surveys. The airborne EM survey provides high resolution resistivity-depth or conductivity-depth maps. It has been in use for subsurface hydrogeological characterization for quite some time. Recently, heliborne TEM with early measurement time has been purposefully developed for groundwater and other near surface investigations. In comparison to FEM, TEM can yield deeper information in a conductive environment. In view of the large area coverage with high data-density acquired in much less time, the airborne FEM or TEM with MAG would be economical and can be conducted under regional groundwater exploration programmes in hard rock areas. The limitation for low altitude high resolution heliborne survey with hooked system is safe terrain clearance, cultural noise and habitation while that for higher altitude flight is primarily the drastic attenuation of field strength.

Besides, plenty of aeromagnetic data collected at different altitudes is available over hard rock areas throughout the world and can be usefully re-interpreted for hydrogeological purposes. According to Astier and Paterson (1987), areas of crystalline basement have quite wide and dense coverage, probably exceeding 75%. A large amount of information on groundwater can be extracted from these aeromagnetic data. These can be interpreted jointly with the photo-lineaments and geological maps as a part of reconnaissance at a smaller scale of 1:250,000 for delineating the regional fractured zones and narrowing down the zone of interest or prioritize the linears at a larger scale of 1:50,000 before taking up ground geophysical surveys. The joint interpretation is done because the photo-lineaments are sometimes misleading.

The area with favourable photo-lineament coincident with aeromagnetic lineament can be confidently taken up for ground geophysical follow-up to pinpoint the drilling site.

The simple linear feature of a photo-lineament is attributed a significant subsurface structural picture by identifying its attitude through aeromagnetic data. According to Astier and Paterson (1987), in comparison to geological mapping and photo-lineaments the aeromagnetics can reveal much more structural features, their regional continuity and also features which are not picked up either by satellite images or geological mapping because of the presence of overburden. In this way the follow-up surface geophysical methods are applied more effectively and economically. Sites drilled through surface geophysical follow-up even in areas picked up from simple qualitative interpretation of aeromagnetic and photo-lineament data tapped high yielding fractured zones (Chandra, 1991 and Joseph and Chandra, 2003). A site drilled away from the aeromagnetic linear was almost dry indicating its great potential in groundwater exploration. The reduction in well yield with distance from aeromagnetic linear is indicated by Astier and Paterson (1987) also. The detailed analysis of aeromagnetic and landsat thematic mapper data from the crystallines of south-central Zimbabwe by Ranganai and Ebinger (2007) establishes the usefulness of joint interpretation in locating high yielding wells.

3.3.2 Surface geophysics

The surface geophysical methods and techniques are used in hard rock area mainly to locate suitable sites for water well drilling. They can also be used for characterization and quantification of subsurface hydrogeological conditions. The basic difference in these is of approach and scale. The type of geophysical investigation through a selected single method or combination of methods is tailored to the individual programme objective, site-specific demands, local geologic conditions, infrastructure including availability of field equipment, field professional, interpretational software and the expertise, so that desired deliverables are obtained and effectively utilized. Because of the cost and time involved in detailed surveys, generally it is carried out at site-scale. If required, a feasibility study for a smaller area can also be carried out before taking up detailed surveys.

The surface geophysical investigation starts with a reconnoitory survey and the inflowing data are continuously reviewed and interpreted so that suitable areas for detailed surveys are identified or narrowed down. A reconnaissance survey is important to make the programme economical, particularly when the area to be covered is quite large and the target has limited spatial extent. A reconnaissance survey should be conducted in such a way that it results conclusively in selection of favourable areas, suitable methods and techniques and instruments for detailed surveys. Such reconnaissance surveys in hard rock may be supported by remote sensing data such as satellite images for identifying the lineaments, deep buried pediments and other significant geomorphological features and the aeromagnetic maps, if available. In hard rock areas reconnaissance is generally initiated by ground magnetic profiling which is fast in operation and quite cost effective. Addition of Very Low Frequency (VLF) electromagnetic profiling can make it more effective. The spatial data obtained from reconnaissance geophysical surveys, can also be integrated through GIS with the data from geological, hydrogeological, lineament and aeromagnetic maps and existing boreholes for selection of favourable areas. To appreciate such integrations the geophysicist should have a good knowledge of hydrogeology and also of satellite imaging data interpretations.

The detailed geophysical surveys follow subsequently covering selected smaller or specific areas for obtaining comprehensive information on specific objectives. These specific areas may require close-grid 1-d or high resolution 2-d along close spaced profiles and also 3-d measurements. For example, to prepare a weathered zone map, there is no need of a reconnaissance survey. The measurement can be either on a close-grid pattern or it can be based on geomorphological conditions. But to delineate a fractured zone it is better to initiate through reconnaissance surveys followed by conventional close-grid 1-d detailing by sounding and parallel profiling or 2-d and 3-d imaging. The area for coverage by reconnaissance survey can be more than a square km, for detailed survey it is within 1000–500 square m and for pinpointing it is done at site-scale of 1 to 10 m.

Besides physical property and target geometry, geophysical response is a function of scale of field measurement. The responses at different scales of measurement with various geophysical methods will be different depending the way the subsurface is sampled. Different types of coverage at different scales and resolution are required to pinpoint the target. To get a proper response and make the application successful measurements at appropriate scale are essential. In many instances unsystematic use of geophysical methods at miss-matched scales led to failures. The sampling-density as well as measurement-scale selection require an understanding of the spatial variations in hydrogeological conditions, scale of inhomogeneities, complexities and the related geophysical anomalies or responses that will be obtained. If the scale of measurement is higher than the scale of variation in macroscopic hydrogeological property it will average out the inhomogeneities and accordingly the resolution of geophysical response required for details will be obscured. For example a thin saturated fractured zone may be missed if the resistivity profiling measurements are made with relatively large electrode separation at a considerably larger station interval. Besides, it may also be missed if the orientation of profiles for field measurement is not appropriate. Conventionally, the sampling interval should not be greater than half the expected width of the target. Of course the target response and the anomaly width will depend on its depth extent also. It becomes quite significant in hard rock where spatial variations in hydrogeologic properties are quite large. For this, the scale of geophysical field measurements should match with the scale of variations in hydrogeological properties that are 'significant'. Also, there may be random, uncertain smaller heterogeneities and if required the specific geophysical data sampling should be sufficient enough to pick out the anomalies adequately and manifest these heterogeneities properly.

To sense the subsurface remotely being the basic perception of surface geophysical investigation, the problem of non-uniqueness and resolution creeps in. Each geophysical method has its limitation in characterizing the subsurface and no single geophysical method delivers the best for the desired information. For effective delineation it is essential to deploy different methods which are sensitive to different physical properties of the target, the surrounding and the overburden and it necessarily demands an integration of interpretations. The objective is to catch the best common anomaly for the target confidently, constrain the interpretation, refine the model and reduce the ambiguity. The electrical and electromagnetic techniques sensitive to conductivity characteristics of the target-fracture zone, could be combined with a seismic survey which is sensitive to the mechanical property.

Geophysical investigation for groundwater is generally a low budget programme. So, the geophysical surveys include the most affordable electrical resistivity method with conventional data acquisition systems which is quite cost-effective (Chandra, 2002). Moreover, electrical resistivity of a geological formation being dependent on the quantity and quality of water in it, the method yields most relevant information measuring lateral as well as vertical variations in resistivity of the subsurface which are interpreted to understand the subsurface groundwater conditions. In hard rock it can be utilised to understand the weathered zone aquifer distribution, fractured zones and quality of water (in terms of salinity) in them, structural control of dykes, quartz reefs and fault zones etc. on groundwater occurrences and the bed rock topography. It has been used extensively. The maximum global expenditure on geophysics for groundwater is on electrical resistivity surveys. This requires a short exposure to carry out field measurements successfully. The interpretation of data and its translation into hydrogeological model of course requires a great deal of expertise and experience.

Interestingly, in granitic terrain without any layering, resistivity sounding measurements succeeded in yielding desired parameters through conventional 1-d layered-earth interpretation, characterising the shallow weathered zone aquifers and the bed rock depth and in places also the in-between thick fractured zone aquifer with intermediate resistivity. The response is affected by the overburden conductance, equivalence and suppression. Because of these inherent limitations causing non-uniqueness in interpretation the conventional sounding technique is not employed for detecting thin targets. The saturated conductive fractured zone at depth impregnated in resistive host rock acting as a micro-level discontinuity in conductivity generally remains undetected in a sounding curve. The electrical resistivity soundings when observed with some modifications in field data acquisition like closer increments in current electrode spacings can be interpreted empirically for shallow fractured zones.

Obtaining a recognizable electrical response from a thin fractured zone depends on its size, shape, depth of occurrence and also whether it is dry or saturated, filled or open, its conductivity contrast with the surrounding and character of the overburden. Prior to induction of resistivity imaging, the technique of electrical resistivity profiling was in practice and was reasonably effective in lateral demarcation of shallow zones holding saturated fractures and thickening of the weathered zone. In resistivity profiling the response is greatly modified by conductivity of the weathered zone/overburden and its thickness in relation to the width of the fractured zone. With so many inherent investigative constraints and the geological variables causing heterogeneities, precise delineation of fractured zones at depths only by profiling becomes difficult and definitely a challenging problem. It may require forward modeling to assess the response and design the survey parameters. The exact definition of the fracture geometry may not be ascertained by resistivity profiling or resistivity imaging alone and a subjective support of magnetic profiling or depending on the overburden conductivity controlled source frequency domain and remote source VLF (very low frequency) electromagnetic profiling can be added to confirm. Resistivity profiling can be even replaced by multi-frequency electromagnetic profiling. Palacky et al. (1981) use electromagnetic profiling in hard rock to locate suitable sites for water well drilling. For VLF profiling the orientation of the target vis-a-vis the location of remote VLF station may create a constraint.

The resistivity sounding yields only 1-d information. Therefore the technique of electrical resistivity imaging has become popular for its fast 2d and 3d data acquisition and lateral inhomogeneity investigation capabilities. The high resolution spatial and time-lapsed information obtained from resistivity imaging has added value to resistivity method and in fact revived it, opening scope for a variety of its applications in near-surface investigation. In hard rock it has been used to decipher lateral variations in weathered zone aquifer character and delineate shallow fractured zones immediately underlying the weathered zone. Also, it is used to monitor the changes in water content in the vadose zone and quality in terms of electrical conductivity.

Though the cost of conventional resistivity equipment is low, the field operation is time consuming. Also, in many areas, particularly in urban agglomerates where straight long open land is not available, the resistivity sounding or imaging cannot be conducted. It is replaced by time-domain electromagnetic (TEM) sounding, with 25 to 50 m side square transmitter loop, which can be conducted in small parks. In this the cost of equipment is higher than that for resistivity equipment. In TEM sounding, the field expenditure is negligible, operation is quite fast and the ratio of depth of investigation to surface spread is reasonably high and is generally more than 1. A disadvantage of TEM is that it can detect the conductive targets better than resistive targets.

Besides, the self potential (SP) method is used to measure the natural potentials of electrochemical and electrokinetic origin. The subsurface groundwater flow either natural or induced by pumping well or artificial recharge and the structures obstructing the flow or facilitating it generate electrokinetic potential. Its measurement can help decipher the direction of groundwater movement. Once a borehole is drilled and a fractured zone is encountered its orientation can be inferred by the mise-a-la-masse method involving placing of an active current source within the borehole against the fractured zone.

The gravity and magnetic methods have also been used in groundwater exploration. The gravity method involves measurement of minor variations in gravitational attraction at a point caused by variations in near surface rock densities and other factors. It is used to delineate the valley fills, palaeo-channels, fault structures, cavities and the depressions in bed rock topography which can act as groundwater repositories. Being a potential field measurement it has inherent ambiguities and may require support from other passive and/or active methods for confirmation. The gravity surveys are mainly used as reconnaissance except the detailed micro-gravity measurements for cavity detection (Hakim et al., 2004) and time-lapse microgravity measurements for monitoring changes in groundwater saturation due to pumping out water or artificial recharge. Murty and Raghavan (2002) conduct gravity surveys over the crystallines of Hyderabad, India and delineate areas of deep weathering holding high yielding wells. Some of the classical gravity studies are by West and Sumner (1972), Ibrahim and Hinze (1972), Carmichael and Henry (1977), Stewart (1980), Hansen (1984), Ghatge and Hall (1989) and Allen and Michel (1996). The magnetic method involves study of lateral changes in magnetic field caused by variation in magnetization due to differing magnetic properties of rock. The magnetic method is used for reconnaissance. Relevant to groundwater exploration in hard rocks, the faults and basic dyke intrusives associated with prominent magnetic signatures and the bed rock topography can be traced (Birch, 1984; Chandra et al., 1994; Kumar et al., 2006; Kayode et al., 2010).

In the seismic method, the propagation of a seismic wave through the earth is studied. The method has developed greatly because of its application in oil exploration. Though the method yields most precise information on lithological interfaces, its application in groundwater is not as common as the resistivity method. It is expensive and time consuming warranting much better planning and specific skill in processing and interpretation. The methods of seismic refraction as well as reflection are used in hard rock groundwater exploration to map the interface between the weathered zone and underlying fractured and compact rock, the lateral variations in material properties and to distinguish the lithology (Lennox and Carlson, 1967; Haeni, 1988; Steeples and Miller, 1990; Robinson et al., 2008). The passive seismic method which uses only ambient noise-field in place of the one created purposely by explosive or other sources can also be deployed effectively in estimating the depth to bedrock (Lane et al., 2008). Ground penetrating radar (GPR) has been used in near-surface investigations for the last 3 decades. It uses electromagnetic waves in the frequency range of MHz to GHz for different depth of investigations and resolutions. Its use in detecting near surface fractured zones and their orientation has been established (Seol et al., 2001, da Silva et al., 2004, Theune et al., 2006). GPR has been used in boreholes also for fracture determination (Hansen and Lane, 1995) and cross-well solute transport through fractures (Day-Lewis et al., 2003). Besides, the magnetic resonance sounding (MRS) that measures the nuclear magnetic resonance signal generated by water molecules present in the subsurface can be conducted to identify aquifers, their water content and hydraulic conductivity within about 100 m depth (Baltassat et al., 2006; Legchenko et al., 1998, 2004 and 2006; Roy and Lubczynski, 2003 and Vouillamoz et al., 2005).

3.3.3 Borehole geophysics

Geophysical logging in the domain of subsurface geophysical methods or borehole geophysics is conducted in boreholes for in situ measurements of physical properties of the geological formations encountered. Geophysical logging for water wells mainly comprises parameters like electrical resistivity, natural gamma radioactivity, borehole diameter variations, borehole fluid conductivity-temperature and groundwater flow. For regular hard rock water well drilling geophysical logging is generally not conducted. It is essentially required for establishing hydraulic properties of individual fractures, spatial connectivity and contaminant movements. Also, in the hydrofracturing process geophysical logging is conducted at pre- and post-hydrofracturing stages. In subsurface characterization required for groundwater modeling and for other purposes like geothermal reservoir definition and radioactive waste dumps, various types of logging and cross-hole imaging including the above are conducted (Keys, 1979, Williams et al., 2002). Also, the surface to borehole and cross-borehole high resolution resistivity, seismic and electromagnetic imaging including borehole radar reflection methods (Robinson et al., 2008) are conducted for detailed studies on fracture characterization and contaminant movement.

3.4 INTEGRATION OF METHODS

At many sites it has been observed that the inferences drawn from geophysical interpretations are not favourable while the local hydrogeological and geomorphological

Table 3.1 Geophysical methods used for groundwater exploration in hard rock.

Method	Major utility in ground water investigations in hard rock	Limitations
Electrical Resistivity	Weathered zone aquifer thickness, bed rock topography, fractured zone demarcation, water quality	Highly conductive and resistive overburden, poor resolution of thin targets due to transition and suppression, non-uniqueness due to equivalence, large surface space for deeper investigation
Frequency domain controlled source electromagnetic	Fractured zone demarcation	Highly conductive overburden, thin targets, equivalence,
VLF electromagnetic	Fractured zone, top layer resistivity	Highly conductive overburden, orientation of fractured zone
Time domain electromagnetic	Conductive layers	Difficult to discriminate resistivity of resistive layers
Airborne TEM	ditto	Cultural noise and high overburden conductivity
Seismic Refraction	Weathered zone thickness, fractured zone and their orientation, dry and saturated fractures	Low velocity zone
Seismic Reflection	ditto	Energy source
Mise-a-la-Masse	Orientation of saturated shallow fractured zone encountered in borehole	Conductivity contrast and depth
Self Potential	Direction of groundwater flow, seepage	Influence of other potentials
Magnetic	Delineation of basic dykes, faults, lithological contacts, bed rock topography	Influence of other magnetic bodies, magnetic contrast, non-uniqueness
Aeromagnetic	ditto	ditto
Gravity	Valley fills, palaeo-channels, cavities	Density contrast, non-uniqueness and a variety of corrections
GPR	Near surface fractures	Saturation and conductivity
Borehole logging	In situ measurements for lithology, quality of formation water, borehole diameter, borehole mud density fluid salinity and turbidity, verticality, radioactivity, flow, temperature, fractured zone-water content, porosity, borehole wall condition	Borehole diameter, borehole mud fluid density and salinity, wall condition, verticality, metal casing etc.

conditions are favourable. Just the opposite has also been observed. The experience is that geophysical favourables may be ambiguous but the geophysical unfavourables are seldom ambiguous. These qualitative aspects and uncertainties of non-uniqueness can be reduced to a large extent if geophysical surveys are carried out through a combination of two or more methods and integrated with geomorphological, hydrogeological and geochemical information. It demands a thorough knowledge of local hydrogeological conditions and associated geophysical responses.

Besides, it essentially requires calibrated test-geophysical measurements under field-conditions and scales and area-specific standardization of different geophysical responses and interpreted parameters either through existing borehole data or through test boreholes drilled. Further, the inflow of lithological, hydrogeological and geophysical log information obtained from subsequent boreholes drilled is effectively used as feedback to refine the interpretations and understand the constraints. It is most essential for getting success in geophysical surveys in hard rocks. Table 3.1 summarizes the application of various geophysical methods in groundwater investigations in hard rocks.

REFERENCES

Ahmed, S. (2014) A new chapter in groundwater geophysics in India: 3D aquifer mapping through heliborne transient electromagnetic investigations. *Journal Geological Society of India*, 84 (4), 501–503.

Allen, D.M. & Michel, F.A. (1996) The successful use of microgravity profiling to delineate faults in buried bedrock valleys. *Ground Water*, 34 (6), 1132–1140.

Astier, J.L. & Paterson, N.R. (1987) Hydrogeological interest of aeromagnetic maps in crystalline and metamorphic areas. In: G.D. Garland (ed.) *Proceedings of Exploration '87: Third Decennial Int. Conf. on Geophysical and Geochemical Exploration for Minerals and Groundwater, 1987, Ontario*. Ontario Geol. Survey, Special Volume 3, pp. 732–745.

Baltassat, J.M., Krishnamurthy, N.S., Girard, J.F., Datta, S., Dewandel, B., Chandra, S., Descloitres, M., Legchenko, A., Robain, H., Ananda Rao, V., & Ahmed, S. (2006) *Proton magnetic resonance technique in weathered-fractured aquifers*. NGRI-BRGM Report, IFCPAR Project 2700-W1, BRGM 2003-ARN-11.

Birch, F.S. (1984) Bedrock depth estimates from ground magnetometer profiles. *Ground Water*, 22 (4), 427–432.

Bromley, J., Mannsrom, B., Nisca, D. & Jamtlid, A. (1994) Airborne geophysics: Application to a ground-water study in Botswana. *Ground Water*, 32 (1), 79–90.

Butler, D.K. (2005) What is near-surface geophysics? In: Butler, D.K. (ed.) *Near-Surface Geophysics, Part 1, Chapter 1*, SEG Investigations in Geophysics Series: No. 13, Tulsa, Oklahoma, USA, Society of Exploration Geophysicists. pp. 1–6.

Carmichael, R.S. & Henry Jr, G. (1977) Gravity exploration for groundwater and bedrock topography in glaciated areas. *Geophysics*, 42 (4), 850–859.

Chandra, P.C. (1991) Study of aeromagnetic map for groundwater exploration: Recent trend in hardrock hydrogeophysics. *Bhujal News, Quarterly Journal of CGWB*, 6 (2), 37.

Chandra, P.C. (2002) Surface geophysical surveys in groundwater exploration: source finding. *Bhujal News, Quarterly Journal of CGWB*, 17 (1 & 2), 35–43.

Chandra, P.C., Reddy, P.H.P. & Singh, S.C. (1994) *Geophysical studies for groundwater exploration in Kasai and Subarnarekha River basins* UNDP Project Central Ground Water Board Min. of Water Resources, Govt. of India Tech. Report

Cresswell, R. & Gibson, D. (2004) *Application of airborne geophysical techniques to groundwater resource issues in the Angas-Bremer Plains, South Australia: A synthesis of research carried out under the south Australia salinity mapping and management support project (SA SMMSP)*. Dept. of Water, Land & Biodiversity Conservation Report DWLBC 2004/35, ISBN 0-9756945-2-9.

da Silva, C.C.N., de Medeiros, W.E., de Sa, E.F.J. & Neto, P.X. (2004) Resistivity and ground-penetrating radar images of fractures in a crystalline aquifer: a case study in Caicara farm-NE Brazil. *Journal of Applied Geophysics*, 56 (4), 295–307.

Day-Lewis, F.D., Lane, J.W., Harris, J.M. & Gorelick, S.M. (2003) Time-lapse imaging of salie-tracer transport in fractured rock using difference-attenuation radar tomography. *Water Resources Research*, 39 (10), 1290–1303.

d'Ozouville, N., Auken, E., Sorensen, K., Violette, S., de Marsily, G., Deffontaines, B. & Merlen, G. (2008) Extensive perched aquifer and structural implications revealed by 3D resistivity mapping in a Galapagos volcano. *Earth and Planetary Science Letters*, 269 (3–4), 518–522.

Ghatge, S.L. & Hall, D.W. (1989) *Geophysical investigations to determine bedrock topography in the East Hanover-Morristown area, Morris County, New Jersey.* New Jersey Geological Survey, Geological Survey Report No. 17.

Haeni, F.P. (1988) *Application of seismic-refraction techniques to hydrologic studies.* U.S. Geological Survey, Techniques of Water Resources Investigations, Book 2, Chapter D2, U.S.G.S.

Hakim, A, Singh, N., Hasan M & Chandra, PC (2004) *A note on micro-gravity survey for detection of voids in limestone in Karwi area, Chitrakoot district, U.P. (INDIA)* Geological Survey of India, Lucknow, Technical Report

Hansen, B.P. & Lane, J.W. (1995) *Use of surface and borehole geophysical surveys to determine fracture orientation and other site characteristics in crystalline bedrock terrain, Millville and Uxbridge, Massachusetts.* U.S. Geological Survey Water-Resources Investigations Report 95-4121.

Hansen, D.S. (1984) Gravity delineation of a buried valley in quartzite. *Ground Water*, 22 (6), 773–779.

Ibrahim, A. & Hinze, W.J. (1972) Mapping buried topography with gravity. *Ground Water*, 10 (3), 18–23.

Jorgensen, F., Scheer, W., Thomsen, S., Sonnenberg, T.O., Hinsby, K., Wiederhold, H., Schamper, C., Burschil, T., Roth, B., Kirsch, R. & Auken, E. (2012) Transboundary geophysical mapping of geological elements and salinity distribution critical for the assessment of future sea water intrusion in response to sea level rise. *Hydrology & Earth System Sciences*, 16 (7), 1845–1862.

Joseph, A.D. & Chandra, P.C. (2003) Integrated interpretation of satellite imagery data and aeromagnetic map for groundwater exploration in hardrocks: case studies. *Bhujal News, Quarterly Journal of CGWB*, 18 (1–4), 66–69.

Kayode, J.S., Nyabese P. & Adelusi, A.O. (2010) Ground magnetic study of Ilesa east, Southwestern Nigeria. *African Journal of Environmental Science & Technology*, 4 (3), 122–131.

Keys, W.S. (1979) Borehole geophysics in igneous and metamorphic rocks, Paper presented at the SPWLA Twentieth Annual Logging Symposium, Tulsa, Oklahoma, USA, June 6, 1979.

Kirsch, R. (ed.) (2009) *Groundwater Geophysics: A Tool for Hydrogeology.* 2nd edition, Berlin Heidelberg, Springer-Verlag Publication.

Kumar, D., Murthy, N.S.K., Nayak, G.K. & Ahmed, S. (2006) Utility of magnetic data in delineation of groundwater potential zones in hard rock terrain. *Current Science*, 91 (11), 1456–1458.

Lane Jr, J.W., White, E.A., Steele, G.V. & Cannia, J.C. (2008) Estimation of bedrock depth using the horizontal-to-vertical (H/V) ambient-noise seismic method. Presented at: *SAGEEP 2008: Symposium on the Application of Geophysics to Engineering and Environmental Problems, April 6–10 2008, Philadelphia, Pennsylvania*, Denver, Colorado, Environmental & Engineering Geophysical Society.

Legchenko, A.V., Baltassat, J.M., Beauce, A., Makki, M.A. & Al-Gaydi, B.A. (1998) Application of the surface proton magnetic resonance method for the detection of fractured granite aquifers. *Proceedings of the IV Meeting of Environmental & Engineering Geophysical Society (EEGS), Barcelona*, pp. 163–166 (abstract).

Legchenko, A., Baltassat, J.-M., Bobachev, A., Martin, C., Robain, H. & Vouillamoz, J.-M. (2004) Magnetic resonance sounding applied to aquifer characterization. *Ground Water*, 42 (3), 363–373.

Legchenko, A., Descloitres, M., Bost, A., Ruiz, L., Reddy, M., Girard, J.-F., Sekhar, M., Mohan Kumar, M.S. & Braun, J.-J. (2006) Resolution of MRS applied to the characterization of hard-rock aquifers. *Ground Water*, 44 (4), 547–554.

Lennox, D.H., & Carlson, V. (1967) Geophysical exploration for buried valleys in an area north of Two Hills, Alberta. *Geophysics*, 32 (2), 331–362.

Murty, B.V.S. & Raghavan, V.K. (2002) The gravity method in groundwater exploration in crystalline rocks: a study in the peninsular granitic region of Hyderabad, India. *Hydrogeology Journal*, 10 (2), 307–321.

Paine, J.G. & Collins, E.W. (2003) Applying airborne electromagnetic induction in groundwater salinization and resource studies, West Texas. *SAGEEP 2003: Proceedings of the Symposium on the Application of Geophysics to Engineering & Environmental Problems, 6–10 April, 2003, San Antonio, Texas, USA*, Environmental & Engineering Geophysical Society, pp. 722–738.

Palacky, G.J., Ritsema, I.L. & De Jong, S.J. (1981) Electromagnetic prospecting for groundwater in Precambrian terrains in the Republic of Upper Volta. *Geophysical Prospecting*, 29 (6), 932–955.

Paterson, N.R. & Bosschart, R.A. (1987) Airborne geophysical exploration for groundwater. *Ground Water*, 25 (1), 41–50.

Paterson, N.R. & Ronka, V. (1971) Five years of surveying with the very low frequency electromagnetic method. *Geoexploration*, 9 (1), 7–26.

Ranganai, R.T. & Ebinger, C.J. (2007) Aeromagnetic and Landsat TM structural interpretation for identifying regional groundwater exploration targets, south-central Zimbabwe Craton. *Journal of Applied Geophysics*, 65 (2), 73–83.

Robinson D.A., Binley, A., Crook, N., Day-Lewis, F.D., Ferré, T.P.A., Grauch, V.J.S., Knight, R., Knoll, M., Lakshmi, V., Miller, R., Nyquist, J., Pellerin, L., Singha, K. & Slater, L. (2008) Advancing process-based watershed hydrological research using near-surface geophysics: a vision for, and review of, electrical and magnetic geophysical methods. *Hydrological Processes*, 22 (18), 3604–3635.

Roy, J. & Lubczynski, M. (2003) The magnetic resonance sounding technique and its use for groundwater investigations. *Hydrogeology Journal*, 11 (4), 455–465.

Rubin, Y. & Hubbard, S.S. (eds.) (2005) *Hydrogeophysics*, Series: Water Science & Technology Library (Book 50), Springer Publication.

Seol, S.J., Kim, J.-H., Song, Y. & Chung, S.-H. (2001) Finding the strike direction of fractures using GPR. *Geophysical Prospecting*, 49 (3), 300–308.

Smith, B.D., Grauch, V.J.S., McCafferty, A.E., Smith, D.V., Rodriguez, B.R., Pool, D.R., D.-Pan, M. & Labson, V.F. (2007) Airborne electromagnetic and magnetic surveys for groundwater resources: A decade of study by the US Geological Survey. In: Milkereit, B. (ed.) *Exploration 07: Proceedings of the Fifth Decennial Int. Conf. on Mineral Exploration, Sept 9 to 12, 2007, Toronto, Canada*. pp. 895–899.

Steeples, D.W. & Miller, R.D. (1990) Seismic reflection methods applied to engineering, environmental and groundwater problems. In: Ward, S.H. (ed.) *Geotechnical and Environmental Geophysics, Vol. I: Review and Tutorial*. Tulsa, Oklahoma, USA, Society of Exploration Geophysicists, pp. 1–30.

Stewart, M.T. (1980) Gravity survey of a deep buried valley. *Ground Water*, 18 (1), 24–30.

Theune, U., Rokosh, D., Sacchi, M.D. & Schmitt, D.R. (2006) Mapping fractures with GPR: A case study from Turtle Mountain. *Geophysics*, 71 (5), B139–B150.

Viezzoli, A., Auken, E. & Munday, T. (2009) Spatially constrained inversion for quasi 3D modelling of airborne electromagnetic data-an application for environmental assessment in the Lower Murray Region of South Australia. *Exploration Geophysics*, 40 (2), 173–183.

Vouillamoz, J.M., Descloitres, M., Toe, G. & Legchenko, A. (2005) Characterization of crystalline basement aquifers with MRS: comparison with boreholes and pumping tests data in Burkina Faso. *Near Surface Geophysics*, 3 (3), 205–213.

Ward, S.H. (ed.) (1990) *Geotechnical and Environmental Geophysics, Vol. I, II and III*. Tulsa, Oklahoma, USA, Society of Exploration Geophysicists Publication.

West, R.E. & Sumner, J.S. (1972) Groundwater volumes from anomalous mass determination for alluvial basins. *Ground Water*, 10 (3), 24–32.

Williams, J.H., Lane, J.W., Singha, K. & Haeni, F.P. (2002) *Application of advanced geophysical logging methods in the characterization of fractured-sedimentary bedrock aquifer, Ventura County, California*. U.S.Geological Survey Water-Resources Investigations Report 00-4083.

Chapter 4

Planning of geophysical surveys

4.1 INTRODUCTION

The planning of a geophysical survey varies with objective, geology, terrain, location, area coverage, accessibility, cultural interference, type of survey, geophysical method, equipment, expertise and time and financial resource available. It is tailored to suit the specific requirements. There could be a variety of objectives for geophysical surveys like locating suitable sites for water well drilling, mapping of aquifers at regional and local scales, identifying fractured zones, defining areas and sites suitable for managed aquifer recharge, delineating groundwater salinity interfaces and contamination, locating sites for monitoring wells etc. and accordingly the planning changes. Incorporating details of planning for all types of geophysical surveys and for various objectives of groundwater exploration is not possible and also beyond the scope. Some relevant aspects of commonly used surface geophysical methods are discussed here.

The purpose of carrying out a geophysical survey for water well drilling is to assess and infer subsurface conditions as accurately as possible so that wells drilled on geophysical recommendations are successful and results are as per expectations. The most important factor to get success in hard rock well drilling either for water supply or for characterization and monitoring well construction and make the programme cost-effective is to allocate adequate time for the geophysical survey and complete it well in advance of drilling and that needs planning. The expenditure on geophysical survey, though only about 5 to 10% of the exploration programme, is of great value. Obviously it increases for a detailed high resolution site characterization programme. The essentiality and advantage of completing a geophysical survey prior to drilling is that, before the drilling rig is placed, survey results provide scope and options for selecting geophysically the most promising sites as per the objective or their locations commensurate with the water demand and supply points and also there is no waiting in point-to-point rig mobilization. The reconnaissance survey initiated much ahead of the exploration programme, is integrated with the hydrogeological survey, remote sensing and geomorphological studies including low altitude aeromagnetic data, if available. It is immediately followed by a detailed survey in selected areas and at sites and which also is completed well in advance of the drilling programme.

The process of geophysical activity commences with translating the hydrogeological objective into a reasonably achievable geophysical objective and after conducting the geophysical survey it ends with translating the interpreted and validated geophysical output into required hydrogeological information; rightly presented by Reeves *et al.* (1997) as taking the physics out of a geophysical map and delivering information that is essentially geological. At times it may not be possible to deliver the required output. For example, if the objective is to detect a thin aquifer layer at depth that cannot be achieved by geophysical survey under prevailing geological and/or anthropogenic conditions, this should be conveyed to the client before taking up the survey or it should be conveyed that the survey is being conducted on an experimental basis. A geophysical survey becomes easier in areas where the general subsurface hydrogeological conditions are known and is used to refine or complement the information. The information about groundwater conditions in bore wells known locally is quite useful in selecting areas or sites for geophysical surveys. Also, in some areas information from earlier surveys may be available which could be used to adopt the best approach. In case bore wells exist in the area a test measurement for standardization of parameters is to be conducted near a few bore well sites which have minimum cultural interference. It is an advantage for near-surface geophysical investigations.

4.2   MODELING GEOPHYSICAL RESPONSE

A significant step is to generate a synthetic geophysical response through forward modeling of model parameters defined through existing borehole data. It is a process of realizing the response for a given hydrogeological condition. This is followed by unconstrained inversion of the synthetic response, to see if the model parameters could be resolved and to what extent the synthetic response conforms to the generated actual test field data. Variations in model parameters in forward modeling for one or more geophysical methods and techniques give an understanding of the effectiveness, limitations and selection of techniques and their integration in achieving the objective or the viable simplified model which can be obtained. Forward modeling generates ideal data and therefore it is necessary that the exercise is carried out with various degrees of noise mixing. It helps standardization of the geophysical response, designing the survey parameters, e.g., optimum electrode spacing with respect to width of the body and ascertains the applicability and efficacy of the method for the output with desired resolution. Like 1D forward modeling mentioned above, 2D and 3D forward modeling can also be carried out, if required, for understanding the geological complexity, but generally this is not done in groundwater exploration.

4.3   TYPES OF SURVEY AND COVERAGE

The surveys can be classified into two categories, the one for 'large area' coverage for aquifer mapping and the other for selecting a suitable site, assessing suitability of a predefined site or characterizing a site in detail. For surface surveys the 'large area' can be a minimum 1 to 4 sq km. Where a 'large area' is available the approach to surveys can be classified as wide-spread and concentrated or detailed and conducted sequentially. While the former is over a 'large area', the latter is for a smaller area with high resolution coverage identified through the former. So, the geophysical survey for the objective can be taken up in two phases. It starts with reconnaissance which is rapid. It is conducted based on geological information at moderate station interval along wide spaced long profiles. Care is taken about selection of scale of measurement and sampling interval in reconnaissance vis-a-vis expected hydrogeological complexity to get the required level of basic details and possibly develop a conceptual understanding of the subsurface. It helps planning the survey parameters for follow-up close-grid detailing. The geophysical methods generally used in reconnaissance are well known. For example, the use of magnetic method in detecting concealed basic dykes and structures, VLF-EM for shallow fractured zones and that of gravity method in detecting a buried valley. They can be employed further in detailing if found useful and necessary by systematically reducing the scale of measurement. Also, the results of the reconnaissance survey may indicate the appropriate geophysical method, combination or alternative methods and the survey design parameters for these alternative methods. Even viability of the whole geophysical programme can be evaluated through a reconnaissance survey. Contrary to it, a geophysical survey for predefined water demand point is generally conducted near habitation and may not provide adequate scope for reconnaissance. Even at times such spot or localized survey may not yield suitable locations and drilling of well is compelled on the local best anomaly which may be a mediocre geophysical anomaly only.

The preparation of a weathered zone aquifer map can be attempted either through reconnaissance or detailing depending on the required scale of information. For such maps surveys can be conducted either on a grid or along a transect for systematic coverage or based on geomorphological conditions. The grid size for geophysical surveys may vary from 50 m × 50 m to 500 m × 500 m or more. The information becomes more of a generalized nature as the size of grid increases, and ultimately represents a random sampling only. On the other extreme, hard rock hydrogeological heterogeneities could be such that even information obtained at a practically feasible close-grid may not suffice selection of a site by extrapolation or interpolation. Except for the selection of favourable area, the investigations are to be made site-specific for such heterogeneous formations. Preparation of a fractured zone map requiring a multiple-method detailed survey is quite tedious, more so, if the area is large. It is time-consuming since narrow-spaced long profiles are to be covered by the methods at short station intervals.

4.4 SELECTION OF METHOD AND EQUIPMENT

The geophysical objective commands the selection of geophysical method or combination of methods and techniques and also the equipment. Besides, the selection of method and equipment depends on surface conditions and availability of land area. For example, in areas with very high resistivity moderately thick sand at the top or compact rock exposures, a conventional electrical resistivity method may not work and the electromagnetic induction method has to be opted. Also in urban areas where only parks are available for surveys or conducting a survey with a large surface spread is not possible, the modified electrode configuration for electrical resistivity or small loop size time domain electromagnetic sounding may have to be employed. Simple resistivity equipment may be adequate for areas with bed rock depth within 100 m, but not for the coastal environment where conductive clays and sand saturated with saline water prevails at depths. The survey methods and equipment are requisitioned on the basis of hydrogeological and geophysical objectives, top layer condition, space available, type of survey-reconnaissance or detail, and the accuracy required. Availability of proper equipment and funding controls the whole geophysical programme including time-schedule and quality of deliverables.

4.5 PLANNING FOR FIELD SURVEY

Hoekstra and Hoekstra (1990) present an outline of general planning and execution of a geophysical survey for groundwater investigation and hazardous waste site evaluation. Arrangement for a geophysical survey for groundwater using only one or two geophysical methods and techniques is not a big one and customized according to the needs. However, it needs a field set-up. The geophysical survey team is led by a team leader or party chief who is the most experienced and communicative person for the work. This person has wide experience relevant to the objective and expected to have expertise for all the methods to be used but not essential. There could be separate team leaders for different methods with a coordinator. The team includes one or more geophysicists/technicians, surveyor and the crew. The size of the team varies with survey type, geophysical method(s) employed, time-duration available and whether the surveys can be conducted sequentially or with two non-interfering methods conducted simultaneously.

The team leader acquires a thorough understanding of the objective of survey, studies the feasibility and scope of work and guides the team geophysicists/technicians. He fixes the responsibilities of the team members. The geophysicist taking up investigation by a specialized technique is expected to be familiar with relevant field processes and requirements. Prior to taking up the geophysical survey, hydrogeological and geophysical objectives are made clear to the geophysical team. Overall, the responsibility of the team leader starts from the initial information collection at planning stage to logistics (field stay, transportation and expenses), data acquisition in field, quality control and finally to presentation stage. Specifically, the team leader covers the pre-, during and post-field processes comprising the following:

- compilation and analysis of existing data and maps,
- collection of local hydrogeological information,
- conceptualization of objectives,
- identification of methods and sequence in approach,
- selection and arrangement of equipment,
- evaluation of accessibility to the area,
- arrangement of proper logistics,
- dialogue with local administration and the people, informing them about the objective of the survey and possible disturbances,
- designing survey parameters for different methods, selection of survey-lines and spots for measurements on priority,
- management of field activities,
- guiding the survey execution,

- day to day planning, monitoring and modification of survey and acquisition parameters,
- optimization of surveys based on progressive accumulation of data and evaluation of findings,
- controlling the quality of data acquisition and the output,
- sorting out the constraints,
- timely completion of surveys,
- ensuring required precautions and safety with accountability including health aspect,
- reviewing the preliminary interpretations done by technicians or geophysicists and
- preparation of report.

The data processing and interpretation require experience. It can be done by the team leader. The drilling feedback from the first few boreholes drilled on geophysical recommendations is also concurrently added to this process so that geophysical results are validated, cause of discrepancies analysed and modification/reinterpretation is incorporated wherever required for the upcoming well drillings. This entire process is to be done jointly by the hydrogeologist and geophysicist. Any shortcoming at any stage in this phased endeavour may become a constraint in getting the optimum output cost-effectively.

4.5.1 General considerations for equipment

It is necessary to have reliable equipment and a consistent approach by different operators. With modernization of instrumentation, the data acquisition process is also changing. The geophysicists are supposed to know pre-operation checking, process of data acquisition and what storage system of equipment is being used. For this, equipment manuals are made available in field. The field team is responsible for field level maintenance of equipment as indicated in the manual. The safety precautions given in the manual are to be observed. The geophysical equipment is quite expensive. Repairing is also expensive and may not be possible immediately. Adequate spare parts are carried to field for minor field level repairs. The accessories are checked regularly. Any deterioration in equipment condition is rectified immediately. All components of the equipment are test checked daily for their proper functioning. Batteries are charged regularly. Maximum care is taken in transportation, handling and safe storage in a stable dust free, clean place. The transportation vehicle should be suitable and have adequate space to accommodate everything. Also, a good communication system is required during field operation.

4.5.2 Access to the area

The restrictions to accessing the land area/site are to be clarified from local administration. Permission from the land owner to work in his land is taken as the survey may cause damage to standing crops and other structures. There can be a provision for paying compensation. Physical constraints like bushes along the profile line are cleared for easy movement and distant visibility along the profiles.

4.5.3   Area details and compilation of data

Maximum possible information is to be collected on local geological and hydrogeological conditions, rock exposures, surface water bodies and anthropogenic interferences prior to moving to the area. Digital topographic map, hydrogeological map and other related maps of the area at 1:50,000 or larger scale, lineament details, existing geophysical survey results, aeromagnetic data (if available), lithological and electrical logs of boreholes and relevant literature are to be collected. Wells located in the area are methodically inventoried to assess the depth of water level and quality of water through chemical analysis. These data are compiled and analyzed. Compilation and analysis are time-consuming, but essentially deserve sincere efforts, since these might give insight into the positive and negative aspects of groundwater conditions in the area, help define the scope and feasibility, survey parameters and make the programme cost-effective.

Depending on the objective and required quantum of surveys, a tentative plan of field measurements may be drawn in the office followed by finalization at field camp after studying the geomorphic features and geological structures in field. The problem/objective can also be discussed with local representatives to plan the surveys and gain their confidence and involvement. Before initiating the survey it is necessary that man-made or cultural noise if any, e.g., iron fencing, power lines, transformers, railway track and particularly the network of exposed or underground metallic pipe lines and cables etc., which can affect the measurements are noted or mapped by the field crew. It helps plan the field operation and lay profiles with minimum interference. At places it may be necessary to shut down the power supply during field operations.

4.5.4   Survey parameter design

The survey parameter is designed in such a way that it helps detect and resolve the target (if present) defined in the programme objectives cost-effectively (Eso *et al.*, 2006). Hard rock areas having much heterogeneity require high density data acquisition for the resolution and accordingly the survey parameters are designed and procedure followed. The profile lines as well as sampling or station interval are kept much closer in hard rock. Also, the orientation of traverse lines is decided according to the strike of geological formation or that of lineament and fault, dyke etc. which may not be necessary for thick alluvial areas. The traverse lines are laid perpendicular/across the lineament or fractured zone. A reconnaissance survey can be conducted at moderate station intervals and wide line spacing with normal acquisition parameters. The station interval in magnetic surveys can be kept closer than that for gravity survey as indicated in the Geological Survey of India document on 'Standard Operational Procedure (SOP) for Geophysical Mapping and Mineral Exploration' (www.portal.gsi.gov.in). The magnetic measurements are made over long profile lines to assess the regional picture.

For conducting resistivity sounding in sedimentary areas, generally the measurement interval is kept in such a way that 6 to 8 almost equi-spaced points are obtained per log cycle of double log graph, while in hard rock measurement the interval is kept much closer to pick up kinks in the sounding curve for interpreting them empirically in terms of fractures. In resistivity imaging (ERT) selection of inter-electrode spacing and configuration are crucial. Detailing can be done by 3-d imaging. In case it is difficult to conduct 3d imaging parallel profiles of 2d imaging can be run across the lineaments. For frequency domain EM profiling minimum two frequencies are to be used. TEM soundings are to be carried out with dual moment arrangement. Since time and cost of survey increase with dense sampling, an optimum density is maintained with minimum data redundancy. The survey parameters may have to be modified during the surveys as an attempt to meet the objective, if initial parameters do not yield much information. Forward modeling can also be used for designing the survey parameters. These are discussed in detail in the concerned chapters.

4.5.5 Surveying work for profile layout

It is necessary to layout straight profile lines of adequate length at defined spacing primarily as per the geological strike requirement. For profiling the lines are laid in such a way that they are perpendicular to the strike but cause minimum disturbance to the standing crops and habitation. Also, the lines are laid in such a way that the high voltage transmission lines, buried cables and pipelines, water bodies, topographic depressions, ditches, rock exposures are avoided or kept at a minimum safe/non-interfering distance. Lines for sounding spread are kept along the strike. Lines are laid precisely in different directions for azimuthal sounding and a square array, required for determining the fractured zone orientation and in perpendicular directions for the square loop of TEM. In seismic refraction surveys the straight line profiles are laid over a distance more than 3 times the expected depth to the bed rock. Surveying work can be carried out by a surveyor using the conventional method of tape and compass measure or using GPS. While laying out several profile lines tie-lines are fixed. In undulating terrain, elevation of measurement points can be recorded using DGPS in a shorter time. The profiles and spot-measurement locations are accurately transferred either on digital maps or conventional toposheets and fixed on the ground by some visible marker and referred to some nearby land marks adequately so that drilling points identified on the basis of interpretations are exactly pinpointed later on the ground. Confusion in location may warrant reinvestigation which could be cost-prohibitive and a mistake may cause well failure. In case data reacquisition is necessary it is done by the same equipment (Yarie Quentin, 2003). To locate the points same GPS is used. The time gap between survey and pinpointing of drilling site is kept at minimum to avoid any slip-up which may crop up due to a variety of activities including construction work or growth of vegetation in the surveyed area.

4.6 GEOPHYSICAL TEAM SIZE AND RESPONSIBILITIES

The requirement of personnel for geophysical surveys varies with the methods. In addition to geophysicist(s), technicians/technical assistants are required. For example total magnetic field measurement by proton precession magnetometer will require only one technician, while electrical and electromagnetic measurements may require two and seismic survey may require more than three. The geophysicists/technicians will take observations; ensure proper alignment of layout, position of electrodes/energy source/receivers/geophone and recording and scrutiny of the data.

4.7   SURVEY COST AND TIME

The cost of a surface geophysical survey programme depends on a number of factors like location and size of area, type of survey, single method or multiple method, survey parameter i.e., line spacing and station interval, type of equipment used, size of the geophysical team, time-schedule, time of survey-winter or summer, data acquisition, processing and interpretation which varies with method, transportation and the form of output required. While planning an approximate quantum of work and the time required are also evaluated. The time schedule includes time required in 'preparation' and 'execution' of project up to 'presentation' of report. The compilation of existing data, formulation of project, contractual agreement, procurement of equipment, permissions and mobilization etc., come under 'preparation' and it takes a considerable time with unforeseen developments and actions. Time taken in 'execution' has to consider weather, interferences, objective and deployment of field parties. A margin time with flexibility in funding is kept for execution of supplementary work involving additional geophysical methods. While estimating the tentative cost of geophysical exploration, provision is made for all these against the time available. Accordingly, excluding the equipment and accessories the survey cost including acquisition, processing and interpretation is minimum for the ground magnetic survey and maximum for the seismic. Some approximate costs and survey time are given in Table 4.1.

4.8   SAFETY AND PRECAUTIONS IN FIELD OPERATIONS

Realistic and cautious planning of field survey especially for inclement weather, remote areas and hostile terrain is essential to acquire meaningful data for the objective envisaged. Common potential hazards include working under high voltage power lines and during lightning, during extremely harsh weather, in extremely remote areas and with high energy sources of electricity and explosives. The surveys are planned for dry weather period, avoiding the rainy season, to set aside current leakage possibility and damage by lightning and extreme summers to safe guard the instrument against high temperature. The operation is discontinued during the rains. In case of expected rains and lightening the current and potential cable connections are removed from the instrument and no one is allowed to touch the terminals. Lightening even at 5 to 6 km distance can cause development of a very high potential and damage the circuit and may cause electrocution. Besides adopting safe operational procedure, hazards are to be anticipated and acted upon to prevent them. It is always obligatory to be prepared in advance for hazards likely to be stumbled upon during transportation and field operation. The vehicle(s) used for transportation has to be kept always close to the area of operation and should not be used for any other purpose during field operation as it may be required in unforeseen situations or emergency. A document on safety code is to be made available for equipment and personnel in operation.

Before starting the operation, the field crew is briefed about the successive operational movements and instruction-words and mode of instruction e.g. verbal, through walkie-talkie, mobile set, different colour flags or hand movement which of course depends on visibility. For distances more than 200 m communication through walkie-talkie/mobile set is desired. Communication with crew is to be clear and unambiguous.

Table 4.1 Approximate expenditure involved in various types of surveys. The costs are subjective and vary with terrain. The processing, interpretation and report writing costs are to be included (modified from Robinson *et al.*, 2008).

Type of survey	Field personnel*	Survey output (per day)**	Approx. cost of equipment	Tentative cost of field survey (per line-km)**
Heliborne TEM with MAG		200 line-km	_____	$150–200
TEM	1 geophysicist 1 tech. assistant	10–15 soundings (in nearby area)	$60–80k	$100–300 per sounding
FEM profiling	1 geophysicist 1 tech. assistant	5 line-km	$60–80k	$100–300
VLF profiling	1 geophysicist 1 tech. assistant	10 line-km	$20k	$50
Terrain Conductivity Mapping	1 geophysicist 1 tech. assistant	10 line-km	$20k	$50
Magnetic Profiling	1 geophysicist 1 tech. assistant	10 line-km	$20k	$50
Electrical Resistivity Imaging	1 geophysicist 2 tech. assistant	2 line-km	$60–100k	$2000–4000
VES	1 geophysicist 1 tech. assistant	2 (AB: 2000 m)	$15–20 k	$300 per VES
Gradient Resistivity Profiling	1 geophysicist 1 tech. assistant	1 line km (with AB: 1000 m)	$15–20k	$600
Self Potential Profiling	1 geophysicist 1 tech. assistant	5 line-km	$15–20 k	$50
Mise-a-la Masse Profiling	1 geophysicist 1 tech. assistant	2 line-km	$15–20 k	$50
Passive Seismic	1 geophysicist 1 tech. assistant	Continuous measurement on grid system	$40k	$50
High Resolution Seismic Profiling	1 geophysicist 3 tech. assistant	1 line-km	$100k	$10k
GPR Profiling	1 geophysicist 1 tech. assistant	5 line-km	$20-30k	$200-400
Well Logging	1 geophysicist 1 tech. assistant	2 (Multi-parametric maximum depth 300 m)	$100k	$ 500–1000 per borehole

*Surveying and field crew additional. Geophysical survey team comprises minimum 1 geophysicist and 1 technical assistant. ** The output and cost varies with transportation, topographic conditions and clearances etc.

Before pressing for current or for shooting, the operator has to ensure that crew has clearly given the 'ready' signal. Measurements are to be made judiciously fast.

Unnecessary use of high voltage input should be avoided. For example, one should be careful while using more than 100 volts or more than 120 milli ampere current. It is made a practice to check the line continuity and readiness irrespective of the voltage range applied. The crew are not supposed to touch the electrodes or the cable till further instruction from the operator. The crew has to restrict animal/human movement near the profile and avoid passage of the cable through water or near a high voltage power line. Also, they are supposed to insulate leakage in the cable and should not stand bare foot on wet ground near the electrode.

In frequency domain EM equipment with a number of frequency selections, frequencies are changed only after switching-off the instrument. In magnetic surveys ferrous objects are kept away from the sensor. In gravity surveys the ground elevation is accurately determined by surveying. Plotting or viewing of data is completed at the site itself, so that data errors are removed by repeating the measurements. The point of observation is recorded by GPS with adequate details and the location is unambiguously marked on the ground by a stone/pillar etc., so that it can be easily identified if required for drilling a borehole.

Seismic surveys being extensively done for oil exploration, several manuals exist on safe survey procedures. The manual on 'Seismic Survey Safety Procedure' prepared by OH & S Management Team of Hydro Tasmania (Albertini, 2010) and the Bulletin on 'Safe Operating Procedures for Seismic Drilling' by Govt. of Alberta (2010) give the details of safety procedure for seismic surveys. As such, the team leader of the seismic survey is responsible for overall safety and ensures that the team strictly follows the safety procedures. He defines the duration of explosive blasting activities, notifies the general public, and puts danger/warning public notice at blasting location and its duration. Explosives required for the survey are procured through proper permit and stored safely and transported by explosive vehicle. Explosives are handled by trained shooters and blasting is done as per permit for the jurisdiction of the area and the rules and regulations. There should not be any overhead power line near the shot hole. At the time of shooting there should not be any movement on the profile line. The persons involved are to use personal protective equipment. The effects of vibrations from traffic movement and wind are to be avoided. The shot hole is damped by water. The charge (explosive) is not kept in the top soil/weathered layer (low velocity) due to dissipation of energy. Detonators are always kept short circuited or shunted (even while carrying). For shallow investigations depth of weathering is estimated by special shooting so that the charge can be placed below the weathered zone. Vibroseis or weight dropping is preferably used.

4.9   QUALITY CONTROL

The quality control is necessary and that of data acquisition is important. It affects the interpretation and final output and hence the decision. Quality control is essential to minimize the uncertainty or increase the reliability and attain the objective confidently and cost-effectively. The basic objective of quality control is that whatever data is collected it must actually be the true response for the subsurface condition and not the misleading one due to any extraneous effects. US EPA (2000) guideline relates data quality and data adequacy with its proposed use. Crumbling *et al.* (2001) indicate dependency of data quality on integrity at every step in the sequence of activities or measurements and stress a careful site-specific consideration for each step. For example, accuracy in surveying is as important as that in data acquisition. The quality and density of data required for a reconnaissance geophysical survey will not be adequate for detailing. In geophysical surveys quality control encompasses the entire sequence,

starting from equipment, field procedures, data acquisition and density to data interpretation, output and up to the preparation of the final report. It is done by the team
leader with the support of team members in the field itself. Because once you come
out of the field area it is difficult to generate exactly same set of readings at the same
point. Recognizing the need of enhancing the data density at a later date and attempt to
fill it becomes impractical or cost-prohibitive as data acquisition is already the costli-
est component of the field programme. Ultimately the inference is compromised. So,
at data acquisition level the team leader has to continuously monitor, evaluate and
ensure the quality as well as density of data required for the objective. For interpretation 'noisy' data are removed judiciously. 'Noisy' data is defined on the basis of
experience, non-repeatability or non-occurrence in the surrounding.

A typical example on quality control could be of conducting a large spread Schlumberger resistivity sounding by a low sensitivity resistivity meter in a coastal environment
holding thick sediments. Because of the low sensitivity, to get measurable potential,
the potential electrode separation is increased quite frequently and a curve is obtained
with several jumped segments. A smoothened curve is only an approximation leading to interpretational uncertainty. Added to this there is the possibility of error in
electrode distance, straightness of profile, current leakages, profile orientation and the
actual number of quality soundings may be limited in an area and taking decisions
based on the results may become difficult.

In hydrogeophysical mapping the density and quality of data controls the quality
of maps, cross-sections and the decision. As regard to optimum data density, a typical
example could be the prevailing practice to correlate lithological and geophysical logs
of boreholes located at considerable distance. It is often observed that another borehole
drilled in between changes the correlations, sometimes totally, illustrating the need for
optimization of data density for reliable subsurface information.

The quality and reliability of data also depend on capability, accuracy and functioning of the equipment used and geological and external noise. It can be improved
by routine check of the equipment and repeat measurement wherever essential; checks
on profile layout, orientation, and spacings, selection of scales and changes made,
avoiding superficial geological inhomogeneities, filtering, stacking, calibrations, base
station selection, infinite electrode position selection and measurements etc. It also
depends on field personnel collecting and processing the data.

Spatial accuracy can be obtained by using GPS. Profile lines should be straight.
Distances of current and potential electrodes, that between transmitter and receiver,
between shot and geophone should be accurate. To measure the spacings, use of marked
string or cable is avoided. Spacings are repeatedly checked or confirmed. In electrical
resistivity survey the effect of lateral inhomogeneities viz., natural boulders in rocky
terrain or man-made buried objects, pipe lines, telephone cables etc. is minimized.

These days mostly the instrument has the facility to store and transfer the data to
a computer. If the instrument has the provision of immediate viewing the data display
as a profile or graph, the data should be viewed. Real-time hard copy conventional
data plotting is adopted to check the data wherever possible, e.g., resistivity sounding
and profiling etc. This helps the operator to maintain the quality of output, check
the spurious measurements and if possible repeat the measurements or reorient the
measurements for proper spacing and check the equipment also. To check the data
quality it is preferred to process the data before closing the field operation every day

or at least after closing the field operations the same day getting scope for repeat measurements next day. It also helps improving or reducing the data density and to plan for further field operations. Finally, the interpretation and output may be allowed for independent review by an expert or opinion sought from an expert for quality of the deliverables against the specific objectives.

4.10 DELIVERABLES

The deliverables in groundwater exploration have two main aspects. The field level deliverable is pinpointing of the drilling site based on geophysical anomaly positions, approachability, availability of land, local conditions, local demands, avoiding physical constraints like, electrical lines, metallic structures, crossing of roads, streams, bridges and topographic depressions etc. It is always better that the sites pin-pointed are physically shown to the drilling crew. The deliverables for planning, execution and record mainly include interim, mid-term and final reports at various pre-defined stages of the programme. The report format and minimum information for the objective it should contain are defined at the beginning of the programme. It may also include the constraints in field investigations, limitations in data acquisition and interpretations etc. The most important is that besides the technicalities, utility maps, tables, location details and map and distinct recommendations, the report is to include the summary of findings in hydrogeological and non-technical languages also.

REFERENCES

Albertini, E. (2010) *Seismic survey safety procedure.* OH&S Management Team, Hydro Tasmania Procedure.

Crumbling, D.M., Lynch, K., Howe, R., Groenjes, C., Shockley, J., Keith, L., Lesnik, B., Van E. J. & McKenna, J. (2001) Managing uncertainty in environmental decisions. *Environmental Science & Technology, American Chemical Society Publication,* 35 (19), 405A–409A.

Eso, R.A., Oldenburg, D.W., & Maxwell, M. (2006) Application of 3D electrical resistivity imaging in an underground potash mine. In: *Proceedings of Society of Exploration Geophysicists International Exposition & 76th Annual Meeting, 1–6 October, 2006, New Orleans, USA,* SEG Publication, pp. 629–632.

Geological survey of India, **n.d.** *Standard Operational Procedure (SOP) for Geophysical Mapping and Mineral Exploration.* [online] Available from: www.portal.gsi.gov.in.

Govt. of Alberta, (2010) *Safe operating procedures for seismic drilling: Workplace Health and Safety Bulletin, IS004-Petroleum industry.*

Hoekstra, B. & Hoekstra, P. (1990) Planning and executing geophysical surveys. In: *Proceedings of Fourth National Outdoor Action Conference on Aquifer Restoration, Ground Water Monitoring and Geophysical Methods. Dublin, Ohio.* National Water Well Association, pp. 1159–1166.

Reeves, C.V., Reford, S.W. & Milligan, P.R. (1997) Airborne geophysics: old methods, new images, (Keynote Session). In: Gubins, A.G. (ed.) *Proceedings of Exploration 97: Fourth Decennial International Conference on Mineral Exploration, 14–18 September 1997, Toronto, Canada,* GEO F/X Div. of AG Inf. Sys. Ltd. Canada, pp. 13–30.

Robinson, D.A., Binley, A., Crook, N., Day-Lewis, F.D., Ferré, T.P.A., Grauch, V.J.S., Knight, R., Knoll, M. Lakshmi, V., Miller, R., Nyquist, J., Pellerin, L., Singha, K. & Slater, L.

(2008) Advancing process-based watershed hydrological research using near-surface geophysics: A vision for, and review of, electrical and magnetic geophysical methods. *Hydrological Processes*, 22(18), 3604–3635.

US EPA (2000) *Guidance for data quality assessment: Practical methods for data analysis EPA QA/G-9 QA00 Update*. U.S. Environmental Protection Agency: EPA 600/R-96/084.

Yarie, Q. (2003) *Geophysical investigations with the Geonics EM61-MK2 and EM61: operational procedures and quality control recommendations*. Geonics Ltd. Ontario, Canada.

Chapter 5

The magnetic method

5.1 INTRODUCTION

The magnetic method is the oldest geophysical method. It is a passive method. It was developed for mineral exploration and still continues as a primary geophysical exploration method. The historical development and various applications are described by Nabighian *et al.* (2005) and Hinze and Frese (1990). Near-surface applications are described by Hansen *et al.* (2005). The method is applied to measure the lateral variations in magnetic field which are magnetic expressions of the subsurface geological conditions. For applying this method it is essential that the subsurface lithology and structural fabric is associated with variations in magnetic property – the magnetic susceptibility. In its diverse applications to groundwater exploration in igneous and metamorphic rocks which, in general, have higher magnetic susceptibilities as compared to sedimentaries, the subsurface geological structures and lithological contacts, such as basic dyke intrusives with high susceptibilities (Roberts and Smith, 1994) and prominent fault and fractured zones with low susceptibilities, can be usefully picked up (Henkel and Guzman, 1977; Cull and Massie, 2002; Chary *et al.*, 2005 and Lopez-Loera *et al.*, 2010). These structures may form the preferential flow path of groundwater, affect groundwater movement and also form groundwater repositories. Gobashy and Al-Garni (2008) use a high resolution survey to identify the location of a subsurface dam to constrain groundwater movement and raise groundwater level in wadis of Makkah Al-Mukarama. It is used to delineate an igneous body concealed under alluvial capping. The depression in unaltered, high susceptibility bed rock leading to thickening of the aquifer in the sedimentary overburden or the thick altered zone of weathering can be detected (Ram Babu *et al.*, 1991). The measurements can also be made to estimate magnetic susceptibility, though the variations in magnetic susceptibility are seldom diagnostic of lithology or rock type because of overlapping ranges. However, the variations can be used as an indicator of lithologic changes (Bordie, 2002). As such, to acquire precise subsurface information magnetic surveys are to be essentially followed by electrical resistivity, electromagnetic or seismic surveys.

The magnetic measurement can be made on the ground as well as airborne. The ground magnetic measurement is generally used as reconnaissance survey to understand the subsurface structures and related hydrogeological set-up to narrow down the zones favourable for groundwater exploration, thereby constraining the time and expenditure on labour-intensive detailed geophysical surveys. The data acquisition in a ground magnetic survey is easy, reasonably fast and cost-effective. The processing of data also requires only a few corrections and interpretations are relatively simple. Preliminary qualitative interpretations can be done even visually by an experienced geophysicist, if the data size is small, of course with the experience of the interpreter. The airborne magnetic measurements known as aeromagnetics cover a large area in a relatively short time and can also be conducted for groundwater exploration to get a regional view, identify structures and tectonics and define favourable localities. The airborne magnetic data is generally acquired together with other airborne geophysical data acquisition to support integrated geological interpretation at a little extra cost on processing and interpretation. The existing low-altitude high resolution airborne magnetic data acquired for mineral surveys and geological investigations in hard rock can also be subjectively reinterpreted for groundwater targeting.

5.2 BASICS

By the magnetic method the magnitude of the magnetic field intensity is measured at specific points. Measurement can be made of the total magnetic field or its vertical and horizontal components and also its vertical and horizontal gradients. The magnetic field at a point of measurement is a vector sum of the Earth's magnetic field, induced field and field of remanent magnetization. The direction of the Earth's magnetic field is vertical at the north and south poles and horizontal at the magnetic equator. Its intensity varies considerably, from about 0.6 Gauss or 60,000 nano-Tesla ($nT = 10^{-9}$ T) or gammas at poles to 0.3 Gauss or 30,000 nT near the equator. The nano-Tesla is the unit of magnetic field intensity commonly used in magnetic field measurements. In response to inducing Earth's magnetic field, rocks and minerals acquire induced magnetization and exhibit an induced magnetic field parallel to the Earth's field. The intensity of induced magnetization is proportional to the inducing field and is related through a proportionality constant known as the magnetic susceptibility (k). The magnitude of induced magnetization varies due to varying amount and nature of magnetic mineral grains in rocks and hence due to their differing bulk magnetic susceptibilities. For example, the magnetic susceptibility of an intruded basic dyke is greater than and differs significantly from that of the granitic country rock which is mostly acidic in nature and hence the concealed dyke is picked up by magnetic survey unambiguously. The magnetic susceptibility of weathered material decreases with higher grades of weathering as it causes chemical transformation of the magnetic minerals. The thick weathered zone or bed rock depression filled with alluvium with less magnetic susceptibilities behave as transparent to magnetic measurements and hence the features in underlying compact bedrock which are primary sources of anomaly can be detected. The remaining component of the measured field is due to the remanent magnetization of rock. It is also induced magnetization in rock but permanent (could be metastable as defined by Clark, 1997) in strength and direction which can be related to the Earth's palaeo-magnetic field, at the time rock cooled through its Curie temperature. It is also known as thermo-remanent magnetization. The direction of remanent magnetization may be different from that of induced magnetization. In a magnetic survey it is difficult to separate the field due to induced magnetization from remanent magnetization and generally the latter is not considered. Besides the above mentioned three main components of the measured magnetic field, there are sources external to the Earth adding variations in the measured field.

The fundamental condition in applying the magnetic method is the lateral variations in magnetic susceptibility of subsurface material causing observed variations in magnetic field intensity – the magnetic anomalies. The local anomalies related to geological structures are the targets for study. The shape of the observed magnetic anomaly, while traversing in an area, depends on the magnetic property and geometry of the local causative body, intensity and the direction i.e., inclination and declination of Earth's magnetic field there and on the location of a measurement point relative to the body. Away from the source the anomaly starts disappearing or overlapping from other sources appears. The local magnetic anomalies superimpose over the regional trend of the Earth's field in the area. The amplitude of the anomaly ranges from as low as a few n T to thousands of nT. While interpreting, the local anomalies are separated from the regional trend of the Earth's field to get the residual anomalies through which the subsurface geology is mapped. In groundwater exploration the local anomalies are significant, could be due to the occurrences of fractured zones, basic dykes and thickening of weathered zone.

5.3 INSTRUMENT

The portable magnetometers are Askania-Schmidt, fluxgate, proton precession and cesium vapour. The Askania-Schmidt magnetometers were earlier used for measuring the vertical and horizontal components of the earth's magnetic field. In hard rock groundwater exploration total magnetic field surveys are conducted using a proton precession magnetometer (PPM) which has sensitivity of the order of 1 nT or 0.1 nT. It is a light-weight, less cumbersome piece of equipment having a sensor and a console. The PPM measures the magnitude of Earth's magnetic field independent of its direction. So, the sensor can be kept in any orientation. The local perturbing field adds vectorially to the Earth's magnetic field. Since the perturbing field is quite small compared to main field, the resulting field is almost parallel to Earth's undisturbed field whose magnitude is measured by PPM. In this way, the component of local perturbation along the main field which is termed as the anomaly can be measured (Breiner, 1973). It is schematically shown in figure 5.1. A proton precession magnetometer gives erratic readings if the gradient of the local field is very high. The other equipment in use with higher resolution is the optically pumped cesium vapour magnetometer. In this, the data acquisition can be either continuous with high sampling rate or discrete at different observation points. Unlike the proton magnetometer, the sensor is orientation sensitive. Its higher cost constrains its regular use particularly in low budget groundwater surveys where the magnetic surveys are mostly of reconnaissance type requiring fast and cost-effective coverage. For this, the 1 nT accuracy total field intensity measurement by proton precession magnetometer is sufficient. If required, an instrument with accuracy of 0.1 nT can be used for a smaller area. To enhance the anomalies from shallow features the vertical gradient measurement with two sensors one above the other can be conducted. This also eliminates the variations due to extraneous field.

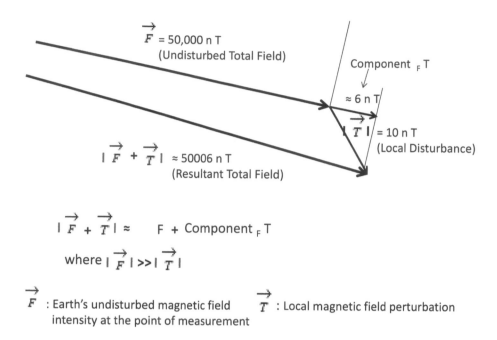

$$|\vec{F} + \vec{T}| \approx F + \text{Component}_F T$$

$$\text{where } |\vec{F}| \gg |\vec{T}|$$

$\vec{F}$: Earth's undisturbed magnetic field intensity at the point of measurement

$\vec{T}$: Local magnetic field perturbation

Figure 5.1 The vector addition of small local magnetic field perturbation (T) to the Earth's magnetic field (F). (source: Breiner, 1973).

5.4 FIELD PROCEDURES

5.4.1 Total magnetic field intensity measurement

The total magnetic field intensity measurement can be made along a single profile or parallel profiles at regular station or sampling intervals. It is not measured at isolated spots as that does not yield any meaningful interpretation. Measurements can also be on a grid spread consistently over the area or on a network of profiles located either through surveying or GPS. Generally, parallel profiles are placed across a geologic structure and lineament whose approximate location is known from local geological information and satellite images or across the trend of the country rock. The orientation of profiles, their spacing and station interval for measurements are decided based on the trend, size and depth of the body or anticipated size and amplitude of the anomaly, details required and availability of space. In groundwater surveys profile spacing could be 50 to 200 m and station interval 25 to 50 m. The selection of station interval is important and can be attempted through a test profile. It should be such that the anomaly due to short lateral variations in magnetic character is not lost and accordingly the station interval and also the profile spacing can be reduced. Closer the spacing the slower is the rate of coverage but the greater is the chance of detecting anomalies and accuracy in interpretation. Reconnaissance survey may be at larger profile spacing and station interval while detailing could be done at close-grid. It is better to cover the area in a grid pattern. The profile must be long enough, 500 m to 1 km, to understand the wavelength of anomalies and differentiate the anomalies from background or regional trend. GPS can be used to correct and control the location information.

Before starting the survey, it is necessary to ensure that operator is totally free from magnetic material on his person (magnetic cleanliness) as these are likely to distort the measurements. The source of magnetic noise may be any item made of iron, like knife, compass, belt-buckle, keys, coins, hammer, wrist watch-strap, umbrella, iron bucket etc. The vehicle used for survey or movement of vehicles near the measurement points affects the readings. Secondly, before starting the normal surveys as well as during the surveys, measurements are repeated at a particular station, to check the repeatability of the reading particularly when high or low values are recorded. If there is any magnetic disturbance the readings will not repeat and show wide variations. In such cases, the cause of such disturbance is identified, whether it is due to an external field or it is related to some cultural noise. The former could be due to the onset of magnetic storms that may persist for a few hours or a day and the measurements are discontinued. While for the latter either the measurement station is skipped or the cause is shifted. As a precaution the profile should not pass near buildings, power or telephone lines, buried pipe lines, near standing vehicles, railway tracks, fencing (metallic) and overhead high tension electric lines which are cultural noise sources of magnetism. Keeping the sensor on the ground may give noisy data due to superficial material of varied magnetic susceptibility lying on it. Therefore, measurements by proton magnetometer are made keeping the sensor on a non-ferrous stand at a fixed height of 1 to 2 m above ground. This helps reduce the spurious anomalies. The sensor height selected is not changed during the survey in an area. Further, the proton magnetometer is tuned by the facility provided in it, to the general value of Earth's magnetic field in the survey area, which can be obtained from the published map. It helps in getting repeatable readings.

5.4.1.1 Correction of data

The Earth's magnetic field varies with time. The temporal variations to the interest of magnetic surveys are secular, diurnal and transient. Secular variations are slow over a long period of time in Earth's magnetic field and are of internal origin related to convections in Earth's outer core. These variations are not of serious concern in the short duration magnetic surveys. The diurnal variations are of external origin caused by ionospheric influences and of periodicity about a day. Diurnal variations are of two types, viz., the 'quiet day' – smooth, low amplitude ($\approx 25\,\text{nT}$) regular variation and 'disturbed day' – rapid, high amplitude ($\approx 1000\,\text{nT}$) irregular transient variation caused by magnetic storms, during which measurements are discontinued. The transient variations are related to magnetic activities in the ionosphere and increased sunspot activities. In magnetic surveys, measurements are affected by these variations superimposing the variations in magnetic field due to geological causes. The effects are nullified or removed applying corrections during data acquisition as well as data processing.

The main objective of applying correction is to extract that part of the data which can be attributed to local geological variations only. Corrections made are mainly for

(a) diurnal and (b) geomagnetic variations. The 'quiet day' diurnal variation is recorded either time-lapsed or continuous through the day at a fixed base station in the survey area and the total field measurements along profiles are corrected by interpolating and distributing the variation linearly over them. Closing or loop error in successive base readings of the order of 5 to 10 nT is also distributed amongst the readings. Diurnal correction should be made for a chosen reference time. The base station should be away from cultural noise and located in such a geological condition that the measured values do not change much by moving the magnetometer sensor in the surrounding few metres. Depending on the length of profile and duration lapsed, measurements can be repeated at the selected base station, by returning to it two to three times at convenient interval (1 to 2 hours) during the operation. For large-scale surveys where repeat measurements at an interval of about 1 or 2 hours at a single base station become cumbersome, more than one base station may be established. The time of observing the repeat readings is recorded. Diurnal correction is more effectively applied if the variation is recorded almost continuously by another magnetometer positioned at the base station and the time of measurement at each station along the profiles is recorded so that observations are corrected as per the time of variation recorded.

The total field measurements are corrected for diurnal or time variations in case surveys continue for a long period over a day or days, along a large profile or over a large area covered in a grid pattern. If the survey is for a short duration, say about an hour or over a short profile with high amplitude anomalies over several hundred nT, diurnal correction is not required. For example, correction may not be required for surveys carried out for groundwater exploration along a single small profile across a dyke or photo-lineament covered within a short period. Breiner (1973) points out that diurnal correction is required for broad anomalies of the order of 20 to 50 nT and essentially for surveys at high magnetic latitudes. For high resolution surveys diurnal correction is necessary. For gradient measurement it is not required. Geomagnetic variation relates the magnetic field variations with geographical location of the observation point. For surveys covering a small area this correction is not required.

5.4.2 Magnetic susceptibility measurement during field survey

The induced magnetization in a material placed in a magnetic field depends on its magnetic susceptibility. Susceptibility is defined as the ratio of magnetization to the magnetic field, given as $\mathbf{M} = k\mathbf{H}$ or $k = \frac{\mathbf{M}}{\mathbf{H}}$ where $\mathbf{M}$ is the induced magnetization of the material, $\mathbf{H}$ is the inducing magnetic field and k is the volumetric susceptibility. The units of $\mathbf{M}$ and $\mathbf{H}$ are the same and hence k is dimensionless. Magnetic susceptibility varies with magnetic field, temperature and direction of measurement and therefore the measurements are made at room temperature and low field – the Earth's magnetic field. The measure of magnetic susceptibility helps magnetic characterization of the geological formation. Also, the measurement can support understanding the anomaly through modeling and geological mapping. Magnetic susceptibility can be measured by susceptibility meter in field as well as laboratory. The proton precession magnetometer deployed for total field measurements can also be used during field operation to get an approximate measurement of magnetic susceptibility (k) of rock specimen (Breiner, 1973). A PPM of 1 nT sensitivity will yield sufficiently accurate data. The sensor of PPM is held above the ground at a place free from magnetic material. The sensor is oriented along the earth's magnetic field (F). The earth's magnetic field direction can be obtained by using a dip needle. The field intensity (T_0) in the absence of rock specimens and near surface features is measured and checked by repeat observation. Then a regular shaped rock specimen whose magnetic susceptibility is to be measured is held at a known distance from the sensor along the earth's magnetic field direction. The distance (r) of the specimen from the sensor is kept more than 5 times its diameter (D) to maintain the dipole approximation. The distance is kept about 20 to 50 cm. Specimens of high magnetic susceptibility can be kept at a distance of 100 cm while low magnetic susceptibility specimen can be kept close to the sensor. Measurements are taken for different positions of the specimen by rotating it about an axis perpendicular to the direction of Earth's field and are repeated turning the specimen by 90 degrees. The minimum (T_{min}) and maximum (T_{max}) values of total field intensity are recorded. The magnetic susceptibility is expressed as

$$k = \frac{3}{2\pi F} \left(\frac{r}{D}\right)^3 (T_{max} + T_{min} - 2T_0)$$

The F, T_{max}, T_{min} and T_0 are in gauss (10^5 nT $= 1$ gauss). The procedure mentioned here is from application manual by Breiner (1973) where it is explained in detail. According to Breiner by this technique, with a PPM of sensitivity 1 nT the smallest k value that can be measured for a large specimen adjacent to the sensor is of the order of 2×10^{-5} cgs unit (dimensionless cgs unit of k is multiplied by 4π to get SI unit). The range of magnetic susceptibility for different rocks (Clark and Emerson, 1991 and Lelievre 2003) is presented through figure 5.2.

5.5 PROCESSING OF DATA

In most of the cases, the magnetic anomaly and noise can be readily recognized without any data processing. However, prior to interpretation, the magnetic measurements are corrected to eliminate the effects which are "non-potential", i.e., not due to the body and have external source and get anomaly which solely manifests the character and geometry of the subsurface formation or body. The correction and normalization is quite simple. The magnetic readings obtained along profile lines are corrected for diurnal variations, if required. The diurnal variation curve (plot of base readings with time) is constructed for each day and correction is linearly distributed among the readings as per the time of taking measurements with the assumption that magnetic field has varied linearly. The corrected data are plotted with magnetic values on the y axis and stations on the x axis to generate magnetic profiles. These profiles plotted at relative positions on graph paper can be combined and contoured to get the total magnetic field variation map of the area surveyed. The other way of preparing a magnetic map from a network of profiles for an area is to compute field values at a grid over the area from these profiles. The drawback of gridded data is that the grid interval is generally larger than the sampling interval and smaller than the profile interval, so it may not represent the actual anomaly trend and also the shallow source anomalies may get obscured (Jain, 1976).

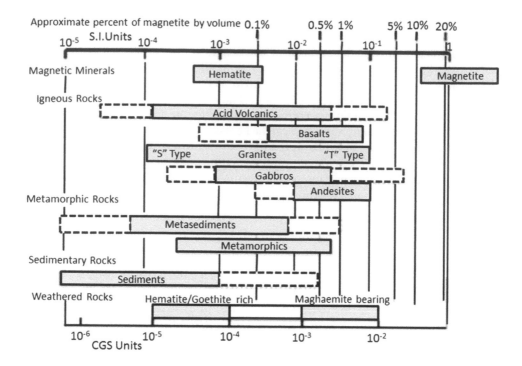

Figure 5.2 Ranges of magnetic susceptibility for different rock types (Source: Clark and Emerson, 1991 and Lelievre 2003).

To get the total field anomaly map or residual map the present International Geomagnetic Reference Field (IGRF), which represents the actual regional field, computed for the latitudes and longitudes of the stations obtained by GPS is subtracted from the respective observed values already corrected for diurnal variations. In this way the correction for secular variation is also incorporated. Except data reduction, as such, the IGRF correction for secular variation, which is quite slow with time, is not important in groundwater exploration studies generally covering small areas. If the recorded total field values are plotted, the plot shows anomalies of highs and lows and subtracting the Earth's field from the anomaly gives positive and negative anomalies.

5.5.1 Regional-residual anomaly separation

The anomalies due to local induced field from shallow subsurface structures-the targets in the study area are separated from the relatively regional trend of the Earth's magnetic field in that area. The separated anomaly is known as the 'residual anomaly'. According to Hinze and Frese (1990) the residual anomaly is a geologically significant problem-specific anomaly in a particular study area. Since the quantitative interpretation is carried out on the residual anomaly, a reliable estimation and removal of the regional field from observed data is desired (Li and Oldenburg, 1998). The simplest empirical

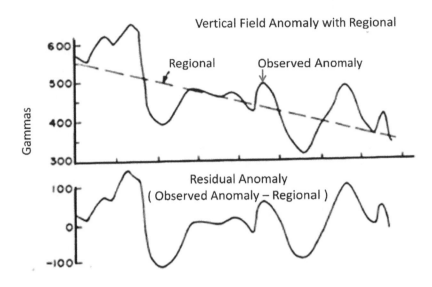

Figure 5.3 Vertical field anomaly and the residual magnetic anomaly (Source: Chary *et al.*, 2005).

method to get the residual anomaly is smoothening of the observed magnetic profile to remove the noise, logically defining and drawing the regional trend manually over it and subtracting the regional trend from the smoothed observed profile data. An example of this graphical procedure presented by Chary *et al.* (2005) for a vertical field profile is shown in figure 5.3. To get rid of the personal bias in estimating the regional trend, there are several techniques and one of them is to get the regional trend through the least-squares method with specified order of polynomial proposed by Agocs (1951). Li and Oldenburg (1998) proposed a different approach for obtaining the residual anomaly from 3-d inversion of data based on magnetic susceptibility distribution. According to them, to get a susceptibility model the inversion is done for two sizes of the area. It is done at 2 to 5 times coarser grid for the large area of dimensions in kilometers which includes the much smaller specific area of interest for detailed inversion at a finer grid. The regional susceptibility distribution obtained from the large area outside the local area forms the base for estimating the regional field which is subtracted from the observed field.

5.5.2 Reduction to pole

The inclination of the Earth's magnetic field changes with latitude. It is $+90°$ at north magnetic pole to $0°$ at equator and $-90°$ at the south magnetic pole. With this, the direction of magnetization in the source body aligned to the Earth's field also changes. The induced field of the source body, mostly dipolar in nature adds to the Earth's local field to produce the anomaly. Therefore, the magnetic anomaly for the same target differs in shape and amplitude with latitude. The shape of anomaly, say, over a vertical magnetic dyke body will vary at different magnetic latitudes (magnetic inclinations) defined through Earth's magnetic field, direction and magnitude of the induced field of the source body magnetization and for different directions of traverse line. The variations in magnetic anomaly with inclination on N-S profile across E-W dyke and on E-W profile across N-S dyke are shown in figure 5.4 (Macleod *et al.*, 1993) and also for different shapes of target are given in Breiner (1973). At the north magnetic pole the Earth's field and induced dipolar field are vertically down. The recorded magnitude of the total field shows a positive symmetrical anomaly over the body and negative anomalies on either side. The negative anomalies are obtained because the direction of the induced field on either side is opposite to the inducing Earth's field. Similarly, the development of an anomaly at the magnetic equator can be visualized. At the magnetic equator, the inclination of Earth's field is zero, i.e., Earth's field as well as the induced field is horizontal. Since above the body the induced field is in the opposite sense to the main field, on a N-S traverse a negative anomaly is recorded, while on either side the fields being in the same direction, positive anomalies are recorded. At mid-latitudes, say in the northern hemisphere, the Earth's field and the induced field in the body both being inclined downward at same inclination, on a N-S traverse across the body, an asymmetrical anomaly is obtained.

Therefore, all magnetic anomalies are reduced to the pole by recalculating the total field data as if the target is located at the pole and the Earth's magnetic field has 90 degree inclination or vertical. In this way dipolar asymmetric magnetic anomalies are transformed to symmetrical anomalies which would have been observed with vertical magnetization. The anomaly is centered over the target and correlation between the anomaly and source is enhanced (Hinze and Frese, 1990). The basic assumption in this reduction process is that induced magnetization is totally parallel to the Earth's field. The direction of remanent magnetization can be different from the direction of Earth's magnetic field and cannot be accommodated in this reduction process. The remanent magnetization being comparatively small it is generally neglected in exploration purposes.

5.6 INTERPRETATION

The magnetic method being a potential field method it has inherent non-uniqueness in interpretation. The measured values can be related to infinite number of causative magnetic sources. Generally, the anomaly is a superposition of contributions from regional and local magnetic sources as well as noise and hence the observed anomaly can be due to any possible combination source model. Therefore, the limitations of the method and possible errors in field measurements and data processing should be understood prior to interpreting the data. To minimize the ambiguities and arrive at a possible geological model, the interpretation is supported by geological and borehole lithological information available for the area and also by subsurface information obtained through other geophysical methods. The selection of supporting geophysical methods depends on the objective and precision desired.

Prior to interpretations corrections are applied to the field data to remove the non-geologic variations and actual anomalies are identified. Interpretation can be attempted at two sequential stages, viz., qualitative and quantitative. In qualitative interpretation, information is obtained on the location and orientation of structures by visual

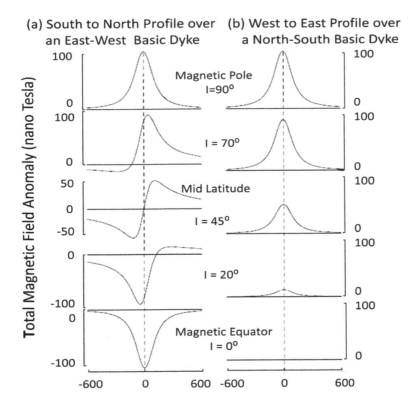

Figure 5.4 Total magnetic field response over (a) east-west striking vertical basic dyke and (b) north-south striking vertical dyke at different latitudes (inclinations) (Source: Macleod *et al.*, 1993).

inspection of the shape of anomaly and the trend on parallel profiles, such as dykes and faults which are associated with typical magnetic signatures at the magnetic latitude of measurements and the convergence or divergence of anomaly contours. In general, a linear magnetic low could be associated with a fault or fractured zone, while a linear magnetic high is associated with a basic igneous dyke intrusion. Dyke may also show a magnetic low if the reverse polarity remanent magnetization dominates (Lee *et al.*, 2012). The anomalies vary widely in shape and they are mostly asymmetrical in nature (Breiner, 1973). The asymmetrical anomaly is due to the dipolar nature of induced field.

The shape and magnitude of the anomaly, and whether it is positive or negative, depend on the inclination of the earth's magnetic field at the point of measurement, orientation of the magnetic body, direction of traverse across the body and relative position of source and point of measurement, i.e., on the depth of the magnetic body. Closer the body to the ground surface, the steeper would be the magnetic anomaly and of smaller extents. The interpretations of total field anomalies are given in Breiner (1973). Also details can be obtained in Hinze *et al.* (2012).

Table 5.1 Approximate maximum depth to magnetic source of different geometries in vertical field based on one-half anomaly width ($X_{1/2}$) at one-half maximum anomaly-amplitude. Z_c and Z_t are depths to centre and top of the ideal magnetic source respectively (after Nettleton, 1940 referred in Hinze and Frase, 1990), figure 5.5a.

Geometric form	Magnetic depth
Sphere (Dipole)	$Z_c \leq 2.0\, X_{1/2}$
Horizontal Cylinder (Line of Dipoles)	$Z_c \leq 2.0\, X_{1/2}$
Vertical Cylinder (Monopole)	$Z_t \leq 1.3\, X_{1/2}$
Edge of Narrow Dyke (Line of Monopoles)	$Z_t \leq 1.0\, X_{1/2}$

5.6.1 Estimation of depth to magnetic source

Several methods are available for approximation of maximum depth to the top or centre of the magnetic source of simplified geometries (Hinze, 1990). These methods are grouped as manual or graphical and computer based or automated. One of the manual methods by Nettleton (1940) referred by Hinze and Frese (1990) is the 'Half-Width' rule. In this the horizontal distance ($X_{1/2}$) between point of maximum anomaly-amplitude and the point of half the maximum anomaly-amplitude is measured. The depth estimations for different simplified source geometries in vertical magnetic field given by Hinze and Frese (1990) are presented in Table 5.1 and in horizontal magnetic field near magnetic equator are presented in Table 5.2 (Breiner, 1973). These are shown in figures 5.5a and 5.5b. For a N-S oriented cylinder and the edge of an E-W striking horizontal sheet in a horizontal field, the half-width is the horizontal distance between the point of maximum or minimum anomaly and the point of 'zero' anomaly cross-over (Breiner, 1973). Another empirical method of depth determination is from the horizontal projection of the slope at the inflection point on the steepest part of the anomaly. It was developed by Vacquier *et al.* (1951) referred by Hinze and Frese (1990) and named as 'slope technique' (Breiner, 1973) shown in figure 5.6. The length of horizontal projection is multiplied by a factor between 0.5 and 1.5 to get the depth to the top of the magnetic source. Gunn (1997) indicates that by this method depth estimates are accurate to ±15% and it is useful for intrusive bodies of varying width and sheet intrusives representing prism and plate models respectively and not for dipping bodies. Stanley (1977) presents a simple procedure for interpretation through differentiation of the magnetic profiles placed perpendicular to strike of a magnetic contact or dyke. This method is applicable to short length profiles also. By this the location, depth and dip along the profile and magnetic susceptibility contrast can be worked out. Al-Rawi (2009) uses Fraser filter and Abdelrahman and Sharafeldin (1996) use iterative least-squares approach to estimate the depth of a dyke like body.

Also, there are several computer based methods for depth estimation. All these methods with their merits and demerits are summarized by Li (2003). The magnetic source location can be obtained by deconvolution using the Euler's homogeneity relation (Thompson, 1982). It is widely used. It can be conveniently applied to reduced-to-pole magnetic data along isolated profiles, usually generated in groundwater exploration, as well as to the gridded data for 3-d interpretation (Reid *et al.*, 1990). It can be used for multiple magnetic sources (Hansen and Suciu, 2002).

Table 5.2 Approximate maximum depth to magnetic source of different geometries in horizontal field based on one-half anomaly width ($X_{1/2}$) at one-half maximum anomaly-amplitude. Z_c and Z_t are depths to centre and top of the ideal magnetic source respectively (after Breiner, 1973), figure 5.5b.

Geometric form	Magnetic depth
Sphere (Dipole)	$Z_c \leq 2.5\, X_{1/2}$
Horizontal E-W Cylinder (Line of Dipoles)	$Z_c \leq 2.0\, X_{1/2}$
Horizontal N-S Cylinder (Monopole)	$Z_c \leq 1.3\, X_{1/2}$
Edge of Sheet (Line of Monopoles)	$Z_t \leq 1.0\, X_{1/2}$

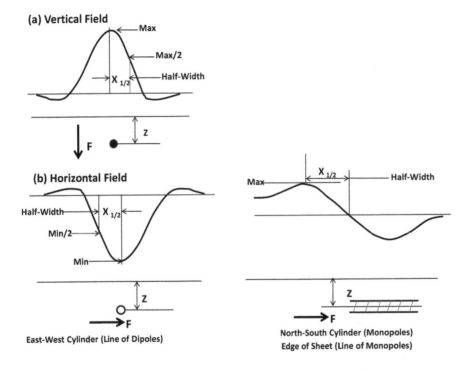

Figure 5.5 Half-width rules for source depth estimation (a) vertical magnetic field and (b) horizontal magnetic field (Source: Breiner, 1973).

The method uses the first order derivative of the total field anomaly and the degree of homogeneity as a structural index which is a measure of the fall-off rate of the field with distance and indicates the type of source structure. The proper selection of structural index is important as it controls the depth estimate. Gobashy and Al-Garni (2008) used one and two-dimensional Euler deconvolution method to estimate the depth to the magnetic source and the magnetic boundaries. Some of the applications of Euler's deconvolution in depth estimation are by Milligan *et al.* (2004) in Australia, by Sultan *et al.* (2009) in central part of Sinai Peninsula, Egypt and by Githiri *et al.* (2011) in Southern Kenya.

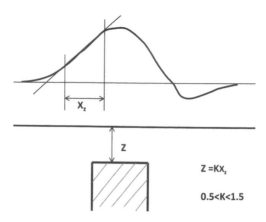

Figure 5.6 Source depth (z) estimation from steepest slope of the anomaly (Source: Breiner, 1973).

The other deconvolution method was developed by S Werner in 1953 (Jain, 1976). The Werner deconvolution method is used for magnetic profile data to estimate the depth to the top of a thin dyke of large (infinite) strike and depth extents, its dip and susceptibility contrast. With the concept that the horizontal gradient of total field anomaly due to the edge of a thick body is equivalent to total field anomaly over a thin dyke, its application gives the edges of a thick body. According to Jain (1976) computation of the susceptibility contrast is reliable when the dip is 40 degrees or more. The interpretation can be erroneous with overlapping of anomalies, noisy data and non-uniqueness in model approximation of the geologic condition (Birch, 1984). To minimize uncertainties and refine parameter estimation Ku and Sharp (1983) present an automated iterative computation for Werner deconvolution. Birch (1984) indicates the use of Fourier spectral analysis in depth estimation. He shows that depth is just the negative slope of the plot of the amplitude spectrum versus wave number. For its useful application the basic geological model should be good, data must be suitable, profile should be perpendicular to the strike of the body, its length at least 10 times the bed rock depth and the spacing interval of measurement should be less than one-tenth the bedrock depth (Birch, 1984). Since due to weathering rock loses its magnetic susceptibility, weathered zone thickness or depth to the compact bed rock can also be estimated using the Fourier amplitude/power spectrum (Ram Babu *et al.*, 1991). They indicate that the Fourier amplitude spectra comprise two linear segments for magnetic sources at two different depths. Compared to depths obtained from drilling, using this method the depths are within rms error of ± 10.2 m.

The parameters of the magnetic source can be estimated through forward modeling and inversion schemes (Hinze, 1990). Computer software is available for forward modeling and inversions. In forward modeling a guess model of the source of simple geometric form is used to generate a theoretical anomaly and it is compared with the observed field anomaly. The discrepancy between these anomalies is reduced iteratively modifying the source model parameters. However, the set of refined model parameters

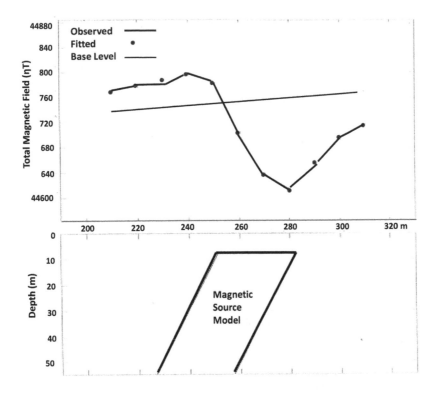

Figure 5.7 Modeling of source model and matching of theoretical and observed anomaly.

thus obtained is not the unique one and there could be a number of equivalent models. An example of modeling is shown in figure 5.7.

5.7 IDENTIFICATION OF FRACTURED ZONE

Henkel and Guzman (1977) from linear magnetic lows infer the fractured zone underlying the alluvial capping. The fractured zone is associated with a decrease in magnetic susceptibility due to martitization and hydration. In this magnetite is transformed to hematite which has much lower order of magnetic susceptibility. Reduction in magnetic susceptibility over fractured zone is also reported by Petersson *et al.* (2007). Lee *et al.* (2012) propose network-extraction method for the identification of linear positive and negative anomalies associated with geological lineaments. It combines the principles of peak-identification algorithms used in geophysical data interpretation and the GIS drainage network-extraction algorithm applied to topographic surface.

5.8 AEROMAGNETICS

The aeromagnetic survey for regional geological mapping, over a wide area and depth range, as an aid to mineral and oil exploration has been in practice for nearly six decades and has advanced remarkably. Hogg (1987), Reeves *et al.* (1997) and Thomson *et al.* (2007) review its comprehensive development and progressive refinement from conventional to the recent high resolution survey. The aeromagnetic data in the form of contour maps or images reveal the variations in total magnetic field intensity. These maps can also be used in drawing inferences related to groundwater conditions in hard rock areas. The litho-structural features being associated with magnetic expressions, a great amount of information on faults, fractured zones, litho-contacts and dykes in hard rock areas influencing groundwater occurrences can be extracted. In areas covered with soil and weathered material, features in the underlying compact rock, having feeble or no surface expressions, are picked up in aeromagnetic maps. Astier and Paterson (1987) demonstrate the increase in productivity of water wells located in proximity of fault zones interpreted from aeromagnetic anomalies and advocate the use of existing data in locating groundwater targets.

While airborne geophysical survey is quite common for mineral and oil prospecting, to date very little survey has been conducted for groundwater. Even the interpretation of the existing airborne data for groundwater targeting has been attempted only sporadically. Therefore, examples of aeromagnetic survey and data interpretation exclusively for groundwater are very few. Major case studies are from Burkina Faso where aeromagnetic survey was carried out during 1972–77 (Astier and Paterson, 1987) and from eastern Botswana where aeromagnetic and airborne electromagnetic surveys were carried out during 1985–88 for groundwater over an area of 3300 sq km (Bromley *et al.*, 1994). It helped reduce ground surveys and guide water well drilling programme. Cresswell and Gibson (2004) conducted aeromagnetic surveys for issues related to groundwater across Angas Bremer Plains in Australia. Smith *et al.* (2007) summarize the advances in airborne geophysical surveys for hydrologic studies within USGS. High resolution aeromagnetic survey was carried out in the southern Espanola basin, New Mexico to understand the geological controls on groundwater flow and storage (Grauch and Bankey, 2003). Wynn (2006) conducted aeromagnetic survey to map crystalline basement topography, structures and aquifers in Upper San Pedro Valley of Southeastern Arizona. Paterson and Bosschart (1987) and Chandra (1991) stress the need of re-examining the available aeromagnetic data and airborne geophysical surveys for hydrogeological investigations. Joseph and Chandra (2003) confirm the lineaments picked up from satellite images in the eastern part of India by studying the existing aeromagnetic data. The regional lineaments picked-up from satellite images and aeromagnetic lineations studied in conjunction to identify their subsurface disposition, result in better reconnaissance on regional basis, enhancing the level of confidence in selecting zones or areas for further hydrogeological and ground geophysical investigations. Astier and Paterson (1987) compare the features on geological and landsat map with those inferred from aeromagnetic data. They opine that the large number of faults and dykes picked up in an aeromagnetic map could be confirmed to some extent with the geological map but not always so with the lineaments traced in the other map. The surface-lineaments possibly with very limited depth dimensions do not correlate because the lineations picked up from the aeromagnetic map include only those which underlie the weathered material. It indicates a significant advantage of an aeromagnetic map in locating prominent deeper fractured zones. Rangarai and Ebinger (2007) also interpreted the aeromagnetic data along with landsat TM data for groundwater exploration in south-central Zimbabwe. Another study by Gettings and Bultman (2005) synthesizes the Landsat TM data and multi-generation aeromagnetic data to delineate deep fractures possibly controlling the recharge to deeper aquifers. Amoush et al. (2013) analyze aeromagnetic data and Landsat images for structural and tectonic interpretation of southern part of Jordan. For groundwater exploration in hard rock, besides confirming a selected lineament as per requirement, it is also imperative to interpret the aeromagnetic map separately, as that may yield hydrogeologically significant structures more in number than what is obtained from the surface expressions through satellite images.

For the purpose of groundwater exploration, to start with, the aeromagnetic data could be studied even qualitatively on the basis of intensity level, contour pattern and extent. The visualization of anomalies as images makes the aeromagnetic data intuitively interpretable at first sight (Reeves et al., 1997). Analysis of maps is preferred at 2 scales, one at smaller scale for the regional features and the other at 1:50,000 scale for smaller selected areas and transects for ground geophysical follow up. While carrying out such analysis, variations in shape of the magnetic anomaly with latitude are considered. For example, the magnetic low in an aeromagnetic map near magnetic equator actually represents a strong magnetization (Astier and Paterson, 1987). The close linear contours or anomalies sharply differentiating the magnetic levels and sudden disruption of contours are the manifestations of structural discontinuities and could be related to faults, basic intrusives or fractured zones. The deep weathering is also picked up from aeromagnetic data (Olesen et al., 2006). Interestingly, Coetzee et al. (2009) indicate the possible interpretation of aeromagnetic data in identifying subsurface acid mine drainage pathways. According to Astier and Paterson (1987) in aeromagnetic map a fault is picked up "(a) as sharp gradients forming a linear boundary between areas of different magnetic level, relief or texture ..., (b) disruptions and/or deflection of magnetic trends ..., (c) linear magnetic lows within country rocks of moderate or high relief ... and (d) narrow linear features with direct magnetic expression"....

The above facts reveal that for groundwater exploration existing aeromagnetic data which may be for different altitudes and spacing can be synthesized as attempted by Ponce and Blakely (2001). Secondly high resolution aeromagnetic surveys along with other methods can be conducted for groundwater studies (Blakely et al., 2000). It is cost-effective in the long term, as far as subsurface detail is concerned, at cost per kilometer (Reeves et al., 1997) or information per sq km per unit cost. The old aeromagnetic data of high altitude and large flight line spacing gives a regional picture of the subsurface. The resolution of the low amplitude anomalies is increased by low altitude flight, higher flight line density with high sensitivity instrument, positional accuracy and sampling rate. Bromley et al. (1994) carried out an aeromagnetic survey in Botswana with a terrain clearance of 20 m only, to maximize the response from the top few hundred metres and compared it with the response obtained earlier from 300 m altitude for the same area, which revealed only broader and deeper structures. High resolution aeromagnetic (HRAM) data resolve better the shallow subsurface structures of smaller wavelength controlling groundwater movement in hard rock within the

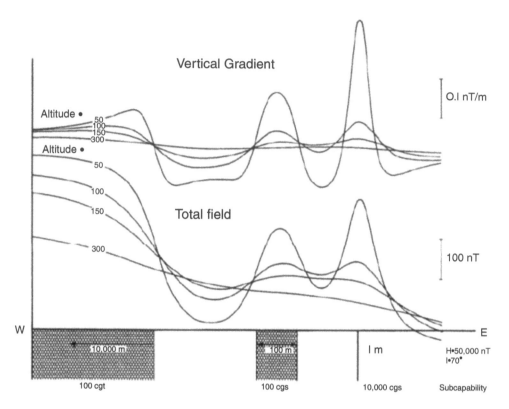

Figure 5.8 The total field and vertical gradient model responses for three vertical prisms of different widths. At 50 and 100 m terrain clearance the prisms are clearly resolved (Source: Hogg, 1987).

depth range of 200 to 300 m. The effect of flight height on amplitude and resolution shown by Hogg (1987) is reproduced here (Fig. 5.8). Obviously, with reduction of flight height the line spacing is also reduced. The line spacing is kept 100 to 200 m in high resolution surveys as compared to 500 m or more in the conventional surveys and flight altitude could be as low as 50 m (Hogg, 1987). The low altitude surveys can be carried out by helicopter. The limitation is in rugged terrain where the sensor altitude or ground clearance varies and needs appropriate correction. In addition to total magnetic field map, residual magnetic map, reduced to pole map and vertical and horizontal gradient maps can also be computed to improve the definition of structures required in groundwater exploration. Hogg (1987) reveals the usefulness of horizontal gradient data in getting information on geological strike and inter-line interpolation of data.

For interpreting existing aeromagnetic data as contour maps, the simplest way is to digitize the map and the values brought to a grid map of convenient grid size and noise removed by filtering. The digitized map can be transformed to a pole-reduced total magnetic field map. The first and second derivative operations on the pole-reduced map yield the near surface features and the geological contacts respectively.

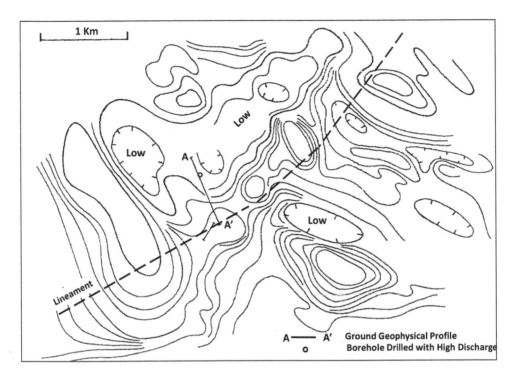

Figure 5.9 Ground geophysical surveys for groundwater in granitic terrain across coincident aeromagnetic and photo-linear (Chandra *et al.*, 1994). Aeromagnetic data source: AMSE, Geological Survey of India.

5.8.1 Case studies

Chandra *et al.* (1994) analyzed the existing aeromagnetic data in parts of eastern India to confirm the lineaments and select profile lines for ground geophysical follow-up. Typical examples are from granitic and metasedimentary terrains of West Bengal and Jharkhand States, India. In the granitic terrain a NE-SW trending regional lineament was picked up from satellite image (Fig. 5.9). An aeromagnetic map of the area shows a linear magnetic low in NE-SW direction confirming the trend of lineament. The trend of magnetic contours indicates fault/contact with a wide shear zone to the west. The wide magnetic low is inferred as due to lateral changes in magnetic property by shearing. On the basis of a ground geophysical follow-up across the aeromagnetic linear, a bore hole was drilled to the depth of 198.76 m. It encountered high yielding saprolite, shallow and deep fractures upto 110.53 m depth. The cumulative discharge of the well was 7.5 litres per second. Similarly, in metasediments the NNE-SSW lineament was picked up from satellite imagery. Aeromagnetic map of the area indicates a magnetic discontinuity and a prominent linear low in the NNE-SSW direction which confirms the lineament (Fig. 5.10). The three wells drilled on the aeromagnetic linear pinpointed through the ground geophysical follow up encountered high yielding

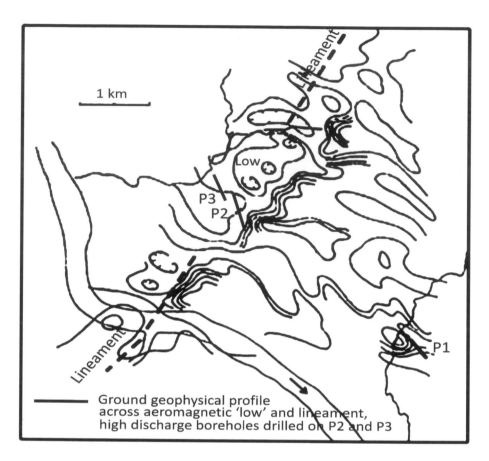

Figure 5.10 Ground geophysical surveys for groundwater in metasedimentary terrain across coincident aeromagnetic and photo linear (Chandra *et al.*, 1994). P1, P2 and P3 are ground survey profiles. Aeromagnetic data source: AMSE, Geological Survey of India.

fractures in the depth range of 27.5–147.0 m with cumulative yields varying from 7.41 to 25.83 lps.

REFERENCES

Abdelrahman, E.M. & Sharafeldin, S.M. (1996) An iterative least-squares approach to depth determination from residual magnetic anomalies due to thin dikes. *Journal of Applied Geophysics*, 34 (3), 213–220.

Agocs, W.B. (1951) Least squares residual anomaly determination. *Geophysics*, 16 (4), 686–696.

Al-Rawi, F.R. (2009) Magnetic depth estimation for dyke-like bodies by using Fraser filter – a new scheme. *Jour. Univ. of Anbar for Pure Science*, 3 (1), 89–97.

Amoush, H.A., Hammouri, N., Farajat, M.A., Salameh, E., Diabat, A., Hassoneh, M. & Adamat, R.A. (2013) Integration of aeromagnetic data and Landsat imagery for structural analysis purposes: a case study in the southern part of Jordan. *Journal of Geographic Information System*, 5 (3), 198–207.

Astier, J.L. & Paterson, N.R. (1987) Hydrogeological interest of aeromagnetic maps in crystalline and metamorphic areas. In: G.D. Garland (ed.) *Proceedings of Exploration'87: Third Decennial Int. Conf. on Geophysical and Geochemical Exploration for Minerals and Groundwater, 1987, Ontario.* Ontario Geol. Survey, Special Volume 3, pp. 732–745.

Birch, F.S. (1984) Bedrock depth estimates from ground magnetic profiles. *Ground Water*, 22 (4), 427–432.

Blakely, R.J., Langenheim, V.E., Ponce, D.A. & Dixon, G.I. (2000) *Aeromagnetic survey of the Amargosadesert, Nevada and California: a tool for understanding near-surface geology and hydrology.* U.S. Gelogical Survey Open-file report 00–188.

Bordie, R.C. (2002) *Airborne and ground magnetic, in 'Geophysical and remote sensing methods for regolith exploration.* In: Papp, E. (ed.) CRC LEME, Australia Open File Report 144, pp. 33–45.

Breiner, S. (1973) *Applications manual for portable magnetometers.* Geometrics, Sunnyvale, California, USA.

Bromley, J., Mannsrom, B., Nisca, D. & Jamtlid, A. (1994) Airborne geophysics: Application to a ground-water study in Botswana. *Ground Water*, 32 (1), 79–90.

Chandra, P.C. (1991) Study of aeromagnetic map for groundwater exploration: Recent trend in hardrock hydrogeophysics. *Bhujal News, Quarterly Journal of CGWB*, 6 (2) 37.

Chandra, P.C., Reddy, P.H.P. & Singh, S.C. (1994) *Geophysical studies for groundwater exploration in Kasai and Subarnarekha River basins.* (UNDP Project) Central Ground Water Board, Min. of Water Resources, Govt. of India, Tech. Report.

Chary, M.N., Srinivas, Y. & Sundararajan, N. (2005) Structural analysis of magnetic anomalies across Gondwana outlier near Tiruvuru, Krishna district, Andhra Pradesh. *Journal of Indian Geophysical Union*, 9 (1), 21–28.

Clark, D.A. (1997) Magnetic petrophysics and magnetic petrology: aids to geological interpretation of magnetic surveys. *AGSO Journal of Australian Geology & Geophysics*, 17 (2), 83–103.

Clark, D.A. & Emerson, D.W. (1991) Notes on rock magnetization characteristics in applied geophysical studies. *Exploration Geophysics*, 22 (3), 547–555.

Coetzee, H., Chirenje, E., Hobbs, P. & Cole, J. (2009) Ground and airborne geophysical surveys identify potential subsurface acid mine drainage pathways in the Krugersdrop game reserve, Gauteng province, South Africa. In: *Proc. 11th SAGA Biennial Tech. Meeting and Exhibition, 13–18 September 2009, Royal Swazi Spa, Swaziland*, pp 461–470.

Cresswell, R. & Gibson, D. (2004) *Application of airborne geophysical techniques to groundwater resource issues in the Angas-Bremer Plains, South Australia: A synthesis of research carried out under the South Australia salinity mapping and management support project (SA SMMSP).* South Australia, Dept. of Water, Land and Biodiversity Conservation Report DWLBC 2004/35, ISBN 0-9756945-2-9.

Cull, J.P. & Massie, D. (2002) Aquifer mapping and groundwater quality in faults and fractures. *Exploration Geophysics*, 33 (2), 122–126.

Gettings, M.E. & Bultman, M.W. (2005) A predictive penetrative fracture mapping method from regional potential field and geologic datasets, southwest Colorado Plateau, U.S.A. *Earth Planets Space*, 57 (8) 701–715.

Githiri, J.G., Patel, J.P., Barongo, J.O. & Karanja, P.K. (2011) Application of Euler deconvolution technique in determining depths to magnetic structures in Magadi area, southern Kenya rift. *JAGST, Jomo Kenyatta Univ. Agri. and Tech.*, 13 (1), 142–156.

Gobashy, M.M. & Al-Garni, M.A. (2008) High resolution ground magnetic Survey (HRGM) for determining the optimum location of subsurface dam in Wadi Nu'man, Makkah Al Mukarammah. KSA, JKAU: *Earth Sciences*, 19, 57–83.

Grauch, V.J.S. & Bankey, V. (2003) *Hydrogeologic framework of the Southern Espanola Basin, New Mexico.* U.S.Geological Survey Open-File Report 03-124.

Gunn, P.J. (1997) Quantitative methods for interpreting aeromagnetic data: a subjective review. *AGSO Jour. Australian Geology and Geophysics*, 17 (2), 105–113.

Hansen, R.O., Racic, L, & Grauch, V.J.S. (2005) Magnetic methods in near-surface geophysics. In. D K Butler, D.K. (ed.) *Near-Surface Geophysics, Chapter 1, Series: Investigations in Geophysics: No. 33*, Tulsa, Oklahoma, U.S.A., Society of Exploration Geophysicists. pp. 151–171.

Hansen, R.O. & Suciu, L. (2002) Multiple-source Euler deconvolution. *Geophysics*, 67 (2), 525–535.

Henkel, H. & Guzman, M. (1977) Magnetic features of fracture zones. *Geoexploration*, 15 (3), 173–181.

Hinze, W.J. (1990) The role of gravity and magnetic methods in engineering and environmental studies. In: Ward, S.H. (ed.) *Geotechnical and Environmental Geophysics, Vol. I: Review and Tutorial*. Tulsa, Oklahoma, USA, Society of Exploration Geophysicists, pp. 75–126.

Hinze, W.J. & von Frese, R.R.B. (1990) Magnetics in geoexploration. *Proceedings of Indian Academy of Sciences (Earth & Planetary Sciences)*, 99 (4), 515–547.

Hinze, W.J., von Frese, R.R.B. & Saad, A.H. (2012) *Gravity and magnetic exploration: principles, practices and applications*. U. K., Cambridge University Press.

Hogg, L.S. (1987) Recent advances in high sensitivity and high resolution aeromagnetics. In: G.D. Garland (ed.) *Proceedings of Exploration'87: Third Decennial Int. Conf. on Geophysical and Geochemical Exploration for Minerals and Groundwater, 1987, Ontario*. Ontario Geol. Survey, Special Volume 3, pp. 153–169.

Jain, S. (1976) An automatic method of direct interpretation of magnetic profiles. *Geophysics*, 41 (3), 531–541.

Joseph, A.D. & Chandra, P.C. (2003) Integrated interpretation of satellite imagery data and aeromagnetic map for groundwater exploration in hardrocks: case studies. *Bhujal News, Quarterly Journal of CGWB*, 18 (1–4), 66–69.

Ku, C.C. & Sharp, J.A. (1983) Werner deconvolution for automated magnetic interpretation and its refinement using Marquardt's inversion modeling. *Geophysics*, 48 (6), 754–774.

Lee, M., Morris, W., Harris, J. & Leblanc, G. (2012) An automatic network-extraction algorithm applied to magnetic survey data for the identification and extraction of geologic lineaments. *The Leading Edge*, 31, 26–31.

Lelievre, P.G. (2003) Forward modeling and inversion of geophysical magnetic data. *MS Thesis, Dept. Earth & Ocean Sciences, Geophysics, The University of British Columbia, Canada*.

Li, X. (2003) On the use of different methods for estimating magnetic depth. *The Leading Edge*, 22, 1090–1099.

Li, Y. & Oldenburg, D.W. (1998) Separation of regional and residual magnetic field data. *Geophysics*, 63 (2), 431–439.

Lopez-Loera, H., U.-Fucugauchi, J. & A-Valdivia, L.M. (2010) Magnetic characteristics of fracture zones and constraints on the subsurface structure of the Colima Volcanic Complex, Western Mexico. *Geosphere*, 6 (1), 35–46.

Macleod, I.N., Jones, K. & Dai, T.F. (1993) 3-D analytic signal in the interpretation of total magnetic field data at low magnetic latitudes. *Exploration Geophysics, ASEG*, 24 (4), 679–688.

Milligan, P.R., Reed, G., Meixner, T. & FitzGerald, D. (2004) Towards automated mapping of depth to magnetic/gravity basement-examples using new extensions to an old method (Extended Abstract). In: Proceedings of the *ASEG 17th Geophysical Conference and Exhibition, 15–19 August 2004, Sydney, Australia*.

Nabighian, M.N., Grauch, V.J.S., Hansen, R.O., LaFehr, T.R., Li, Y., Peirce, J.W., Phillips, J.D. & Ruder, M.E. (2005) The historical development of the magnetic method in exploration. *Geophysics*, 70 (6), pp. 33ND–61ND.

Nettleton, L.L. (1940) *Geophysical Prospecting for Oil*. New York, USA, McGraw-Hill Publication.

Olesen, O., Dehls, J.F., Ebbing, J., Henriksen, H., Kihle, O. & Lundin, E. (2006) Aeromagnetic mapping of deep-weathered fracture zones in the Oslo Region – a new tool for improved planning of tunnels. *Norwegian Journal of Geology*, 87, 253–267.

Paterson, N.R. & Bosschart, R.A. (1987) Airborne geophysical exploration for groundwater. *Ground Water*, 25 (1), 41–50.

Petersson, J., Skogsmo, G., Vestgård, J., Albrecht, J., Hedenström, A. & Gustavsson, J. (2007) *Bedrock mapping and magnetic susceptibility measurements, Quaternary investigations and GPR measurements in trench AFM001265, Forsmark site investigation.* SKB Sweden, P-06-136, ISSN 1651-4416.

Ponce, D.A. & Blakely, R.J. (2001) *Aeromagnetic map of the Death Valley groundwater model area, Nevada and California.* U.S. Geological Survey Miscellaneous Field Studies Report MF-2381-D.

Ram Babu, H.V., Rao, N.K. & Vijay Kumar, V. (1991) Bedrock topography from magnetic anomalies-an aid for groundwater exploration in hard-rock terrains. *Geophysics*, 56 (7), 1051–1054.

Ranganai, R.T. & Ebinger, C.J. (2007) Aeromagnetic and Landsat TM structural interpretation for identifying regional groundwater exploration targets, south-central Zimbabwe Craton. *Journal of Applied Geophysics*, 65 (2), 73–83.

Reeves, C.V., Reford, S.W. & Milligan, P.R. (1997) Airborne geophysics: old methods, new images, *(Keynote Session).* In: Gubins, A.G. (ed.) *Proceedings of Exploration 97: Fourth Decennial International Conference on Mineral Exploration, 14–18 September, 1997, Toronto, Canada,* GEO F/X Div. of AG Inf. Sys. Ltd. Canada, pp. 13–30.

Reid, A.B., Allsop, J.M., Granser, H., Millett, A.J. & Somerton, I.W. (1990) Magnetic interpretation in three dimensions using Euler deconvolution. *Geophysics*, 55 (1), 80–91.

Roberts, C.L. & Smith, S.G. (1994) A new magnetic survey of Lundy Island, Bristol Channel. *Proceedings of the Ussher Society*, 8 (3), pp. 293–297.

Smith, B.D., Grauch, V.J.S., McCafferty, A.E., Smith, D.V., Rodriguez, B.R., Pool, D.R., D.-Pan, M. & Labson, V.F. (2007) Airborne electromagnetic and magnetic surveys for groundwater resources: A decade of study by the US Geological Survey. In: Milkereit, B. (ed.) *Exploration 07: Proceedings of the Fifth Decennial Int. Conf. on Mineral Exploration, Sept 9 to 12, 2007, Toronto, Canada.* pp 895–899.

Stanley, J.M. (1977) Simplified magnetic interpretation of the geologic contact and thin dike. *Geophysics*, 42 (6), 1236–1240.

Sultan, S.A., Mekhemer, H.M., Santos, F.A.M. & Abd Alla, M. (2009) Geophysical measurements for subsurface mapping and groundwater exploration at the central part of the Sinai peninsula, Egypt. *The Arabian Journal for Science & Engineering*, 34 (1A), 103–119.

Thompson, D.T. (1982) EULDPH: A new technique for making computer-assisted depth estimates from magnetic data. *Geophysics*, 47 (1), 31–37.

Thomson, S., Fountain, D. & Watts, T. (2007) Airborne geophysics – evolution and revolution. In: Milkereit, B. (ed.) *Exploration 07: Proceedings of the Fifth Decennial Int. Conf. on Mineral Exploration, Sept 9 to 12, 2007, Toronto, Canada,* pp, 19–37.

Vacquier, V., Steenland, N.C. Henderson, R.G. & Zietz, I. (1951) Interpretation of aeromagnetic maps. *Geological Society of America Memoir 47.*

Wynn, J. (2006) *Mapping groundwater in three dimensions – an analysis of airborne geophysical surveys of the Upper San Pedro River Basin, Cochise County, Southeastern Arizona.* U.S. Geological Survey Professional Paper – 1674.

Chapter 6

The electrical resistivity method

6.1 INTRODUCTION

Amongst all the geophysical methods, electrical resistivity has been used extensively in ground water exploration. It has addressed a variety of groundwater exploration and development issues in hard rock terrain the world over. In hard rock the method is used to identify the weathered zone, its thickness and groundwater yielding character, bed rock topography, saturated fractured zones and their depths, lateral extension and orientation, other structures like basic dykes and quartz reefs controlling groundwater movement and the quality of groundwater in terms of electrical conductivity. Also, it has been used to characterize the vadose zone for managed aquifer recharge and to monitor groundwater movement.

Electrical resistivity of geological formations having direct bearing on the presence of groundwater, characterizes the variations in groundwater conditions. The electrical conduction in near surface geologic formations with relevance to groundwater is mainly electrolytic. It is through the charges carried by ions resulting from dissociation of salts dissolved in water present in the pores. The bulk resistivity of a formation depends only a little on its matrix (without electronic conduction) and mostly on its ability to conduct electric current by ionic movement through connected primary inter-granular pores and secondary inter-connected pore spaces created by structural discontinuities at different scales like joints, fractured zones, faults and shear zones.

With an increase in saturated pore volume, resistivity decreases, but the relation is not direct, as it varies with lithology as well as total dissolved salts in saturating pore water. The higher the conductivity of water the lesser is the bulk resistivity. In formations with clay, the surface conductivity of clay particles having high surface area prevails where as in clay free sand formation the relation between formation resistivity and pore water resistivity is linear for low resistivity saturating water. The ratio of bulk resistivity of formation (ρ_t) to the resistivity of saturating water (ρ_w) is known as the 'formation factor' (F) or formation resistivity factor. The relationship known as Archie's Law is expressed as

$$F = \frac{\rho_t}{\rho_w} = a\phi^{-m}$$

where ϕ is porosity, a and m are empirical constants and known as the coefficient of saturation and cementation factor respectively. Generally for sediments 'a' varies from 0.6 to 1 and 'm' varies from 1.5 for poorly cemented to 2 for well cemented

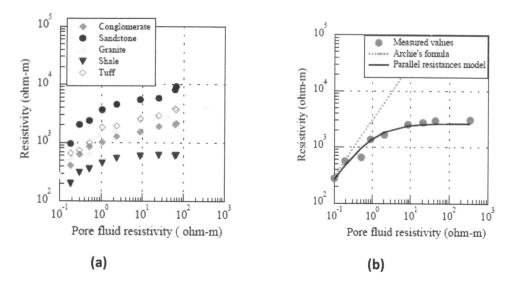

Figure 6.1 Relationship between formation and pore fluid resistivity: (a) resistivity of different rock specimens with saturating fluid of NaCl concentration 20 to 32500 ppm and (b) comparison between measured and calculated resistivities (Source: Matsui et al., 2000)

formations. For very low resistivity formation water the bulk resistivity depends mainly on formation water resistivity and it increases with water resistivity. Matsui *et al.* (2000) reveal that when saturating water resistivity is high, between 20 and 80 ohm.m, the variation in its resistivity has little effect on bulk resistivity (Fig. 6.1).

6.2 RANGES OF ELECTRICAL RESISTIVITY IN HARD ROCK

As already mentioned, bulk resistivity of any water bearing formation manifests the variations in quantity and quality of groundwater present and the lithological character. Therefore, amongst all the physical properties of rocks, electrical resistivity varies most, roughly over a range of 10^{-1} ohm.m for highly saline water saturated sand to the order of 10^4 to 10^7 ohm.m for dry and compact metamorphic and igneous rocks. There are generalized ranges of resistivity for different formations saturated with fresh water, e.g., sands of various grain size, clays, weathered and fractured granites and gneisses, sandstones, cavernous limestones, vesicular basalts etc. As a result of the combined effect of quality of saturating formation water, degree of water saturation, lithology and mineralogy there are overlaps in resistivity ranges which require area-specific standardization to help characterize the formations. The significant contrast in resistivity between dry and water saturated formations, fractured and compact formations and formations with fresh and brackish/saline water makes the electrical resistivity measurements effective in groundwater prospecting.

Compact igneous rock – a hard rock without any porosity has very high resistivity, generally much higher than 1000 ohm.m. The metamorphic rocks have lesser

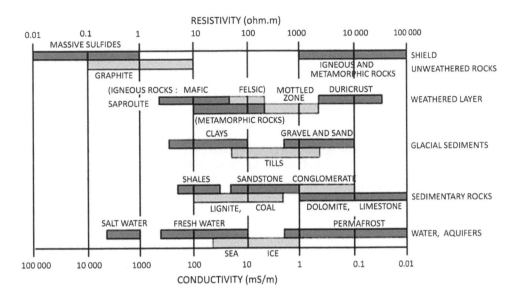

Figure 6.2 Ranges of electrical resistivity for various rock formations (Source: Palacky, 1991).

resistivities. The weathered portion of these rocks generally holds water and has resistivity less than the compact rock. The alteration of compact igneous and metamorphic rocks to weathered material, saprolite and transformed clay, presence of fractured zones/joints and water saturation and deterioration in quality of groundwater drastically reduce the bulk resistivity of these materials from thousands of ohm.m to hundreds or even tens of ohm.m. A very small presence of water reduces the resistivity considerably as shown by Telford *et al.* (1990). The resistivity of the order of 10^{10} ohm.m for dry granite sample reduces to 1.8×10^6 ohm.m with 0.19% water content and to 4.4×10^3 ohm.m with 0.31% water content whereas the resistivity of 1.3×10^8 ohm.m for dry basalt reduces to 4×10^4 ohm.m with 0.95% water content. Also, with higher clay content the resistivity is reduced drastically. The generalized resistivity ranges for different geological formations (Palacky, 1991) are shown in figure 6.2. As such there is no characteristic resistivity value for the weathered, saprolite and fractured material and they are generally associated with overlapping ranges of resistivity. In general, there is a decreasing order of resistivity (from more than 1000 ohm.m to about 50 ohm.m) from compact to fractured rock followed by saprolite and weathered material. Palacky *et al.* (1981) summarize the resistivity ranges obtained from West Africa for weathered and fresh rocks (Table 6.1). Balakrishna and Ramanujachary (1985) summarize the resistivity ranges for basalts and granites in central and southern parts of India (Table 6.2). Based on surface resistivity measurements a detailed range (Table 6.3) for basalt under various conditions is given by Chaturvedi *et al.* (1979) and CGWB (1982). As compared to basaltic terrain the resistivity ranges obtained from granitic terrain are distinct with less overlapping. The resistivity of weathered zone in granites can range from 10 to 100 ohm.m and at the most up to 250 ohm.m. This is observed in the crystallines in India and by Palacky

Table 6.1   Formation resistivity ranges in ohm.m obtained from hard rocks of West Africa (Palacky, 1981 based on Engalenc, 1978).

	Granite	Schist	Amphibolite
Weathered layer	25–50	10–30	5–15
Transition zone	40–200	250–400	10–80
Fresh parent rock	>1500	>1000	>500

Table 6.2   Resistivity ranges in ohm.m for basalts and granites in central and southern parts of India (Balakrishna and Ramanujachary, 1985).

Weathered Basalt	15–25
Weathered Basalt with clay	<15
Compact Basalt	>50
Jointed Basalt	15–40 up to 100
Intertrappean	<40
Highly weathered granite	20–50
Semi-weathered layer (Saprolite)	50–120
Fractured and jointed granite	120–200
Hard granite	>200

Table 6.3   Resistivity ranges in ohm.m for Deccan Trap basalts, Maharashtra, India (Chaturvedi *et al.*, 1979, CGWB, 1982).

Red Bole (clayey material)	5–15
Highly weathered, fractured/or vesicular, saturated with water	15–30
Moderately weathered, fractured or vesicular, saturated with water or partly dry	30–50
Slightly weathered, fractured or vesicular that may contain some water or highly weathered or fractured devoid of water	50–70
Weathered or vesicular dry	70–100
Massive	>100

and Kadekaru (1979) in parts of Brazil. The resistivity of the fractured zone starts from about 200 ohm.m and can be as high as 500 ohm.m depending on the degree of fracturing and saturation. These resistivities are obtained from resistivity sounding interpretations standardized through borehole lithological and geophysical logs. The bulk resistivities determined in situ from field investigations under natural environment differ from those determined in the laboratory. The laboratory determined resistivities are, in general, higher and cannot be used for characterization at field scale (Angenheister, 1982 referred in Palacky, 1991).

6.3   BASICS

The electrical resistivity of any homogeneous material is defined as numerically equal to the resistance offered to the flow of direct current between two opposite faces of

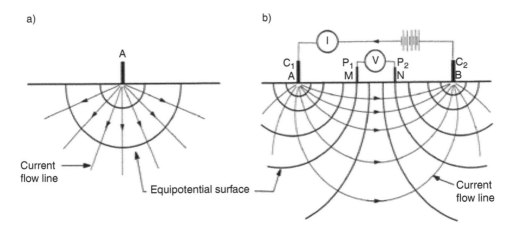

Figure 6.3 Current and equipotential lines in homogeneous subsurface (a) single current source and (b) current and potential electrode pairs.

a unit cube of the material. Resistivity of any material sample can be determined by measuring its resistance and the dimensions- length and area of cross-section. If R is the resistance of the material sample measured in ohm, L its length in metres and A its area of cross-section in square metres, then its resistivity ρ in ohm.metre is defined as ρ = R A/ L. Resistivity is a bulk property, independent of shape and size, i.e., geometry of the material sample.

To measure the depth-wise and lateral variations in resistivity of geologic formations, an electrical resistivity survey is conducted on the ground surface by injecting a known amount of direct current into the ground between a pair of electrodes and measuring the potential developed across another pair of electrodes on ground due to the current (Fig. 6.3b). The basic objective is to study the response of current injected into the ground. The electrodes can be planted on the ground in a variety of arrays or configurations.

The current entering from a point electrode in a homogeneous and isotropic whole space of resistivity ρ flows radially. Considering a sphere with electrode at the centre, the current density, electric field and potential on a spherical surface at a distance r from the centre are

$$J_r = \frac{I}{4\pi r^2}, \quad E_r = J_r \rho = \frac{I\rho}{4\pi r^2} \quad \text{and} \quad V_r = \frac{I\rho}{4\pi r}$$

and for a homogeneous isotropic half space (subsurface), figure 6.3a

$$J_r = \frac{I}{2\pi r^2}, \quad E_r = J_r \rho = \frac{I\rho}{2\pi r^2} \quad \text{and} \quad V_r = \frac{I\rho}{2\pi r}$$

Two electrodes (the source and sink) are required for current to flow in the half space (Fig. 6.3b). The potential V_{P1} developed due to these current electrodes, say C_1 and C_2

at a point P_1 located at distances r_1 and r_2 from these electrodes is obtained by adding the individual potentials due to C_1 and C_2. It is given as

$$V_{P1} = \frac{I\rho}{2\pi} \left[\frac{1}{r_1} - \frac{1}{r_2} \right]$$

To obtain a potential difference between 2 points P_1 and P_2 developed due to current I between the current electrodes, potential V_{P2} at P_2 is calculated as above and then by the method of superposition, the difference in potentials (ΔV) between V_{P1} and V_{P2} is obtained, which can be written as

$$\Delta V = \frac{I\rho}{2\pi} \left[\frac{1}{r_1} - \frac{1}{r_2} - \frac{1}{r_3} + \frac{1}{r_4} \right]$$

where r_3 and r_4 are the distances of point P_2 from current electrodes C_1 and C_2. Then the resistivity of the homogeneous half space is given as

$$\rho = \frac{\Delta V}{I} \cdot K$$

$$K = \frac{2\pi}{\left[\frac{1}{r_1} - \frac{1}{r_2} - \frac{1}{r_3} + \frac{1}{r_4} \right]}$$

K is known as the 'Geometric Factor' and depends on the geometry of electrode array or configuration.

In case the subsurface is layered or inhomogeneous which is, in general observed, multiplication of the ratio of developed potential difference to current input with 'geometric factor' for the electrode geometry gives 'apparent' resistivity (ρ_a) of the 'inhomogeneous' ground. It varies with electrode position. In case the resistivity varies with direction, i.e., in an anisotropic subsurface, the apparent resistivity varies with orientation of electrode arrangement. To explain further, for homogeneous isotropic ground the plot of calculated resistivities vs electrode positions with the latter along the abscissa will be a straight line parallel to the abscissa while in the case of inhomogeneous ground the plot may take any shape based on depth-wise resistivity distribution. Apparent resistivity is not the average of true layer-resistivities and may not also represent the true resistivities. However, values mostly fall within the range of true resistivities of subsurface layers over which the measurements are made. In other words, apparent resistivity of inhomogeneous ground will be equivalent to true resistivity if ground becomes homogeneous. There are several approaches to computing apparent resistivities over layered earth and details can be obtained in Keller and Frischknecht (1966) and Bhattacharya and Patra (1968). Besides, readers may refer to other classical publications by Van Nostrand and Cook (1966), Kunetz (1966), Zohdy et al. (1974) and Koefoed (1979a).

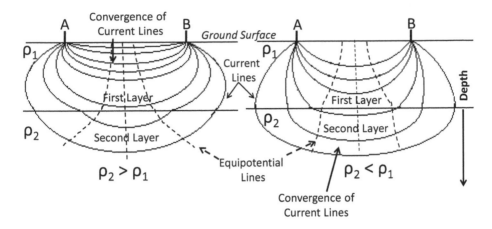

Figure 6.4 Generalized distortion in current and equipotential lines in a two-layered earth of different resistivities.

6.4 VERTICAL ELECTRICAL SOUNDING

An electrical resistivity survey is carried out in three ways, viz., sounding or Vertical Electrical Sounding (VES) for 1-d information on depth-wise resistivity distribution, profiling for information on lateral resistivity variation and imaging which combines sounding and profiling for 2-d and 3-d information on subsurface resistivity distributions. To study resistivity variation with depth, i.e. acquiring information deeper and deeper, the separation between current electrodes is increased successively keeping the central point of the electrode array fixed. As the current electrode distance is increased the current penetrates deeper and deeper. In homogeneous earth the current lines form a hemisphere. But in an inhomogeneous, i.e., layered earth the shape of current and equipotential lines depends on depth-wise resistivity distribution (Fig. 6.4). A graphical plot of observed apparent resistivity values for current electrode spacings is the apparent resistivity field curve at the point of observation (Fig. 6.5). Since, 'depth of investigation' to be discussed latter, is increased by increasing the distance between current and potential electrodes, the variations in apparent resistivity with increasing current electrode distance can be interpreted for subsurface depth-wise variation in resistivity in terms of distinct geoelectrical layers and hence the hydrogeological conditions at the point of measurement.

6.4.1 Electrode arrays

For sounding the current and potential electrodes can be placed on the ground in a variety of configurations or arrays. Szalai and Szarka (2008) identify as many as 102 independent array types and classify them on the basis of superposition of measurements, focusing of current and collinearity of the array. However, the conventional arrays are, Schlumberger, Wenner, dipole-dipole, two-electrode, half-Schlumberger (Keller and Frischknecht, 1966; Yadav, 1988), Lee-partitioning etc. Out of these, the

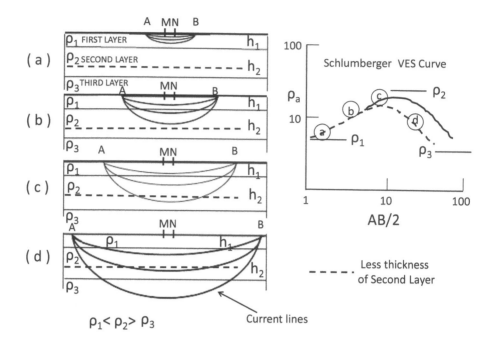

Figure 6.5 Development of a sounding curve; (a) at smaller current electrode spacing $\rho_a \approx \rho_1$, (b) at intermediate spacing ρ_a attains an intermediate value depending on ρ_2, (c) by further increasing the electrode spacing the effect of ρ_2 is pronounced and with greater thickness of second layer ρ_a becomes closer to ρ_2 and (d) effect of ρ_3 is seen and $\rho_a \approx \rho_3$ at very large current electrode spacing.

most common electrode arrays are Schlumberger, Wenner and dipole-dipole (Fig. 6.6a and 6.6b). The 'Schlumberger' and 'Wenner' array of electrodes are named after French and American physicists respectively who developed the technique of earth resistivity measurements in the first two decades of 20th century. These two arrays have active four electrodes, collinear and symmetrical, i.e., electrodes are placed in a straight line and locations of current electrodes are symmetrical with the potential electrodes conventionally positioned in between the current electrodes. The current electrodes are referred as A and B electrodes and the separation or spacing between them as AB and potential electrodes as M and N electrodes and the spacing as MN.

In principle, for sounding with the Schlumberger array the potential gradient is measured at the centre of the array. For this, theoretically, MN should be infinitesimal. Keller and Frischknecht (1966) show that by keeping MN less than 0.2175 times AB, the error is within 5%. In practice, MN is kept not more than 0.2 (1/5th) AB. In field operation both the current electrodes are moved outward simultaneously in steps along a straight line keeping the closely spaced collinear potential electrodes fixed at the centre of the sounding point as long as a measurable (determined by the sensitivity of instrument) potential difference is obtained. The MN can be increased under the above mentioned condition or whenever the potential value is too low (may be at MN = 1/10

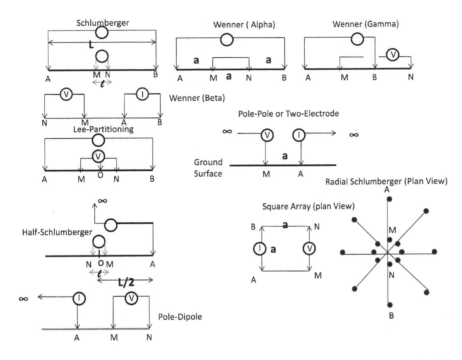

Figure 6.6a Different electrode arrays, current and potential electrodes on the ground surface.

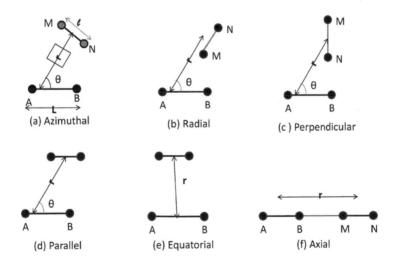

Figure 6.6b Various dipole-dipole electrode arrays (plan view).

or 1/20 AB) for the sensitivity of instrument used. In hard rock areas, formation resistivity generally increases with depth and therefore very low potentials are rarely recorded even for larger separations between current and potential electrodes and therefore MN can be increased only 2 to 3 times upto AB: 200 or 300 m. Since the sounding has to be devoid of any effect of changes in MN, even the closely spaced potential electrodes with which the sounding is initiated may be kept fixed throughout the sequential increase of current electrode spacing, if a measureable potential difference exists. At larger current electrode spacing the geometric factor is quite large and measured potential is too small so to keep noise at the minimum Stummer (2003) gives a formula relating current injection (I_{max}), expected average subsurface layer resistivity (ρ_m), minimum potential (ΔV_{min}) which can be recorded and the geometric factor (K_{max}), which is $K_{max} = \frac{I_{max}\rho_m}{\Delta V_{min}}$. This equation can be used to decide upon the maximum spread of AB and MN for a particular equipment used in an area. For example, if $I_{max} = 100$ mA, $\rho_m = 50$ ohm.m, $\Delta V_{min} = 1$ mV, then $K_{max} = 5000$ m; so for maximum AB of 400 m, keeping K_{max} around 5000 m the MN can be around 15 to 20 m i.e., 1/20 AB.

In Wenner array sounding, MN is kept fixed at 1/3rd of AB. That is, all the electrodes are always placed equi-spaced and for each outward movement of current electrode, unlike the Schlumberger array, the potential electrodes are also moved outward keeping the inter-electrode distance equal so that 1:3 spacing ratio is maintained. The Wenner array is sometimes preferred over the Schlumberger as it gives higher potential differences, but its major demerit is increased probability of encountering near surface lateral inhomogeneity at potential electrodes because of shifting of potential electrodes for each measurement and its effect on apparent resistivity values. Also, there is no way to remove the effect of inhomogeneity under the potential electrode as done in Schlumberger sounding by shifting the curve segments. The presence of inhomogeneity in proximity of the electrodes affects the shape of sounding curve and misleads the interpretation. The effect is prominent in Wenner sounding as compared to Schlumberger sounding shown by numerical computation (Loke, 2001). As such the noise due to inhomogeneities in Schlumberger sounding being the minimum it is preferred for soundings in hard rocks.

In dipole-dipole arrays, the electrode spacings for current as well as potential electrode pairs are considerably small as compared to the distance between these two pairs. Depending on the relative orientations of these electrode pairs, dipole-dipole arrays can be azimuthal, equatorial, radial, parallel, axial and perpendicular (Fig. 6.6b). These offer the possibility of conducting sounding in uneven topographic conditions as well as in areas where collinear arrangement of electrodes is not feasible due to non-availability of long straight stretches of open space. Also the problems of cable leakages during field operation can be avoided. Sounding for deeper investigations is generally conducted by dipole-dipole array, for which Schlumberger or Wenner arrays will require large spreading of cables and also the field surveys will be quite tedious. The demerit of the dipole-dipole array is that it requires higher energizing current than that for Wenner and Schlumberger arrays. This is because the electric field of a dipole falls of inversely with cubic power of distance. Also, dipole array sounding is more sensitive to near surface inhomogeneities and produces scattered data (Ward, 1990). In hard rock areas soundings for groundwater exploration are rarely conducted by this array.

For two-electrode or pole-pole array, out of the current and potential electrode pairs, one current and one potential electrode are kept at infinity, off the profile line along which the active electrodes are moved, i.e., at a distance more than 10 times the distance between other two active electrodes. In the half-Schlumberger array, one of the current electrodes is kept at infinity (i.e., distant current electrode is located about 5

Table 6.4 Geometric Factors for different electrode arrays shown in Fig. 6.6 a and 6.6 b.

Electrode array	Geometric factor
Schlumberger	$\pi\{(L/2)^2 - (\ell/2)^2\}/\ell$; L and ℓ are the current (AB) and potential electrode (MN) spacings respectively
Wenner	α: $2\pi a$; β: $6\pi a$; γ: $3\pi a$, a is the inter electrode spacing
Dipole-dipole: azimuthal	$2\pi r^3/L\,\ell\,\sin\theta$; L and ℓ are the lengths of current and potential dipoles, r is distance between the centres of current and potential dipoles, θ is the angle between the two dipole axes
radial	$\pi r^3/L\,\ell\,\cos\theta$
axial or polar	$\pi r^3/L\,\ell$
equatorial	$2\pi\ r^3/L\,\ell$
parallel	$2\pi\ r^3/L\,\ell(3\cos^2\theta - 1)$
perpendicular	$2\pi\ r^3/3L\ l\ (\sin\theta\cos\theta)$.
Two-Electrode/Pole-Pole	$2\pi a$; a is the distance between active electrodes
Half Schlumberger/ Pole-Dipole	$2\pi\{(L/2)^2 - (l/2)^2\}/\ell$; L and ℓ are the current (AB) and potential electrode (MN) spacings respectively, i.e., twice the geometric factor for Schlumberger array
Lee-partitioning	$4\pi a$; a is the inter electrode spacing
Square	$2\pi A/\{2 - (2)^{1/2}\}$, where A is the length of a side of the square

times the distance to near current electrode) and it need not be collinear with the other three electrodes and generally kept in a perpendicular direction. This is quite useful where sufficient straight space is not available for conducting Schlumberger sounding (Anjorin and Olorunfemi, 2011). Lee-partitioning is same as Wenner array, except that one additional electrode is placed at the centre in between the potential electrodes. The potential difference between the central electrode and the other two potential electrodes is sequentially measured. The geometric factors for the arrays shown in figures 6.6a and 6.6b are given in Table 6.4.

6.4.2 Depth of investigation

The 'depth of current penetration' and 'depth of investigation' (DOI) are commonly used terms. When an electrical resistivity sounding is conducted a question is generally asked, up to what depth we get the information for a certain electrode array and electrode spacing. That is, to relate the resistivity sounding measurement up to a particular electrode spacing to a certain depth. Though there are rules of thumb for the DOI, they are not valid always and the answer is quite complex.

The depth of current penetration is a function of spacing between the current electrodes. For homogeneous earth, Van Nostrand and Cook (1966) show that only 37% of total current penetrates to a depth equal to 1/3rd of the current electrode spacing, 50% penetrates to a depth equal to half the spacing and 70.5% penetrates to a depth equal to the current electrode spacing. Depth of current penetration depends on current electrode separation only, whereas, DOI depends on the separation between current and potential electrodes.

The DOI is explained in a different way. In electrical resistivity measurement the potential difference measured between potential electrodes is the sum of contributions from different depths, and contributions from different depths are not the same. So,

DOI could be defined as that depth at which a thin horizontal layer contributes maximum to the total signal measured at the ground surface. This definition has been used by Roy and Apparao (1971, referred Evjen, 1938), Banerjee and Pal (1986), Bhattacharya and Sen (1981) and Barker (1989). The DOI has been computed from depth of investigation characteristic (DIC) function response to a certain array of electrodes for a homogeneous, isotropic horizontal thin layer placed at different depths in a homogeneous subsurface. The DIC curve is presented as a function of depth of thin layer. According to Roy and Apparao (1971) for a particular electrode arrangement the depth of the thin layer at which DIC curve attains maximum response is the DOI for that electrode arrangement. For Wenner it is 0.11L and for Schlumberger it is 0.125L, where L is the distance between current electrodes. Edwards (1977), instead of maximum response uses the median of DIC function which he refers to as effective DOI. Median depth is that depth at which the integral of DIC function from surface to that depth is equal to that from that depth to infinity. Accordingly the DOI calculated by Edwards is greater than that calculated by Roy and Apparao. It is 0.17L and 0.19L for Wenner and Schlumberger arrays respectively. The computation of DOI described above is for homogeneous earth. For layered-earth it is affected by subsurface layer resistivities and thicknesses and different DOIs can be obtained and it cannot be defined in a straightforward way based on current-potential electrode separation. Bhattacharya and Sen (1981) indicate a decrease in DOI with increasing anisotropy, i.e., where resistivity varies with direction. DOI is related to vertical resolution. With increase in DOI the vertical resolution decreases. Szalai et al. (2009) indicate that though the pole-pole array has a large DOI, almost 3 times as compared to Wenner and Schlumberger arrays, its vertical resolution is the least. The DOI is defined in another way by Oldenburg and Li (1999). They define DOI as the depth beyond which any perturbation in layer parameter does not affect the curve shape or in other words the depth beyond which data are not sensitive to variations in physical property. Olderburg and Li define DOI index ranging between 0 and 1.

6.5   RESISTIVITY PROFILING

Sequential measurement of apparent resistivity at equally spaced observation points along a straight profile line with an array of fixed current and potential electrode spacing, i.e., with almost a fixed DOI, is known as resistivity profiling. This gives lateral variation in apparent resistivity along the profile for a certain depth. It is interpreted in terms of lithologic variation/contact and/or geological structure like fault and fractured zone in the compact bed rock, dyke, quartz reef etc representing a vertical to sub-vertical resistive or conductive body and its association with presence or absence of groundwater. Though the resistivity profile across these structures is supposed to present highs and lows in apparent resistivity anomaly, the sharp features get obscured or subdued by the presence of conductive overburden of varying thickness and also get modified with the orientation of profile with respect to the structure. Profiling is conducted across structures and orientation of the resistivity profile is kept perpendicular to the strike of geological structure to be investigated. Profiling can be conducted by all the electrode arrays mentioned above for sounding. However, it is, in general, conducted using Wenner, axial dipole-dipole, and gradient arrays (modification of

Schlumberger array). For deeper lateral scanning the inter-electrode spacing in Wenner arrays, distance between current and potential electrodes in Schlumberger and distance between centres of current and potential dipoles in dipole-dipole arrays is increased. Profiling with gradient array is a simpler technique, in which the array of current and potential electrodes is not moved from one station to other. The current electrodes are planted at a fixed large separation and only the potential electrode pair is moved. The gradient resistivity profiling (GRP) technique is useful in fracture detection. The field operation is easy and economical. The technique is discussed below in detail.

Prior to conducting profiling by any array a few VES are conducted in the area to qualitatively decide upon the electrode spacing(s) of profiling required for a certain DOI. The station interval for measurement is fixed on the basis of the width of structure and is kept less than anticipated width of the structure. For drawing inference from resistivity profiling data the effects of top layer resistivity and near surface inhomogeneities are considered as these complicate the response. A reduction in top layer resistivity alone may reflect a resistivity low in the profile (Chandra *et al.*, 1983). This is illustrated by an example of 3-layer VES curve commonly observed in hard rocks where the substratum is highly resistive (Fig. 6.7). In this case out of 5 parameters (3 layer resistivities: ρ_1, ρ_2, ρ_3, and 2 layer thicknesses: h_1 and h_2) only the top

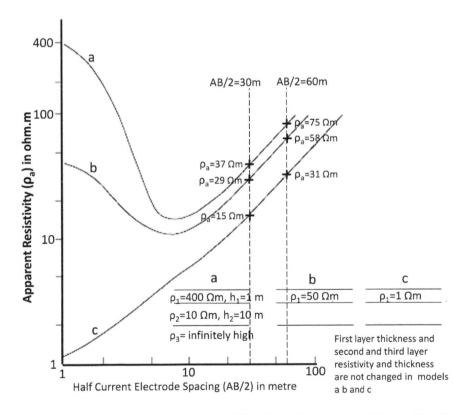

Figure 6.7 Effect of top layer resistivity as seen in VES and its influence on resistivity profiling (Source: Chandra *et al.*, 1983).

layer resistivity is varied from 1 to 400 ohm.m which affects the measured apparent resistivities at all the current electrode spacings. For a particular electrode separation, say 30 or 60 m, which may be used for profiling, three apparent resistivity values can be obtained by just varying the top layer resistivity. It is a practice to add a resistivity profiling with very small electrode spacing (5 to 10 m) to the multi-spacing profiling planned for deeper information and check the value of top layer resistivity which is more or less equal to the apparent resistivity values obtained at smaller spacings. A caution is to be observed if the apparent resistivity values at small electrode spacings are low. For details of resistivity profiling and interpretations the classical works of Van Nostrand and Cook (1966) and Apparao (1991) may be referred.

6.5.1   Gradient array resistivity profiling

The technique of Gradient Array Resistivity Profiling or Gradient Resistivity Profiling (GRP) also known as Brant array (Zohdy et al., 1974) is found to be fairly effective in delineating relatively conductive fractured zones in compact rock immediately underlying the weathered zone. The technique is discussed in detail as the 'Rectangle Method' by Kunetz (1966). It has been used for near surface investigation by Miele et al. (1996) to delineate shallow targets using smaller current electrode separations (AB) of 60 to 130 m. Shettigara and Adams (1989) conducted gradient array resistivity measurements for deeper targets using current electrode separation as large as 4000 m. Its effectiveness in delineating saturated fractured zones in crystalline and metamorphic rocks for groundwater exploration was explored by Chandra et al. (1994) and has been in practice for more than three decades in India. The advantage of gradient array techniques is that unlike other resistivity profiling the current electrodes are fixed and except for the laying of the profile line it requires relatively less labour in potential measurements. The field operation is less cumbersome, faster and economical in comparison to other resistivity profiling.

The concept of potential gradient measurement was introduced by Hedstrom in 1932 (Van Nostrand & Cook, 1966). While traversing for gradient measurements current electrodes are kept fixed at a separation as large as possible, depending on the availability of space, say 500 to 1000 m across the suspected zone or structure, foliation trend or lineament traced from satellite imagery and aeromagnetic map (Fig. 6.8). The orientation of GRP with respect to strike of lineament or geological structure is important as it controls the measured resistivity values (Whiteley, 1973). The current electrodes at large separations produce a quasi-homogeneous electric field in the central part of the array. That is, in a homogeneous earth current lines will be almost parallel and uniform in density. The central one third stretch between the current electrodes, where DOI is high, attaining maximum at the centre, is discretely scanned by the potential dipole. Large current electrode separation provides a larger central one third stretch for potential gradient measurement. It measures potential gradient in the direction of the electric field. The observed quantity approximates the electric field. The measurements are made at a regular station interval which can vary from 2 to 10 m, depending on the scale of target. The potential dipole length could be 10 to 20 m. It can be kept as 1/30 of the current electrode spacing (Ward, 1990). It depends on the range of potential developed due to current electrode separation for a normal current input. According to Middleton (1974) for constant field assumption the required minimum

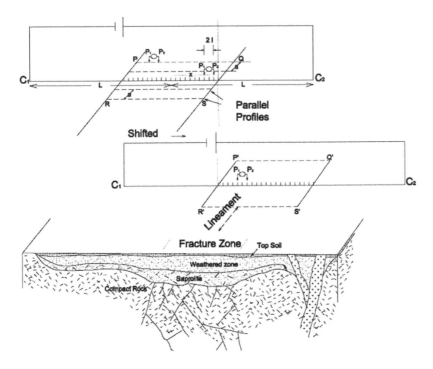

Figure 6.8 Layout of Gradient Array Resistivity Profiling; C_1C_2: fixed current electrodes; P_1P_2: moving potential dipole; PQRS, P'Q'R'S': central area between the current electrodes with parallel profiles for scanning by the potential dipole; RS or R'S' = 1/3 C_1C_2; PR or P'R' = 1/2 C_1C_2, profiles are shifted for regional coverage.

potential electrode to current electrode separation should be 1/100 as the anomaly gets altered when the value of potential electrode separation approaches the depth of the target. The measurement being an average gradient between the potential electrodes, it is better to keep the potential dipole as small as possible to record the variation due to all subsurface heterogeneities. Pratt and Whiteley (1974) based on computer simulation concluded that potential electrode spacing should be at most equal to the width of the target for optimum geometric coupling and lesser volume averaging. In practice while deciding the length of potential dipole, the accuracy of instrument used is also considered as very small potentials will be obtained with such large current electrode spacing. The potential dipole length is kept as a multiple of station interval for easy operation. The measured potential is divided by current input and multiplied by geometric factor to get apparent resistivity. The formula for apparent resistivity (ρ_a) computation is

$$\rho_a = \frac{\pi}{2l} \cdot \frac{L^2 \left(1 - \dfrac{x^2}{L^2}\right)^2}{1 + \dfrac{x^2}{L^2}} \cdot \frac{\Delta V}{I} = K \cdot \frac{\Delta V}{I}$$

where 2l is distance between the potential electrodes (shown in figure 6.8)
2L is distance between the current electrodes
x is distance between the centre of the current electrodes and centre of potential
   electrodes.
Δ V is potential observed, I is the current input
K is geometric factor

The plot of apparent resistivity values on a linear scale at the centre of potential
dipole locations gives apparent resistivity profile across the desired zone (Fig. 6.9).
Longitudinal continuation of profiling along the profile line can be attempted by

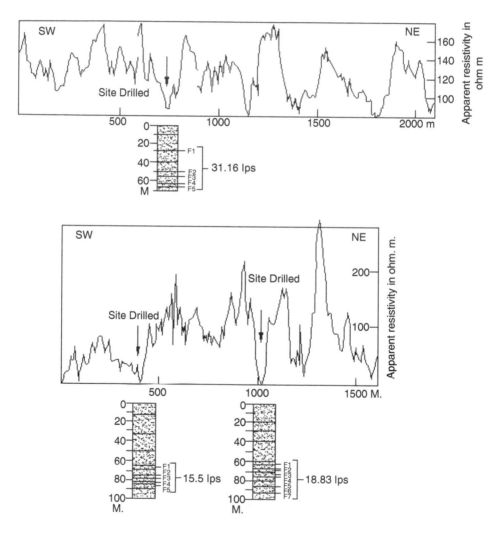

Figure 6.9 Regional Gradient Resistivity Profile across lineament to locate site for water well drilling in
compact sandstones and quartzites, U.P., India The boreholes drilled encountered a number
of fractured zones (F1 to F7) with cumulative discharge ranging from 15.5 to 31.16 lts per
second (lps).

roll-on. For example, with 500 m current electrode separation, at a time only 170 m (≈one thirds of 500 m) stretch can be covered, so, for covering a 300 m stretch, current electrodes are shifted along the profile with overlapping in such a way that a few common measuring points on profile ends are available. The resistivity values at the common end points may not match and a shift in apparent resistivity value can be observed (Fig. 6.9). To locate lateral extensions of the anomaly orthogonal to profile line, parallel profiles can be placed on either side of the central profile keeping the position of the current electrode stationary (Fig. 6.8). An example of such parallel profiles is shown in figure 6.10. In this fashion, the central one-thirds area can be covered through parallel profiles. The area covered is also named as 'Rectangle of Resistivity' (Zohdy *et al.*, 1974).

Since the longer the fractured zone the better is the probability of getting higher discharge of groundwater, conducting parallel profiles is quite useful in delineating lateral extents of a low resistivity fractured zone. The parallel-profile spacing depends on the separation of current electrodes. It is to be noted that 40% of total current

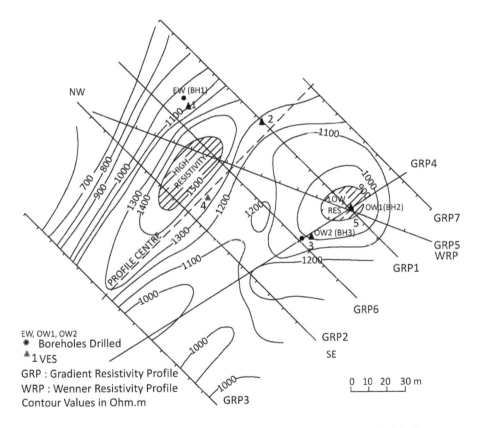

Figure 6.10 Parallel gradient array resistivity profiles (GRP) to ascertain the trend of the low resistivity fractured zone in granites, Jharkhand, India. The borehole BH1 and BH2 drilled on the relatively low resistivity zones had higher discharge than the BH3 drilled on the relatively higher resistivity anomaly in the south eastern part. Wenner resistivity profile (WRP) was conducted to check the over burden resistivity (Chandra *et al.*, 1994).

remains within a cylinder of radius equal to 66% of current electrode spacing (Van Nostrand & Cook, 1966). Kunetz (1966) as well as Bhattacharya and Dutta (1982) however restricted it to one-fourth of the current electrode separation. The apparent resistivity values for profiles other than the central collinear one are corrected for profile off-set by a graph. In the case of parallel profiles apparent resistivity is computed with the geometric factor (K′). Here

$$\rho_a = \frac{\pi L^2}{2l} \cdot K' \cdot \frac{\Delta V}{I}$$

The value of K′ is obtained from the graph between Z and D, where Z = a/L and D = x/L, a is spacing of parallel profile from central collinear profile. The graph of Z vs D (Sumner, 1976) is given in figure 6.11.

The sequence in field practice is to first conduct measurements along a profile line joining the current electrodes. This is followed by similar measurements made along profile lines parallel to main profile on either side of it within the central 1/3 of

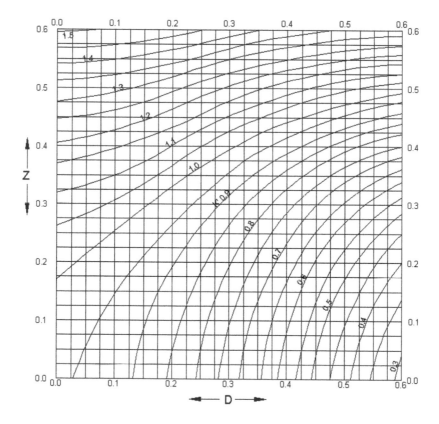

Figure 6.11 Graph between Z = a/L and D = x/L, for computation of the geometric factor for parallel profiles (after Sumner, 1976) where a, x and L are separation of parallel profile from the central profile of collinear electrodes, separation of potential dipole centre from the centre of the configuration, and L is half of the current electrode separation as shown in figure 6.8.

the area keeping current electrode position fixed. The spacing between parallel lines is kept 30 to 100 m depending on the separation between current electrodes. The furthest parallel profile on either side can be kept at a distance equal to 1/4 of the separation between current electrodes. Such 5 to 6 parallel profiles, are sufficient to trace the fractured zone. The resistivity lows in parallel profiles are logically connected to get the trend of resistivity low zone (Fig. 6.10). In the case where the area of interest exceeds 1/3 of the central area, either current electrodes are fixed at a large separation, so that the area gets covered automatically, or gradient profiling array is shifted along the line with same current electrode separation in such a way that latter 1/3 of the central area overlaps the former. In general, gradient profiling should be carried out for stretches larger than the requirement so that background resistivity of country rock and anomaly 'shoulders' are clearly picked up.

Schulz (1985), Eloranta (1986) and Furness (1993) computed gradient array response for different geometries and attitudes of target. For a symmetrical conductor, the anomaly expected is symmetrical with a central low resistivity over the conductor and high resistivity shoulders on either side falling to host rock background resistivity. With increasing depth to the conductor the anomaly flattens. The asymmetry in anomaly is observed obviously due to asymmetry in geometry and dip of conductor. Across a dipping conductive vein i.e., a dipping thin saturated fractured zone, asymmetrical resistivity 'low' with 'high' shoulders on either side coming back to background resistivity is obtained. The effect of dip shifts the shoulder up and also the magnitude of the 'low' varies. Schulz (1985) points out that a dip up to 30° and anomalies from two closely located conductors cannot be distinguished. This was mentioned by Ward (1990) also. Since in detection of water saturated conductive fractured zones mostly the anomalies are qualitatively analyzed, these conditions do not seriously affect the results. The gradient profiling anomalies are simple to interpret. It provides reasonable information on the dip of 2-d targets (Whiteley, 1973). Middleton (1974) describes a thumb rule for interpretation of gradient data. For a shallow sphere the depth to its centre is $2.3X_{1/2}$, for a cylinder it is $2.2X_{1/2}$, where $X_{1/2}$ is the 'half-width' of anomaly at half of anomaly amplitude (Fig. 6.12).

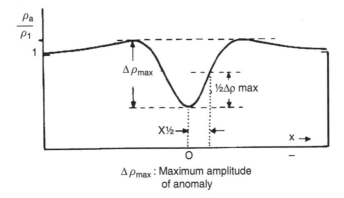

Figure 6.12 Parameters in 'rule of thumb' for spherical and cylindrical bodies. The background or surrounding resistivity is ρ_1 and measured apparent resistivity is ρ_a (Source: Middleton, 1974).

In gradient profiling for detection of the saturated fractured zone, generally, the ratio of background value to the resistivity 'low' is considered. According to Kunetz (1966) the absolute value of the low is not important as the data is used to map the trend of the low resistivity zone. This is true within an area of consistent rock type. A resistivity low of 500 ohm.m with a background of 5000 ohm.m may not have a comparable hydrogeological situation for a low of 50 ohm.m with a background of 500 ohm.m. This could also be due to the presence of conductive overburden. The ratio of resistivity 'low' to the background, steepness of profile 'low', shape of 'low' anomaly, value of resistivity 'low' and stretch of the 'low' zone picked up through adjacent parallel profiles are important considerations. At least 100 m wide 'low' resistivity anomaly with values half the background is preferred (Bernard and Valla, 1991). On a GRP a single point 'low' is not considered and checked by measurements at a closer station interval. Theoretically, the array gives 'least' response to thin vertical conductive body or thin resistive horizontal sheet (Furness, 1993). Corbett (1992) compared GRP with dipole-dipole and concluded that while response of the former is strong over a resistive dyke, that of the latter is for a conductive dyke. However, it has been used effectively, to demarcate dipping conductive bodies/veins/saturated fractured zones.

Variation in conductivity and thickness of top layer or overburden affects measurements and an anomaly may be misleading. An anomaly is produced even by varying the thickness of conductive overburden, but its shape is gentle and different from a fractured zone. A conductive overburden obscures the anomaly as most of the current remains in it. Coggon (1973) studied the effect of conductive overburden on GRP anomalies due to conductive vertical dyke, horizontal slab and dipping beds. The effect of overburden is more pronounced for a vertical dyke. The percentage of current remaining in the conductive overburden resting on a resistive substratum depends on its resistivity contrast and thickness ratio with the current electrode separation. Edwards and Howell (1976) present a generalized derivation for this through a dimensionless parameter 'α'. It is defined as

$$\alpha = 2S\rho_2/AB = 2(\rho_2/\rho_1).(h_1/AB),$$

where S ($=h_1/\rho_1$) is longitudinal conductance and ρ_1 and h_1 are resistivity and thickness of overburden and ρ_2 and AB are resistivity of substratum and current electrode spacing respectively. The percentage $f(\alpha)$ of current remaining in the overburden, a function of α, is shown in figure 6.13. The percentage of current penetrating the bedrock is $100 - f(\alpha)$. This indicates that conductance (h_1/ρ_1) of the overburden should be as small as possible. That is, the GRP technique can work with a thin conductive overburden. The parameter which can be controlled to make α much less than 1 is current electrode spacing AB which is to be kept large. It is to be noted that in this derivation only the ratio of bedrock and overburden resistivity is considered and not the actual resistivities which also play a significant role.

The GRP technique is best suited in areas with a resistive weathered zone. It is better to conduct shallow Wenner profiling with a = 5 or 10 m along the gradient profiles to assess top layer conductivity and differentiate the desired 'low resistivity' anomalies due to deeper zones from those caused by conductive overburden. Though the resistivity lows could be due to thickening of the conductive overburden, the resistivity highs are essentially negative zones without any saturated fracture.

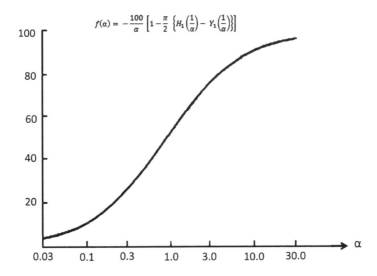

$$f(\alpha) = -\frac{100}{\alpha}\left[1-\frac{\pi}{2}\left\{H_1\left(\tfrac{1}{\alpha}\right)-Y_1\left(\tfrac{1}{\alpha}\right)\right\}\right]$$

Figure 6.13 The function f(α) determines the percentage of current remaining in a conductive thin overburden (top or surface layer) capping the resistive substratum, where $f(\alpha) = \frac{100}{\alpha}\left[\frac{\pi}{2}\left\{H_1\left(\tfrac{1}{\alpha}\right)-Y_1\left(\tfrac{1}{\alpha}\right)\right\}-1\right]$ H_1 and Y_1 are Struve function and Bessel function of the second kind respectively, of order 1 (Source: Edwards and Howell, 1976).

A big disadvantage of the technique is its insensitivity to vertical resolution as DOI cannot be changed effectively and therefore target depth cannot be inferred. The expected steepening of anomaly with a shallow target and flattening with a deeper one being obvious, a qualitative comparison of target depth may be attempted as shown by Pratt and Whiteley (1974). As such, qualitatively GRP reveals the presence of deeper targets with horizontal resolution adequate for visually demarcating the fractured zones. The same profile with at least two different current electrode separations could be conducted to confirm the anomaly as shown in figure 6.14 (Chandra *et al.*, 2002). The GRP 1 and 6 are parallel and with current electrode spacing 1200 m. The central resistivity low picked up in GRP 6 is not seen in GRP1. While it is clearly picked up in GRP 2, 3, 4 and 5 conducted with current electrode spacing of 800 m. Mostly, steepness of the anomaly qualitatively indicates the depth of the target. The DOI is at maximum at the centre and reduces marginally as the pair of potential electrodes moves towards either of the current electrodes. Bhattacharya and Dutta (1982) based on the concept of Evjen (1938) and Roy and Apparao (1971), analyzed DOI in gradient array. The DOI is normally taken as 0.125 times the current electrode separation. Schulz (1985) analyzed depth of detection and effect of the overburden conductivity on it. Like other resistivity technique, the gradient array suffers from equivalence and it is more effective for detecting 2-d targets as compared to 3-d ones (Schulz, 1985; Whiteley, 1973).

The GRP has several advantages, like, less cumbersome field layout, fast operation because of only two moving electrodes, fewer technicians, less operational hazards, symmetrical anomaly, easy and intuitive interpretation and presentation of data as

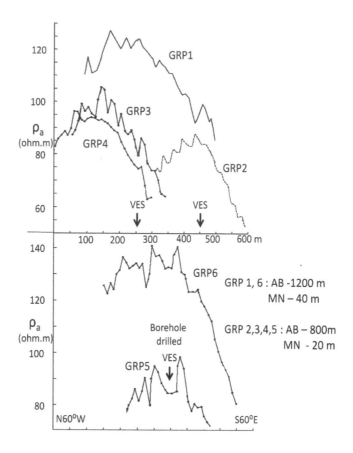

Figure 6.14 Parallel Gradient Resistivity Profiles with two current electrode spacing of 1200 m and 800 m in compact sandstones/quartzite, U.P., India (Chandra *et al.*, 2002).

apparent resistivity maps. Sumner (1972) adds the advantage of less topographic effect. The requirement of a higher current input is an operational disadvantage. It has been in regular practice in India over more than 3 decades for locating water wells in hard rock and several high yielding zones have been identified in most challenging areas. Some of the case studies from eastern part of India are discussed below (Chandra *et al.*, 1994).

The resistivity soundings (Fig. 6.15) at a site in Pre-Cambrian granite gneisses of Chhotanagpur plateau, Jharkhand state, India reveals a maximum 20 m thick weathered zone associated with resistivity in the range of 112 to 240 ohm.m. It is underlain by highly resistive bed rock. As such, there was no favourable indication for groundwater except the possibility of fractures in the depth range of 60 to 70 m. The N-S trend of fracture orientation and foliation anisotropy observed from a polar diagram constructed from radial soundings helped orient the GRP profiles lay-out. Accordingly, 5 parallel and 2 cross gradient array resistivity profiles with AB: 500 m and MN: 20 m were conducted in the area (Fig. 6.10). Resistivity contouring revealed a closed low of

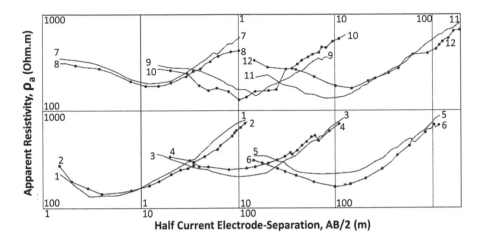

Figure 6.15   Schlumberger radial soundings from an area in Jharkhand State, India traversed by parallel GRP shown in Fig. 6.10.

the order of 800 to 900 ohm.m as well as a linear resistivity low of 500 to 900 ohm.m in a surrounding resistivity of 1300 to 1500 ohm.m. The boreholes BH1 and BH2 drilled on resistivity lows encountered 2 to 2.4 lps yielding fractures in the depth range of 65 to 93 m below ground level (Fig. 6.10). The borehole BH3 drilled at a location with a relatively high resistivity did not encounter any fracture and had a little discharge of 0.66 lps from saprolite in the depth range of 17 to 23 m.

At another site in the granite gneisses of the Chhotanagpur plateau, along with gradient profiling (AB: 500 m, MN: 20 m) Schlumberger (AB: 300 m) as well as Wenner (a: 20 m and 100 m) profiling were conducted on two profiles. These profiles revealed the presence of two low resistivity (about 300 ohm.m) zones separated by a resistive (about 900 ohm.m) block (Fig. 6.16 Site A). The extensions of resistivity low zones were demarcated by parallel profiles which indicated the presence of two parallel shear zones separated by a compact block. Three boreholes were drilled, out of which two (a: drilled depth 100 m and c: drilled depth 209 m) located on southwestern and northeastern low resistivity zones encountered yielding (4.6 and 3.8 lps) saturated fractures in the depth ranges 177 to 179 m and 33 to 35 m respectively. The third borehole (b) located in between on a relatively high resistivity block drilled up to 107 m was almost dry. It is observed that resistivities in the northeastern part where only shallow fractures were encountered are less than that in southern part with much deeper presence of saturated fractures. Borehole drilled (f) on high resistivity at another site (Fig. 6.16 site B) also had very little discharge. At another site in granitic terrain resistivity soundings (Fig. 6.17a) did not show any favourable layer resistivity indicative of groundwater occurrence except the possible presence of a fractured zone. The GRP at this site indicated prominent resistivity lows (Fig. 6.17b). The borehole drilled on one of the lows encountered saturated fractured zones. These case studies reveal that GRP can be used effectively for 'demarcating' shallow fractured zones and resistivity highs

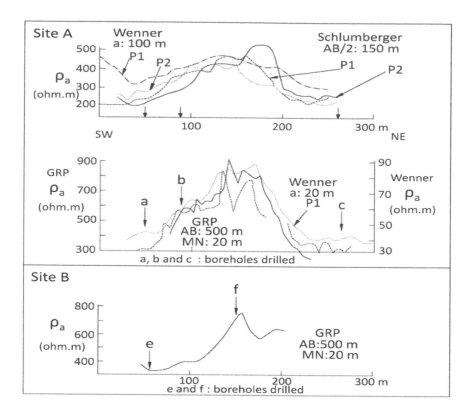

Figure 6.16 The gradient, Schumberger and Wenner profiles from granitic terrain. (A) The boreholes a and c drilled on GRP low had high discharges and borehole b was dry. (B) The borehole e encountered a saturated fractured zone at 44.8 m and a saturated weathered zone and saprolite up to 17 m and the f drilled on gradient resistivity high had negligible discharge. Wenner profile confirms the moderate resistivity of the over burden.

are surely negative areas. In the case where more than one fractured zone is present at depth, the response is mainly due to the shallow prominent one.

6.6   RESISTIVITY IMAGING

The resistivity imaging technique has been developed to investigate complex subsurface geological conditions. Resistivity sounding yields 1-d information at a point. Under normal field conditions, the sounding is affected by lateral changes but it cannot provide distinct information on lateral resistivity variations and these are not expected from it, being based on the assumption that layers are homogeneous and laterally extensive. Lateral changes affect parameter estimation and therefore the sounding may not be effective in areas with high frequency lateral variations in resistivity. This may be observed in hard rock areas where rapid lateral changes in resistivity due to complex geological conditions and subsurface structures are common (Loke, 2000;

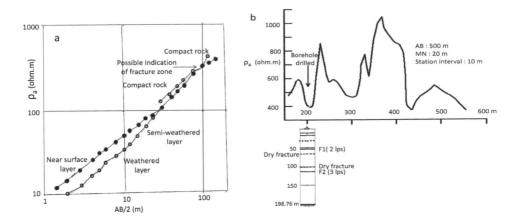

Figure 6.17 VES and GRP at a site in granitic terrain, West Bengal, India (a) VES indicate compact rock at depth, (b) fractured zone location demarcated on the ground surface by GRP and the yielding fractured zones confirmed through drilling of a borehole (Source: Chandra *et al.*, 1994).

Stummer *et al.*, 2004). Similarly, profiling cannot yield vertical resistivity variations. To assess aquifer disposition comprehensively it is necessary to get a close-grid 2-d or 3-d view of the subsurface resistivity distribution which cannot be obtained either from sounding or profiling individually. To get this information, combining conventional sounding and profiling, multi spacing profiling or very close-grid sounding is tedious, labour-intensive as well as time consuming, uneconomical and would prove to be an inefficient way. These are overcome by the technique of simultaneous measurement of vertical and lateral resistivity distributions in the subsurface through multiple-electrode arrangement. This is known as electrical resistivity imaging (ERI) or resistivity tomography (ERT). When conducted in a borehole it is known as tomography. But, in practice ERT is a common term used for imaging on the surface as well as in a borehole. It generates high density near- continuous data involving thousands of measurements as compared to sounding with only 20 to 25 measurements. The multi-electrode measurements were initiated about 25 to 30 yrs back by innovative development of manually operative systems through modification in the existing process of resistivity data acquisition. It was time consuming and prone to acquisition errors. The details of early development of imaging are given in Griffiths and Turnbull (1985), van Overmeeren and Ritsema (1988), Griffiths *et al.* (1990) and Griffiths and Barker (1993). The early application of imaging in hard rocks for siting borehole locations using the Wenner configuration was made by Olayinka and Barker (1990). By now, with the development of fast, automatic, multi-channel data acquisition systems, related processing and available inversion software the entire process has become significantly faster. The resistivity method has experienced resurgence with wide applicability of imaging as a significant near-surface geophysical approach (Stummer *et al.*, 2004) and it has almost replaced conventional resistivity surveys. It has also been used for infiltration studies in the vadose zone (Daily *et al.*, 1992). Obviously the cost of imaging equipment is quite

Figure 6.18 Field lay-out of multiple electrodes along a profile-line (Courtesy: Dr Subash Chandra, NGRI, India).

high compared to even the most modern 1-d resistivity data acquisition system. However, the operational cost of a single image is comparable with conventional sounding. The operational cost increases for a continuous subsurface coverage by roll-on imaging.

Resistivity imaging requires planting of a large number of electrodes at equal spacing on the ground surface along a straight profile line. The electrode spacing could be in centimetres to metres depending on the objective of the survey. Generally, a maximum spacing of 10 m is used. For a system of 48 electrodes 470 m long profile is covered. Field set-up of electrodes is shown in figure 6.18. These electrodes are connected to the resistivity meter through a multi-core cable having many conductors and take-outs and the switching-over system for selection of electrode combinations. The instrument is kept either at the centre of profile and electrodes are planted on both sides (for 48 electrodes, 24 on each side) or shifted one or two times depending on the sequence of data acquisition and number of electrodes.

To acquire a good image it is necessary that the data be without noise. For this, primarily the electrodes are planted firmly with watering if required, so that electrode contact resistance is minimum and within the permissible range of a few kilo-ohms. There should not be any negative resistivity value recorded. There should not be any cultural noise in the vicinity of the profile, like metal wire fencing, power transmission line, underground pipelines and built structure, concrete roads and buildings. Even roots of big trees in the vicinity may cause disturbance. Besides these, it is necessary to clean the connectors and takeouts of the cable. As mentioned by Loke (2000), the time spent in field operation is mostly in laying out the profile, planting of electrodes and once measurement starts waiting for completion of thousands of measurements. The measurement time increases with number of stackings for each reading. The time is now reduced through multi-channel measurements wherein several measurements are carried out simultaneously.

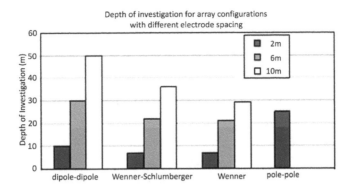

Figure 6.19 Depth of investigation for different arrays and electrode spacings (Source: Seaton & Burbey, 2002).

According to the electrode array selected, viz., Wenner (W), Wenner-Schlumberger (WS), Dipole-Dipole (DD), Gradient (G) etc. electrodes are automatically selected for current and potential and planned shifting of electrode addresses or change of array along the profile line is done through an automatic electronic switching system. The data are automatically stored in the computer. The information obtained is up to a depth around 20% of the total profile length in case of WS, and DD arrays and increases to 90% for Pole-Pole (PP) array. Pole-pole array requires planting of one of the current and potential electrodes at infinitely large distance and therefore it is not commonly used. Besides, the resolution at depth is also reduced in PP array. The details of imaging can be obtained in Loke (2000), Dahlin (2001), Seaton and Burbey (2002) and Dahlin and Zhou (2004).

The arrays used for resistivity imaging have varied resolution, sensitivity and information-depth capabilities. Seaton and Burbey (2002) carried out a detailed study on the applicability of different electrode configurations in crystalline terrain. According to them W and WS have lesser DOI, sensitivity and resolution as compared to DD. The PP has much higher DOI but with less resolution and sensitivity (Fig. 6.19). Seaton and Burbey recommend the DD array as being sensitive to small subsurface pockets of high and low resistivities and as such DD is suitable for imaging fractured zones in hard rocks. The obvious disadvantages of DD pointed out by Searon and Burbey are fast diminishing signal strength with increasing separation between current and potential electrode pairs and DD cannot be used in areas with highly conductive zones which add to the diminishing signal strength.

The general procedure for imaging by Wenner (α) array is that the first four equidistant (a) electrode positions on the profile are considered as C_1 (electrode 1), P_1 (electrode 2), P_2 (electrode 3) and C_2 (electrode 4) and subsequently through a switching-over system the current and potential electrode positions go on shifting automatically from one end of the profile to the other as $C_1(2)$, $P_1(3)$, $P_2(4)$, $C_2(5)$... $C_1(n-3)$, $P_1(n-2)$, $P_2(n-1)$, $C_2(n)$, where n is the total number of electrodes and nth electrode is the last electrode on the profile. In this way, $C_1(1)$, $P_1(2)$, $P_2(3)$, $C_2(4)$

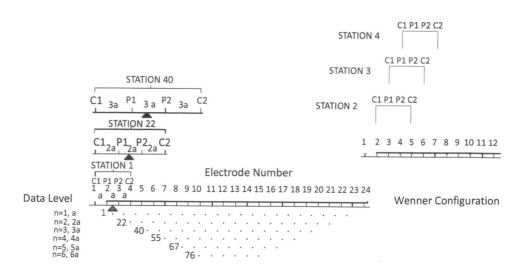

Figure 6.20 Wenner (α) electrode array movement and subsurface coverage in imaging.

represents first measurement, $C_1(2)$, $P_1(3)$, $P_2(4)$, $C_2(5)$ represents second measurement and $C_1(n-3)$, $P_1(n-2)$, $P_2(n-1)$, $C_2(n)$ represents n-3th measurement. That is, n number of electrodes planted will yield n-3 readings in the case of the Wenner array (Fig. 6.20). The distance between current and potential electrodes is increased to 2a in the second set of measurements by skipping alternate electrode positions. That is, the first measurement will be $C_1(1)$, $P_1(3)$, $P_2(5)$, $C_2(7)$ and switching-over will continue till the array occupies $C_1(n-6)$, $P_1(n-4)$, $P_2(n-2)$, $C_2(n)$ position. This set up will yield $n-6$ readings. Measurement with increased current-potential electrode distances and switching over continues till C_2 occupies the nth electrode position. The other arrays are also used in a similar fashion. As observed, in figure 6.20 with increasing electrode separation the data plotting point shifts to deeper levels. Though increasing CP separation gives higher DOI, the number of measurements decreases with increasing separation and therefore with increasing depth the coverage is reduced. Larger lateral as well as depth coverage can be achieved with roll-on measurements and a large number of electrodes. Dahlin (2000) indicates effects of electrode charge-up by making potential measurements with the electrode which was just used as a current electrode and suggests a measurement sequence to minimize this effect (Fig. 6.21).

Conventionally, apparent resistivity data is plotted at the mid-point of the array. In case topographic variations exist along the profile line, it is incorporated using differential GPS measurements. As such the locations of all the electrodes or that of central and end electrodes are obtained using GPS. Contouring of values yields a pseudo-resistivity depth section. It is inverted to actual resistivity depth section by 2d inversion code. The details of interpretation procedure are given in Loke (2000). The interpretation may include a laterally constrained inversion (LCI) scheme. Resistivity imaging is mostly used for shallow investigation within 100 m depth. As shown in figure 6.22 it is quite effective in weathered zone mapping and could prove useful in delineating concealed basic dykes (Chandra *et al.*, 2006) and relatively conductive saturated shallow

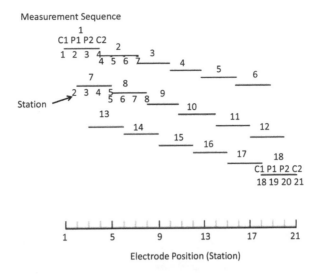

Figure 6.21 Wenner (α) array measurement sequence that avoids reading potentials by an electrode immediately after it has been used as current electrode (Dahlin, 2000).

fractured zones in compact rock underlying the weathered zone (Bernard *et al.*, 2006). Forward modeling simulating shallow and deep fractured zones has been attempted by several researchers. Seaton and Burbey (2002) advocate dipole-dipole configuration for fractured zone detection. Modeling by Louis *et al.* (2002) reveals that a fractured zone of large depth extent underlying a moderately thick overburden can be mapped by resistivity imaging, while thick overburden and limited depth extent of the fractured zone reduce the resolution. Chandra *et al.* (2012), after corroborating modeling and borehole results opine that PP electrode configuration can be used for delineating deeper fractured zones. Further, they indicate that distant electrodes can be placed transverse to the imaging profile. The advantage of placing distant electrodes transversely is that these can be placed nearer, i.e., 5 times instead of 10 times required in collinear distant electrodes. Over all, colourful presentation of the images is quite attractive but before conducting imaging it is necessary to carry out forward modeling for different electrode configurations with different resistivities and geometries of fractured zone, overburden and host rock to get an idea of 2d resistivity image and assess the capabilities and limitations.

6.7 ELECTRODE ARRAYS FOR INVESTIGATING FRACTURE/STRUCTURE-INDUCED ANISOTROPY

For detecting subsurface anisotropy created by dipping or vertical structures like dykes, quartz reefs, joints or fractures, which are considered electrical as well as hydraulic conductivity discontinuities in hard rock, soundings at a point within the area of influence of such structures can be conducted in several radial directions, viz.,

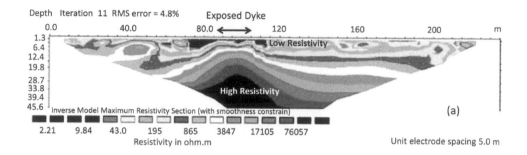

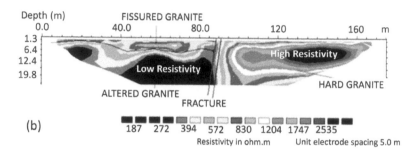

Figure 6.22   Electrical resistivity imaging in granitic terrain across (a) an exposed dyke (Source: Chandra et al., 2006) and (b) shallow fractured zone (Source: Bernard et al., 2006).

E-W, N-S, NE-SW, NW-SE and/or any other directions at 10° to 20° interval covering 360°. The apparent resistivity values for different current electrode spacings obtained from the soundings are plotted in these directions and connected. These are known as polar diagrams. In a case where the subsurface is isotropic, i.e., devoid of such structures, a polar diagram is a circle, i.e., an equipotential line at the surface will be a circle (Taylor & Fleming, 1988, Chandra et al., 1994). Otherwise, for anisotropic subsurface it is an ellipse whose major axis reveals the strike direction of the structure or fractured zone (Fig. 6.23A). Polar diagrams are compared with Rosette diagrams which are radial histograms of strike density or frequency of structure prepared for the area. The ratio of major axis to minor axis of apparent resistivity ellipse yields an apparent anisotropy. The plot of coefficient of apparent anisotropy with increasing current electrode spacings apparently indicates variation in anisotropy with depth as shown in figure 6.23B (Chandra et al., 1987).

The variation in resistivity with direction can be obtained for dipping beds also. Barker (1981) and Watson & Barker (1999) proposed offset Wenner sounding measurements to differentiate the effects of dipping beds from that of structures mentioned above. In an off-set Wenner array measurements are sequentially made by the left four and right four electrodes and the array is rotated at a point in radial directions. The azimuthal square-array is also used to study anisotropy. Habberjam (1972, 1975) and Lane et al. (1995) showed that a square array of electrodes is more sensitive to subsurface anisotropy and also, it requires less surface stretch in comparison to

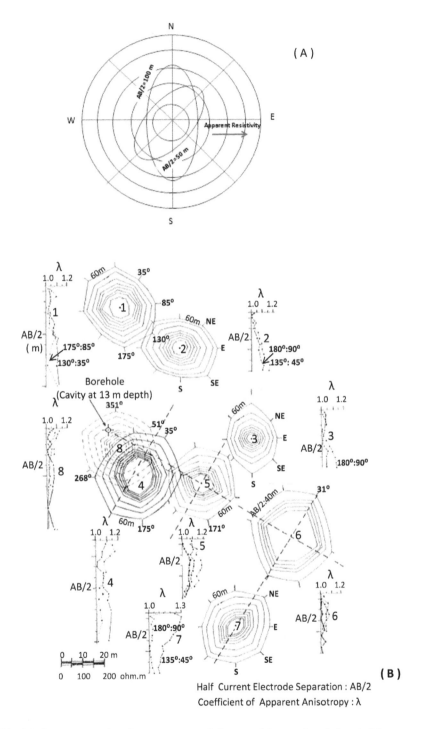

Figure 6.23 (A) Resistivity polar diagram prepared from radial or azimuthal resistivity soundings, (B) Polar diagrams from karstic terrain showing variation in apparent anisotropy with AB/2 (≈depth), (Chandra *et al.*, 1987).

collinear Schlumberger and Wenner arrays mentioned above. Falco *et al.* (2013) suggest null-array to detect fracture location and orientation. In null-array the potential electrodes MN are kept perpendicular to AB. On homogeneous isotropic ground it should record zero potential, otherwise variations in potential due to the presence of fractures are recorded. The square array is discussed in detail later.

6.8   SITE SELECTION IN HARD ROCK AREAS

Because of the intense heterogeneities in subsurface geological conditions and wide variations in groundwater availability, in hard rock areas the geophysical information or results are generally site specific with little scope of extending them to nearby sites. There are several cases where within a distance of 10–20 m, wells have different yields tapping fractured zones at different depths. Since electrical response in a surface survey is primarily a function of scale of measurement and the scale of target, planning of the survey and scales of study and measurement are to be such that they serve the objective envisaged. Identification of the proper site or area based on geophysical results is very important and crucial to achieving success. Site selection is of 2 types. One in which a large area is available for reconnaissance survey with enough space-flexibility in laying out profile-lines in different directions over varied distances and detailed surveys are carried out to pinpoint the most favourable drilling site or identify a patch or alignment for construction of artificial recharge structure. The other is where the surveys are conducted at predefined points of demand. In the former, it is necessary to have a prior study of the area through satellite images and hydrogeological mapping to assess geomorphological and topographic suitability, pick up lineaments, drainage patterns and general lithological and groundwater conditions obtained through existing wells. In a case where a favourable geophysical anomaly falls at a point which is not approachable for drilling, its extension is identified/demarcated by observing additional profiles so that the drilling rig can be placed. For a pre-defined point of demand the best possible profile lines, type of survey and array etc. are considered within the available space to get maximum vertical and lateral information and evaluate suitability of the site.

6.9   INSTRUMENT AND FIELD ACCESSORIES

A resistivity survey is carried out by an instrument, known as resistivity meter that basically comprises a current generator and a voltmeter. Accuracy in generation of field data depends on the equipment used so it has to be of standard quality and sensitivity. Theoretically a direct current (DC) source is required, but it creates polarization at the electrodes and also natural potentials are imposed on the signal. To overcome this, the polarity of the DC source is reversed to filter the parasitic effects. Otherwise, a very low frequency square pulse alternating current (AC) type resistivity meter of suitable wattage can be used which offers the advantages of DC. By using an AC source the aforesaid undesirable aspect of DC is effectively eliminated.

The instrument should preferably have a measurement accuracy in the microvolt range. It should be equipped with a microprocessor based stacking facility to enhance the signal to noise ratio by averaging out spurious signals and digital display of direct resistivity/resistance or current input and voltage measured, built in noise rejection and line-continuity checking. Also, a facility of displaying standard deviation will help reject noisy data. It should have multi-selection constant voltage or constant current input. In some of the instruments standard deviation is also displayed for each measurement, which gives statistical confidence in the data set. Rechargeable or non-rechargeable direct-current power source or source with converter or power generator with converter is required for constant rated supply with multiple range selection.

The minimum accessories required are: 2 nos. of small diameter, one end sharp pointed stainless steel rods/stakes of 0.5 to 0.75 m length, 2 numbers of 2 to 3 kg hammer, two nos. of non-polarizing electrodes – either conventional type comprising copper rod in porous pot with saturated copper sulfate solution and a packet of copper sulphate powder or simple 2 stainless steel rods, 4 nos. of rugged winches/reels with insulated base each holding 100 to 500 m PVC insulated cable of single conductor comprising multi-strand thin wires (length of cable depends on the maximum spread of sounding/profiling to be observed), compatible crocodile clips (different sizes), banana pin connectors and insulation tapes, hand-held walkie-talkie sets, measuring tapes or arrangement for measuring distances at a stretch for 100 m, surveying compass with tripod and ranging rods for fixing profile layout direction or any other arrangement, GPS, topographic sheets (1: 50,000 scale) for locating the survey sites, survey umbrella and arrangement for watering the electrode locations, if required.

6.10 FIELD LAYOUT, OPERATION AND DATA ACQUISITION

Using the surveying equipment straight profile-lines for sounding or profiling are placed in the area or at the site observing that the ground elevation along the profile line is consistent and topographically electrodes are placed approximately at the same elevation. The current and potential electrode distances and station intervals along these profile lines are fixed as per the electrode array used and profile alignments are checked, otherwise the geometric factor would be erroneous leading to incorrect determination of apparent resistivity. In hard rock areas resistivity profiling is conducted across a hydrogeologically favourable geological structure like fault zones or fractured zones to demarcate or narrow down the zone which is locally associated with the best or most favourable anomaly. So, once such structures are identified on the ground, through satellite images and geological mapping, conventionally surveys are initiated through Wenner and/or Gradient resistivity profiling along 2 to 3 parallel lines placed across the structure. For, fixing the electrode spacing in Wenner profiling a few test soundings may be required. The standard approach would be to include a Wenner profiling with very small electrode spacing of say, 5 m or 10 m, that will yield resistivity variations in top soil which affects apparent resistivity values obtained for larger spacings and thus the spurious effects of top soil can be identified. In Wenner profiling effect of near surface inhomogeneities can be reduced by conducting measurements with off-set arrangement of electrodes (Barker, 1981) and taking average of the measurements. While carrying out gradient resistivity profiling across highly resistive structures, negative resistivity values may be obtained. It could be due to the effect of strong anisotropy and lateral inhomogeneity and is overcome by changing the profile orientation.

For gradient resistivity profiling the current electrode spacing can be kept as large as possible (AB: 500 to 800 m), so that one third of the central space is sufficient to cover the zone of interest. Two to 3 parallel profiles located 50 to 100 m apart or as per requirement can be conducted to demarcate the lateral extensions of fracture or anomalous zone by tracing resistivity 'lows' on the profiles. A resistivity low in a single profile may be erroneous and misleading. The profile results can be combined either as an apparent resistivity map or simply apparent resistivity profiles are stacked and compared and extensions of the zones are demarcated.

Once such zones are demarcated Schlumberger soundings spread along or ERT along and across these zones are conducted. Otherwise, the sounding or ERT is located and orientated with respect to the geologic structures, dip of the bed etc, such that they have minimum spurious effect on sounding measurements and the 2d picture of fractured zone is obtained from ERT. Conducting soundings in hard rock is time consuming. For conventional measurements, 6 to 8 measurements or data points per log cycle are sufficient, but for empirical assessment of depth of saturated fractured zones sounding is conducted with shorter increments in current electrode spacing say 1, 2 or 5 m. While carrying out such measurements the conventional increase in potential electrode spacing is kept at minimum to avoid shifting in the curve segments. Potential electrodes are shifted only when it becomes essential to enhance the potential values.

It is generally observed that Schlumberger sounding measurements are conducted as per pre-defined combinations of current and potential electrode distances with tabulated geometric factors. This practice is not useful, particularly in hard rock areas where sporadic rock exposures at the surface are quite common and pre-defined current or potential electrode locations may fall on the exposure warranting their shifting to facilitate current penetration or reduce contact resistance or even skip the electrode position. But many such skippings may lead to erroneous interpretations. The required shifting should be such that the electrode is moved perpendicular to the profile by a maximum of 5% of the electrode distance. Flathe and Leibold (1976) limit this perpendicular shifting up to 20% of the current electrode distance. Szalai et al. (2007) suggest a minimum movement of electrode in perpendicular direction of the profile. It is desired that current and potential electrode locations are accurately measured and kept aligned on the straight profile line.

6.10.1   Checks in field operations

It is to be ensured that current electrodes driven in the ground by hammering are in proper contact with the ground. In a case where the electrode location falls on dry/compact soil/sand sufficient water is put in the hole by removing the electrode and then placing again to minimize contact resistance at the electrode. The contact resistance can be reduced by driving current electrodes deeper (increasing the contact area), and putting saline water in the electrode pit, if required. At large electrode distances in soundings if required a few more electrodes are planted near the current electrode, about a metre apart and then connected in parallel to the main electrode. By doing so overall resistance would be approximately the average contact resistance of one electrode divided by the number of electrodes connected. Since the steel rod electrode for current injection behaves as a point electrode at a distance about 10 times the length of rod driven in the ground (van Nostrand & Cook, 1966), for short current electrode separations rods are driven only about 4 to 5 cm to maintain them as point electrodes.

The potential electrodes are placed on natural earth and its moisture condition is maintained. In dry soil conditions, sufficient water is put into pits/holes well before placing the electrodes. Water need not be added after measurements start as water percolation may create electrokinetic potentials and change the measured potential. Zohdy (1968) indicates that error in current electrode distance affects the apparent resistivity values more than that in potential electrodes. The rock-pieces and metallic objects lying in the immediate vicinity of potential electrodes that can disturb the potentials are removed. In case of Schlumberger sounding, whenever the potential electrode position is changed repeat measurements for at least two preceding current electrode readings with new potential electrode position are obtained for overlapping curve segments. In instruments where measured potentials and current inputs are separately displayed, the minimum potential value is generally kept beyond 5 times the least count of instrument as even a minor variation in potential brings in noise in the sounding curve. Such small values of potential are generally obtained at large electrode separations for which geometric factors are quite large and the product of even a small potential variation with geometric factor gives a large variation in apparent resistivity.

Szalai *et al.* (2007) explain the effect of electrode positioning error in resistivity imaging. The rugged topography, rock exposures or thick bushes etc., which are unavoidable, bring in errors in locating the electrode on ground at its theoretical position on the image profile. This inaccuracy in electrode geometry leads to false anomalies in the near surface region. The effect reduces with depth. It may be argued to skip the electrodes where location error is expected. Such skipping though may help avoid the false anomalies, but affects the overall interpretation.

The leakages in current and potential cables are to be necessarily checked. Zohdy (1968) mathematically analyzes the distortions in sounding curves due to current leakages in current carrying cables and suggests placing of current cable winches at the centre of array where the instrument is kept. Also, the instrument may use high voltages so it is advisable to stay away from cables and electrodes while measurements are going on. Besides, personal precautions of wearing insulating boots and gloves may be necessary particularly while working in areas with high moisture content in surface soil or water on the ground surface. There should be a watch on the total stretch so that presence of stray animals and unauthorized persons near the cables or crossing the cables by vehicles etc. is avoided.

6.11 PROCESSING OF DATA

For soundings, apparent resistivity values are plotted against half current electrode separations (AB/2-Schlumberger, OA-Half Schlumberger), inter-electrode spacings (a-Wenner) or distances (na) between the current and potential dipoles (dipole-dipole) on a double-logarithmic graph paper. The apparent resistivity is on the ordinate and electrode separation on the abscissa. The plotting of VES data as small dots is preferred over joining the dots with continuous lines. Logarithmic plotting has advantages over arithmetic plotting as it can accommodate a wide range of resistivity values in a single 3-cycle logarithmic graph of 62.5 mm modulus in A4 size paper. Also, the VES curve shape is preserved in logarithmic plotting and interpretation by curve matching is feasible. The field-curve is matched with the theoretical curve directly by superposition and shifting vertically or horizontally.

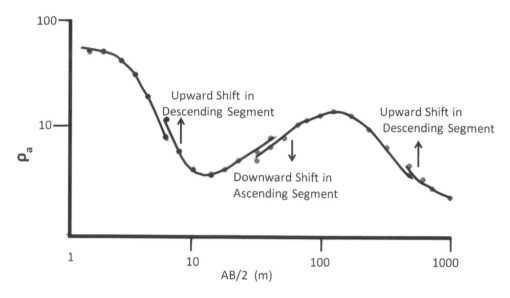

Figure 6.24 Shift in Schlumberger VES curve segments with change in potential electrode position.

In profiling, the apparent resistivity values are plotted at points in between the potential electrode positions for each station on arithmetic graph paper. In dipole-dipole profiling, the intersecting point of two lines drawn from centres of dipoles at 45° is taken as the plotting point.

Sounding curves obtained by Schulmberger array are generally discontinuous, with upward or downward shifting of curve segments, because of the shifting of potential electrodes. There are two types of displacements on account of the shifting of potential electrodes (MN), viz., (a) due to change in geometric position of potential electrodes in relation to current electrodes and (b) due to lateral inhomogeneities near potential electrodes. The 'geometric' shifting of a curve segment will be in a prescribed manner if there is no lateral inhomogeneity (Fig. 6.24). The sounding curve can be smoothened by shifting the curve-segments up or down depending on type of curve (ascending or descending). Conventional shifting of curve depends on relative resistivities of the layer sequence. When MN is increased, the DOI is somewhat reduced and therefore theoretically the apparent resistivity tends towards the preceding value i.e., while the ascending curve shows a downward shift, the descending curve shifts upward. In some of the curves, it may be difficult to decide which segment is to be shifted, the first one or the subsequent one. Difficulty in shifting the curve segments can be overcome by observing the trends of nearby soundings. Shifting of the curve-segments due to lateral surface inhomogeneity near potential electrodes is due to accumulation of charge around the inhomogeneity (Everett and Meju, 2005) and can be differentiated from the shifting due to conventional change in MN. The latter is regular and shifting is not much, of the order of a few ohm.m only. However, the increase of MN should not be as a big jump, e.g. continuing with MN: 1 m upto AB: 50 m and then increasing to MN: 10 m. In such cases MN should be increased minimally in such a way that a considerable potential is developed. For example, with AB: 50 m MN can be 2 or 3 m which may give twice the potential value observed with MN: 1 m.

Inhomogeneities near current electrodes can also be recognized by distortion in the sounding curve. A sharp curvature of maximum value in a sounding curve does not indicate any resistive layer of regional extent but a lateral inhomogeneity. Curves with sudden rise or fall, for change in current electrode position only, indicate presence of lithological contact of varied resistivity, i.e., sounding profile is expanded across strike of the bed or a fault and the curve cannot be smoothened. In such areas, other nearby sounding curves can help smoothen the distorted curve and identify which current electrode has caused the shifting.

6.12 INTERPRETATION

6.12.1 Manual interpretation of vertical electrical sounding curve for layered-earth

The resistivity soundings are interpreted for a 1-d information of geoelectrical layered-earth at a point. The layering varies in its character laterally as well as vertically. However, the basic assumptions for layered-earth interpretations are that layers are i) horizontal, ii) extend to infinite distance in the x and y directions, iii) considerably thick and uniform, iv) homogeneous and isotropic in resistivity and v) with appreciable resistivity contrast. All these conditions are seldom met with. However, interpretation of sounding yields depth-wise interfaces of resistivity contrasts and it is assumed that there is no variation in resistivity at the point within a pair of interfaces. These are geoelectrical interfaces and need not necessarily match with geological interfaces. The rapid variations in electrical resistivity as observed in the logging of boreholes, suggest that geoelectrical layers comprise several thin layers and interpreted resistivity is weighted average resistivity or mean resistivity (discussed latter) of a group of thin layers. It is the skill and experience of the interpreter to define geoelectrical layerings and convert them precisely to match with local hydrogeological conditions.

Qualitative interpretation of sounding curves includes, visual inspection to identify "type" of curve and demarcate areas with similar types of curve, e.g. 2-layer ascending ($\rho_1 < \rho_2$), descending ($\rho_1 > \rho_2$) type or 3-layer H ($\rho_1 > \rho_2 < \rho_3$), A ($\rho_1 < \rho_2 < \rho_3$), K ($\rho_1 < \rho_2 > \rho_3$) or Q ($\rho_1 > \rho_2 > \rho_3$) type curves for various combination of resistivities of first layer (ρ_1), second layer (ρ_2) and third layer (ρ_3) or multi-layered subsurface resistivity variations (Fig. 6.25). Occurrence of highly resistive (more than 40 times the resistivity of overlying layer) bed rock (thick, bottom most geoelectrical layer) is reflected as 45° slope of last segment of the curve. Unlike this, highly conductive bottommost layer shows a very steeply descending last segment. Quantitative interpretations of sounding curves are done manually by empirical or semi-empirical methods and also by computer based techniques.

6.12.1.1 Curve matching technique

Manual interpretation of Vertical Electrical Sounding (VES) is carried out by curve matching technique (CMT) using sets of theoretical sounding curves known as master curves. It is a graphical method. The field curves can be matched fully with a master

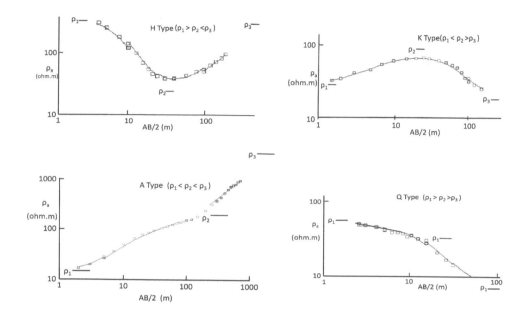

Figure 6.25 Types of 3-layer Vertical Electrical Sounding (VES) curves.

curve or in parts/segments through a sequence of partial curve-matching depending on the layer thickness-resistivity combination and number of layers picked up from the field curve. The concept is that if the field VES curve matches well with theoretical VES curve, the model parameters of the theoretical curve are valid for the field curve also. Two-layer ascending and descending types Schlumberger theoretical curves are plotted for dimensionless parameters ρ_a/ρ_1 vs $AB/2h_1$ for different values of ρ_2/ρ_1 (Fig. 6.26a and 6.26b). Similarly H, A, K, Q types 3-layer theoretical curves are plotted for dimensionless parameters ρ_a/ρ_1 vs $AB/2h_1$ for different values of ρ_2/ρ_1, ρ_3/ρ_1 and h_2/h_1 and are available as published master curve sets. The most commonly used are Schlumberger curves by Orellana & Mooney (1966) and Rijkswaterstaat (1969).

To start with, the Schlumberger apparent resistivity sounding curves (ρ_a vs $AB/2$) are plotted on double-logarithmic transparent graph paper of 62.5 mm modulus or of modulus for which master curves are available. The plotted field curves are smoothened to remove the effect of inhomogeneities and noise and visually matched with a variety of 2- and 3-layer master curves for partial matching along with the auxiliary point charts (Zohdy, 1965). Because of the logarithmic plotting of field curves and matching with master curves generated from dimensionless parameters, the values of top layer resistivity and thickness are directly obtained. Subsequent layer parameters are obtained using the principle of theoretically combining two or more homogeneous and isotropic (assumed) layers in a single anisotropic (introduced) layer, which in turn is equivalent to another fictitious single homogeneous and isotropic layer. This brings into the phenomenon of pseudo-anisotropy.

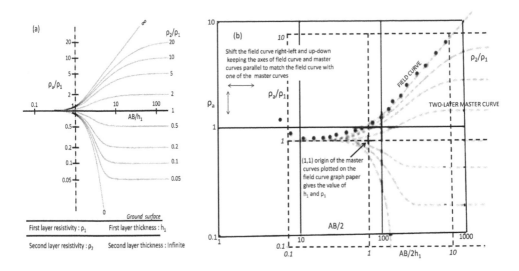

Figure 6.26 Two-layer master curve set for interpreting Schlumberger VES curves (Keller and Frischknecht, 1966). (a) Two-layer master curves, (b) interpreting an ascending type field curve by placing it on the master curve (dotted curves).

The transparent sheet of field curves is placed on 2-layer ascending or descending type master curves (Fig. 6.26b). For A and K type field curves with ascending first segment, the ascending 2-layer curve is selected and for H and Q type the descending 2-layer master curve is selected. By shifting the field curve left-right and backward-forward and keeping the axes parallel, its first segment is matched with one of the master curves. If it is falling in between two master curves, the value is interpolated. The cross ($\rho_2/\rho_1 = 1$ and $AB/2h_1 = 1$) of the master curve is plotted on field curve transparent sheet. The ordinate and abscissa values of the cross are noted from field curve transparent sheet. This gives the value of resistivity and thickness (ρ_1 and h_1) for the first layer. The ρ_2/ρ_1 ratio of the selected master curve being known a guess of ρ_2 is obtained. The second step is to place the field curve transparent sheet on relevant auxiliary chart (A, H K or Q type) and the plotted cross is kept on the origin of the auxiliary chart. Keeping the axes of the field curve and auxiliary chart parallel the relevant resistivity ratio (ρ_2/ρ_1) line on the auxiliary chart is traced from the auxiliary chart on the field curve. The third step is to interpret the next segment of the field curve. For this field curve is again placed on a 2-layer theoretical curve: ascending (for A and H type) or descending (for K and Q type). To match with the master curve the field curve is moved in such a way that the origin of master curve moves on the line traced from the auxiliary chart and the axes are parallel. On matching of the second segment of the field curve with one of the master curves the origin of the master curve (second cross) is plotted and value of ρ_2/ρ_1 of the master curve is noted. The fourth step is to place the field curve again on the used auxiliary chart superimposing the first cross ($\rho_2/\rho_1 = 1$ and $AB/2h_1 = 1$) on the origin of the auxiliary chart and the axes are kept parallel. The second cross either falls on one of the thickness ratio lines in the auxiliary chart or in between two thickness ratio lines. The value of the thickness ratio line for the second cross is noted. This thickness ratio value when multiplied by h_1 already obtained from the first cross gives value of h_2. The ordinate value of second cross multiplied by resistivity ratio (ρ_2/ρ_1) of the 2-layer master curve with which the second segment of field curve matched gives the value of the third layer resistivity (ρ_3). Thus, for a 3-layer curve the values of ρ_1, ρ_2, ρ_3 and h_1 and h_2 are obtained. The process is continued till the last segment of the curve is matched. Of late, with the easy availability of computers and software, manual interpretation is done only to generate a guess model with local geological/hydrogeological bias for further refinement through computer interpretations.

As mentioned above in manual interpretation a parameter- pseudo-anisotropy gets involved. It is a theoretical term resulting from mathematically combining 2 or more geoelectrical layers in a single layer. To understand, let us take a sequence of layers which are homogeneous and isotropic and have different resistivities. The current passing through this sequence between two current electrodes at the ground surface will have a preferred path leading to the concept of resistors in parallel or in series. That is whether overall current flow is parallel or perpendicular to the layers. That is, there exists a resistivity value parallel to the bedding ($\rho_{parallel}$) different from the one perpendicular to the bedding ($\rho_{perpendicular}$) and thus the sequence of layers can be replaced by a single anisotropic layer with resistivities $\rho_{parallel}$ or $\rho_{logitudial}$ (ρ_L) and $\rho_{perpendicular}$ or $\rho_{trasverse}$ (ρ_T). It is the pseudo-anisotropy which is introduced by combining the layers. To convert it further into an isotropic layer the thickness of the single layer no more remains as the arithmetic sum of the thicknesses of individual layers but is changed and termed as pseudo-thickness. The combining of layers can be done through the Dar Zarrouk parameters T (transverse resistance) and S (longitudinal conductance) introduced by Maillet (1947). When the flow of current is along the bedding plane longitudinal conductance (ratio of layer thickness to its resistivity) is considered and when the flow of current is across the bedding plane transverse resistance (product of layer thickness and resistivity) is considered. Through this it is observed that pseudo-anisotropy λ is equal to the square root of the ratio of transverse resistivity to longitudinal resistivity and pseudo-thickness is equal to the product of pseudo-anisotropy and actual thickness. The essence of the auxiliary point method is to establish a proper mathematical expression based on the above concept for combining two layers into a single homogeneous and isotropic fictitious layer (Zohdy, 1965). The method of combining the layers will depend on the layer resistivity distributions.

In the curve matching technique, correct estimation of resistivity and thickness of the top layer is very important, as subsequent layer parameters are controlled by the "first cross" (ρ_1 and h_1) obtained from matching the first segment of the curve. Error introduced in estimation of top layer resistivity, would introduce similar order of error in top layer thickness estimation if the substratum is resistive, while it would be less in the case where the substratum is conductive. The curve matching technique is effective for 3- to 4-geoelectrical layering, as curves can be interpreted either by partial curve matching or complete curve matching. When layering exceeds, errors are introduced in partial curve matching and other computer based techniques should be attempted. Besides these, interpreting thin (effective relative thickness) layers is difficult in curve matching and to some extent depends on skill, experience and local geological information available with the interpreter. In the case where the sounding is observed in an area with lateral variation in resistivity, sounding perpendicular to resistivity discontinuity would reflect a break in the curve. Sounding conducted parallel to the discontinuity would not show any break, and instead a smooth curve is obtained including the effect of resistivity variations which would be difficult to resolve, and may be misinterpreted. The effect of a gentle dip of layers is not significant on sounding curves except that layer resistivity contrasts are apparently reduced.

6.12.1.2   Inverse slope method

The inverse slope method (ISM) is a semi-empirical method of interpreting resistivity sounding curves developed by Sanker Narayan and Ramanujachary (1967) for Wenner sounding. It can be used for Schlumbrger sounding also by applying a correction. The method is effective for H-type curves where the flow of current is parallel to the layering and the layers are treated as resistances in parallel (Keller, 1968). At first, the field VES curve is smoothened if shifts are present for the changes in potential electrode position (MN). The values of apparent conductance: $(AB/2)/\rho_a$ for all AB/2 spacings of the field curve are calculated. The calculated apparent conductance values against AB/2 values are plotted on a linear scale graph paper, keeping $(AB/2)/\rho_a$ on the Y axis and AB/2 on the X axis. The points with the best fitting straight lines are joined such that a minimum of 3 points fall on each line segment. The number of straight line segments indicates the number of layers. The abscissa values (X coordinates) of the intersection points of line segments are multiplied by 2/3 to get depths to the layers. The inverse slope ($\Delta X/\Delta Y$: cotangent) for each line segment on the graph represents the resistivity of the corresponding layer (Fig. 6.27). The near surface thin layers remain undetected in ISM.

6.12.2   Interpretation of sounding curve for bedrock depth

The VES curves over hard rock areas with ascending last segments may be interpreted for bed rock depth estimation without going through the process of curve matching. If the bottom most layer representing bed rock is highly resistive and overlying layers are relatively conductive, the last segment of VES curve approaches asymptotically a line rising at 45°. In such a situation most of the current will flow through the overburden layers and at large distance from the current electrode the current must flow parallel to the bedding plane (Keller & Frischknecht, 1966) and in case of hard rock it flows parallel to the interface between bed rock and overburden. At a large distance 'a' the current density can be assumed to be uniform in the overburden of thickness 't'. The electric field intensity can be related to total current from an electrode (source) by integrating current density over a complete circular equipotential surface of radius a and height t as

$$E = \frac{\rho_1}{2\pi a t} I$$

For measurements made in between two widely spaced source and sink electrodes, it becomes

$$E = \frac{\rho_1}{\pi a t} I$$

The apparent resistivity for Schlumberger array will be

$$\rho_{aS} = \frac{\pi a^2}{I} \frac{\partial U}{\partial a} = \pi a^2 \frac{E}{I} = \rho_1 \frac{a}{t} \quad \text{or} \quad \frac{\rho a S}{\rho_1} = \frac{a}{t} \quad \text{(Keller and Frischknecht, 1966).}$$

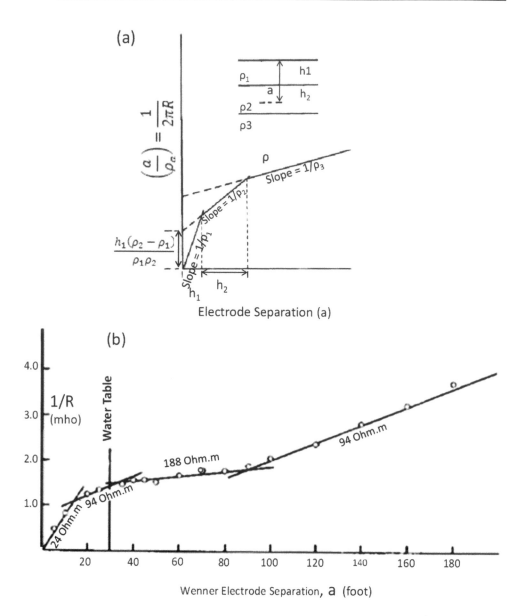

<figure>**Figure 6.27** Inverse slope method of interpreting Wenner VES data (a) and example (b) (Source: Sankar Narayan and Ramanujachary, 1967).</figure>

That is, the relation between the ratio of apparent resistivity to top layer resistivity and electrode spacing to top layer thickness is linear and the slope is 1 or 45°. When a is small the apparent resistivity measured will be equal to the layer resistivity i.e., $\rho_{aS} = \rho_1$ and hence $\frac{a}{t} = 1$. The curve passes through the point $\frac{\rho_{aS}}{\rho_1} = 1$, $\frac{a}{t} = 1$. In hard rock areas if the right hand segment of the sounding curve attains a slope of 45° it is an indication of encountering highly resistive bed rock within the depth probed by the sounding.

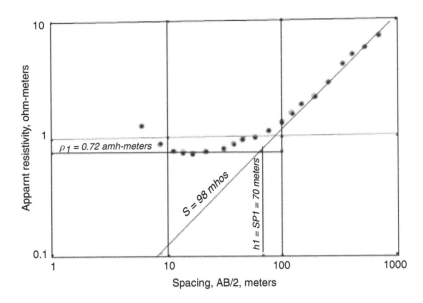

Figure 6.28 Graphical estimation of longitudinal conductance and depth to bed rock (Source: Keller and Frischknecht, 1966).

The relation $\frac{a}{\rho_a S} = \frac{t}{\rho_1}$ is quite important as the ratio t/ρ_1 is the total longitudinal conductance of overburden or layers overlying resistive bed rock. The depth to bed rock (t) can be obtained from this relation. The ratio of spacing (AB/2) to apparent resistivity for any point on the line with 45° slope will be exactly the total longitudinal conductance (S) of all the layers overlying the bedrock. For a geoelectrical layer sequence with highly resistive substratum (bed rock), curve showing 45° slope in last segment, total longitudinal conductance (S) for the layers overlying the substratum could be obtained graphically by drawing a 45° sloping line through the last segment and reading the value of S on the abscissa for the ordinate value of 1 ohm.m. Now, if the resistivity of top layer (ρ_1) estimated from apparent resistivities for small electrode separations is known, the thickness of the top layer can be calculated by multiplying the longitudinal conductance (S) by this resistivity (Fig. 6.28). The asymptotic method of interpretation may be extended to any number of horizontal layers, provided the bottom layer is highly resistive. Since total longitudinal conductance is obtained by this method, the thickness of individual layers cannot be determined.

6.12.3   Interpretation of sounding curve for fractures detection

One of the objectives of electrical resistivity surveys for groundwater in hard rock is to detect saturated fractured zones. The surface electrical resistivity surveys may not yield precise information on individual fractures and their lateral continuity and connectivity, but the persisting fractured zones which are interconnected and with consistent orientation over a large area can be delineated and the orientations identified.

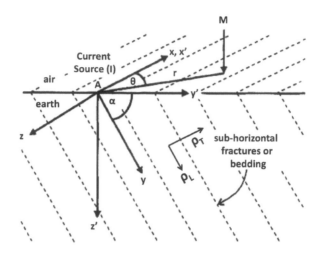

Figure 6.29 Sub-horizontal fractures or bedding creating a 2-dimensional anisotropic half-space and the coordinate system (Source: Watson and Barker, 1999).

The fractures and joints develop in certain directions and form a preferential path for groundwater flows and accordingly control the physical properties – the electrical resistivity. The presence of fractures in hard rock makes it electrically anisotropic and the measured apparent resistivity becomes a function of the orientation of the electrode array with respect to the fracture direction. The anisotropic behavior of resistivity is studied to ascertain the presence and orientation of fractured zones, joints etc. through radial soundings and constructing polar diagrams (Figs. 6.23A and 6.23B).

The idealized case is of vertical to sub-vertical fractures as shown in figure 6.29 (Watson and Barker, 1999). It shows 2-dimensional anisotropy. The fractures dip at an angle α from the horizontal. In such a situation the true longitudinal resistivity ρ_L parallel to the plane of fracturing is less than the true transverse resistivity ρ_T perpendicular to it. The coefficient of anisotropy λ is defined as $\lambda = \sqrt{\frac{\rho_T}{\rho_L}}$ and there is one more term mean resistivity ρ_m where $\rho_m = \sqrt{\rho_T \rho_L}$. That is, $\rho_T = \lambda \rho_m$ and $\rho_L = \frac{\rho_m}{\lambda}$. For isotropic earth the value of λ is 1. The value of λ ranges from 1 to 2 in anisotropic earth.

When fractures are long enough compared to current electrode spacing, the measured apparent resistivity (ρ_a) on the ground surface over such fracture systems is a function of orientation of electrode array with respect to the fracture orientation, type of array used and dip of fractures. The potential V at a point M on ground surface at a distance r from current source I at A (Fig. 6.29) is

$$V = \frac{I\rho_m}{2\pi r[1 + (\lambda^2 - 1)\sin^2\theta \sin^2\alpha]^{1/2}} \quad \text{or} \quad \rho_a = \frac{V}{I}2\pi r = \frac{\rho_m}{[1 + (\lambda^2 - 1)\sin^2\theta \sin^2\alpha]^{1/2}}$$

where θ is the angle between the electrode array alignment and the strike of fractures and r is the distance between current source and potential measurement point

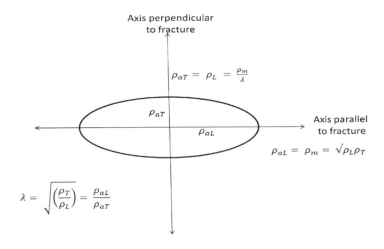

Figure 6.30 Resistivity anisotropy ellipse.

(Watson and Barker, 1999). When $\alpha = 0$, $\rho_a = \rho_m = \lambda \rho_L$. With $\alpha = 90°$, there will be two extreme conditions, viz., $\theta = 0$ and $\theta = 90°$. When $\theta = 0$, again $\rho_a = \rho_m$. and for $\theta = 90°$ $\rho_a = \rho_m / \lambda = \rho_L$. Apparent resistivity for $\theta = 0$ is termed as ρ_{aL} (longitudinal apparent resistivity) and that with $\theta = 90°$ is termed as ρ_{aT} (transverse apparent resistivity). This shows that measurements made across the strike of the fracture will yield apparent resistivity equal to ρ_L and those along the fracture strike will yield $\lambda \rho_L$ (Fig. 6.30). That is, for vertical fractures apparent resistivity measured along the fracture is more than that across it, while the true resistivity along the fracture is less than that across it. It is known as 'paradox of anisotropy'.

Sounding measurements for azimuthal variations in apparent resistivity due to steeply dipping or vertical fractures are made in several directions by rotating the collinear electrode array about its fixed centre at a point. The directions of measurements are kept to 4 at the minimum, e.g., E-W, N-S, NE-SW and NW-SE or any other 4 orthogonal directions. Measurements can also be made at 10° to 20° interval depending on the geological situation and space available. The apparent resistivities for certain electrode spacings for all the azimuthal soundings are plotted on a graph paper along respective directions to generate the 'polar diagram' as shown in figure 6.23. If the polar diagrams are ellipses then the major axes of the ellipses indicate the orientation of fractures. As far as the sensitivity of various electrode arrays is concerned, according to Ehirim and Essien (2009), offset Wenner array, which is described below, is the most sensitive and Schlumberger array is the least sensitive.

Another electrode array for detecting anisotropy is the square array. Habberjam (1972, 1975), Hagrey (1994) and Lane et al. (1995) indicate the square array to be sensitive to anisotropy while it requires relatively less space on the ground surface for measurements. In this array electrodes are positioned at the vertices of a square

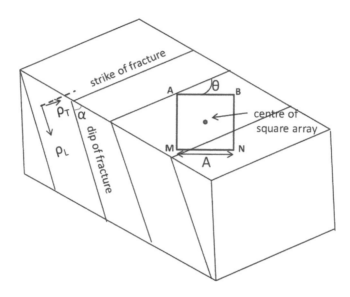

Figure 6.31 Square array of current (AB) and potential (MN) electrodes over fractures dipping at an angle α from horizontal. The side of the square array of length A makes an angle θ with the strike of fracture. The resistivities along and across the fracture are ρ_L and ρ_T respectively (modified after Lane, *et al.*, 1995).

(Fig. 6.31) and the side of the square is taken as the length of the array. For sounding measurements the square array is expanded by increasing the square side length in steps of $\sqrt{2}$. If the length of a side of the square is A, then the successive increments in side length are A$\sqrt{2}$. The geometric factor is given as $K = \frac{2\pi A}{2 - \sqrt{2}}$. The centre of square is the point of measurement and remains fixed. Lane *et al.* (1995) described the square array and its interpretation in detail. By interchanging the positions of four electrodes three measurements can be made. They are α, β and γ measurements. The arrays are shown in figure 6.32. The α and β arrays with current electrode alignments perpendicular to each other measure apparent resistivities in two orthogonal directions. The relation between α, β and γ apparent resistivities is $\rho_{a\gamma} = \rho_{a\alpha} - \rho_{a\beta}$ and therefore measurement with the γ array is made to check the accuracy of measurements with the α and β arrays. For isotropic medium and $\rho_{a\alpha} = \rho_{a\beta}$ hence $\rho_{a\gamma} = 0$

To decipher the orientation of fracture, measurements are made with several square arrays having different orientations and side lengths keeping the centre of square fixed at a point. Lane *et al.* (1995) proposed measurements with six such orientations separated by 15°. This yields 12 apparent resistivity data sets (6 α and 6 β) for a particular square side length adequately covering 360° for graphical display by 24 data points. For example α measurements at 0, 15, 30, 45, 60 and, 75° (accordingly at 180, 195, 210, 225, 240 and 255°) have associated β measurements at 90, 105, 120, 135, 150 and 165° (accordingly at 270, 285, 300, 315, 330 and 345°). The data for different square-side lengths are plotted on a rosette diagram as shown by Lane *et al.* (1995) in

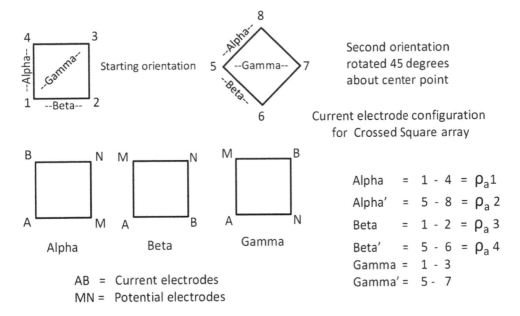

Alpha   = 1 - 4 = $\rho_a 1$

Alpha′ = 5 - 8 = $\rho_a 2$

Beta   = 1 - 2 = $\rho_a 3$

Beta′   = 5 - 6 = $\rho_a 4$

Gamma = 1 - 3

Gamma′ = 5 - 7

AB = Current electrodes
MN = Potential electrodes

Figure 6.32 Square array electrode arrangements (Source: Lane *et al.*, 1995).

figure 6.33. The direction of minimum apparent resistivity (ρ_{aL}) in rosette diagram is the direction of fracture strike and direction of maximum apparent resistivity (ρ_{aT}) is the direction perpendicular to it. While in collinear electrode arrays the direction of maximum apparent resistivity is the strike of fracture. That is, the apparent anisotropy (λ_a) for different arrays is defined as

$$\lambda_{a\,square} = \frac{\rho_{aT}}{\rho_{aL}} \quad \text{and} \quad \lambda_{aSchlum} = \frac{\rho_{aL}}{\rho_{aT}} \quad \text{or} \quad \lambda_{a\,Wenner} = \frac{\rho_{aL}}{\rho_{aT}}$$

The variations in resistivity with direction can be due to dipping beds as well as lateral variations in lithology and moisture/water content. Watson and Barker (1999) proposed a field technique with offset Wenner array to differentiate the effects of dipping bed and lateral variations from that of fractured zones. In this array a collinear arrangement of 5 equally spaced electrodes is used (Fig. 6.34). Azimuthal measurements are sequentially made with the left 4 electrodes as well as the right 4 electrodes. The variations in these two resistances measured with azimuth and the inter-electrode separations help differentiate anisotropy due to lateral changes from dipping bed effects. For horizontal bedding resistances measured by left 4 electrodes (R_{a1}) and right 4 electrodes (R_{a2}) will be same for all azimuths. For measurements over anisotropic rock due to the presence of vertical fractures, the resistances R_{a1} and R_{a2} will vary with azimuth but the rise and fall will be concurrent. As discussed above, maximum resistance will be recoded along the fracture direction and minimum across it. In the

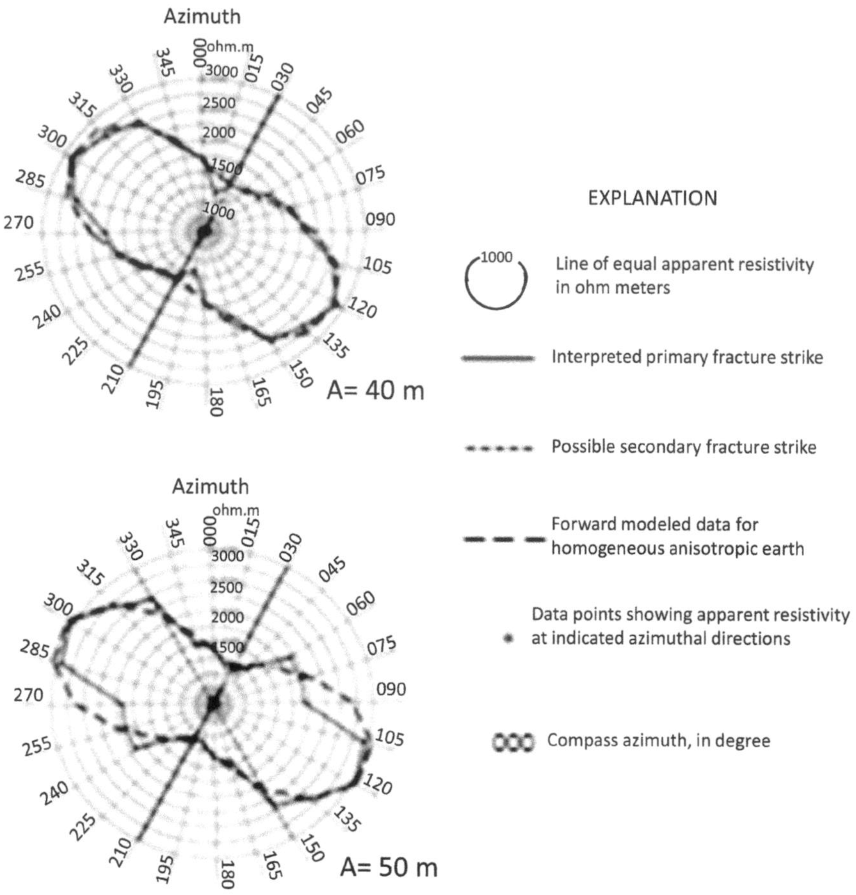
Azimuth
ohm.m
3000
2500
2000
1500
1000
000
015
030
045
060
075
090
105
120
135
150
165
180
195
210
225
240
255
270
285
300
315
330
345
A= 40 m
Azimuth
ohm.m
A= 50 m
EXPLANATION
1000
Line of equal apparent resistivity in ohm meters
Interpreted primary fracture strike
Possible secondary fracture strike
Forward modeled data for homogeneous anisotropic earth
Data points showing apparent resistivity at indicated azimuthal directions
000 Compass azimuth, in degree

Figure 6.33 Square-array apparent resistivity plotted against azimuth for 40 and 50 m A-spacings (Source: Lane *et al.*, 1995).

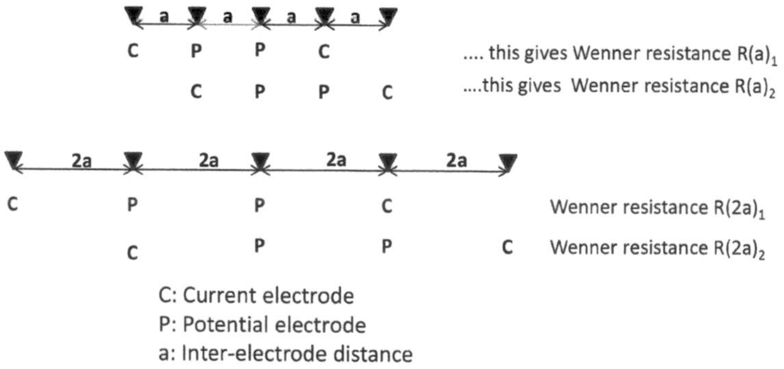
a a a a
C P P C
.... this gives Wenner resistance R(a)1
C P P C
....this gives Wenner resistance R(a)2
2a 2a 2a 2a
C P P C
Wenner resistance R(2a)1
C P P C
Wenner resistance R(2a)2
C: Current electrode
P: Potential electrode
a: Inter-electrode distance

Figure 6.34 Offset Wenner measurement using five-electrode array (after Watson and Barker, 1999).

case of dipping beds the maximum difference in R_{a1} and R_{a2} is observed when the electrode array is aligned in the direction of true dip. For resistivity variations the maximum difference in R_{a1} and R_{a2} is observed in the direction of greatest lateral resistivity variation (Watson and Barker, 1999).

6.12.4   Detecting fractures from sounding curve by empirical methods

It is not possible to delineate thin fractured zones from layered-earth interpretation of VES curves. However, two empirical methods – the curve break method (Ballukraya and Sakthivadivel, 1983) and factor method (Muralidharan, 1976) described below are in use and their application has yielded favourable results. The curve-break method is supported by an analysis (Sharma and Biswas, 2013). Generally for layered-earth interpretation, the VES curve is smoothened to avoid the kinks due to noise. Contrary to it, for detecting fractured zones, the VES curve is not smoothened and kinks are analyzed to relate them to presence of fractures. In a Schlumberger array, current electrodes can be moved outward symmetrically at a smaller increment of 2 to 5 m without changing the potential electrode positions (as long as a measurable potential is obtained) to record minor 'breaks' in the curve. The breaks in the VES curve if not associated with cultural noise, are minor sags or flattening obtained when there is a drop in potential and are selectively linked with underlying fractures. For this purpose constant current or constant voltage source equipment can be used.

6.12.4.1   Curve-break method

For some of the current electrode spacings, input current becomes automatically high (when constant current source is not used) and VES curve shows a descending kink for that spacing. Van Nostrand & Cook (1966) report an empirical method of examining the sounding curves for 'breaks' which can be attributed "to some horizontal discontinuity in resistivity at a depth corresponding to the electrode separation at which the break occurs". Ballukraya et al. (1983) studying a large number of sounding curves and borehole results reveal that in hard rock fractures on a local scale are horizontal discontinuities and cause the resistivity breaks in sounding curve. Sharma and Biswas (2013) analyze the horizontal current density revealing the magnitude of current flow through normalized apparent conductance for different current electrode spacings and indicate an increase in it with the presence of a thin conductive layer at depth.

A statistical correlation between kinks observed in VES curves and depths of saturated fractures encountered in boreholes drilled, revealed a linear relation. In the Schlumberger VES curve the current electrode spacing (AB/2) where a break with increased current occurs, is taken as the depth to fracture. For using this technique the VES is observed with close increment of 2 to 5 m in current electrode spacings. The objective is to precisely pick-up the kinks (minor flattening or drops) in the curve which are associated with inherent increase in current input "reading", in case the equipment used has the facility of recording current and potentials separately. So, the current input should not be externally increased unless it is very much essential

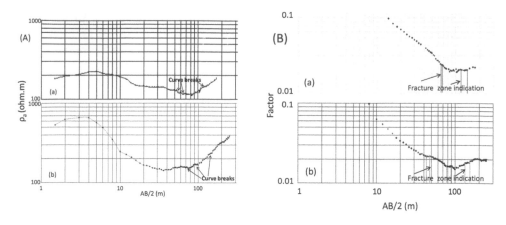

Figure 6.35 Delineating fractured zones from VES curve by empirical methods: (A) Curve-Break method and (B) Factor method.

to pick the voltage. In such cases Sharma and Biswas (2013) suggest for computing normalized apparent conductance (C_{Ai}) which is the ratio of current input at the ith current electrode spacing to the potential measured at that spacing. Increasing the potential electrode spacing should be avoided as change in potential electrode spacing also induces a shift in the curve and may hinder picking the actual breaks. The breaks in the curve may also be associated with near surface lateral resistivity changes. They are to be ascertained and separated from the breaks of interest. The graphical presentation of variations in apparent conductance for different current electrode separations (AB/2) can be plotted as vertical bars (Sharma and Biswas, 2013). For constant current–source equipment, it may not be possible to study the inherent current variations except the potential drops or resistance drops indicating the kinks. Figure 6.35A presents field VES curves from which the breaks have been picked up for fracture indications and confirmed subsequently by drilling of boreholes. The technique gives satisfactory results within 50 to 60 m depth and maximum upto 100 m.

6.12.4.2 *Factor method*

This is another semi-empirical method developed by Muralidharan (1996) to identify thin productive layer-the fractures at depth from the ascending segment of A or H type curve. A parameter called 'factor' (F) is estimated. The process is to calculate the ratio of apparent resistivity for an electrode spacing n ($\rho_{a\,n}$) to the sum of all apparent resistivities for earlier spacings ($\rho_{a\,n-1} \ldots \rho_{a1}$). The factor for AB/2 = 10 m is given as

$$F\rho_{a10} = \rho_{a10}/\text{sum of } \rho_{a1.5} + \rho_{a2} + \rho_{a3} + \rho_{a4} + \rho_{a5} + \rho_{a6} + \rho_{a8}.$$

The factor values for all the AB/2 values are plotted on the same double-log graph used for VES. The flat portions of the factor curve indicate the possible presence of fractured zones and their depths are taken to be equal to the relevant AB/2 (Fig. 6.35B). The factor analysis has been found effective within 100 m depth only.

6.12.5   Computer based interpretations of sounding curves

The development of computer based inversion techniques has made interpretation easier and induced appreciable confidence. Computation of theoretical curves and layer parameters is simplified by the use of a digital filter technique developed by Ghosh (1971) and further developed by others. For interpreting the sounding curves there are two methods of inversion, viz., direct and indirect. The direct method of inversion involves computation of subsurface geoelectrical layer parameters directly from field data without any initial guess model of layer resistivity and thickness. It is carried out in two steps, viz., computing kernel or raised kernel function from the field data and obtaining layer parameters from the resistivity transform. Ghosh (1971) used filter theory to obtain resistivity transform from the linear relation between apparent resistivity and the resistivity transform. In the indirect method of inversion an initial guess of layer parameters derived from manual curve matching or visual approximation is required. From the initial guess model a theoretical apparent resistivity curve is generated. The guess parameters of the model are modified or adjusted through an iterative process to get a satisfactory match between the field curve and the generated theoretical curve under a given error criterion.

The kernel function forms the base of resistivity sounding interpretation. In the direct method a kernel function is computed from field data and from the kernel function other functions are computed to yield the layer parameters. In indirect method, through kernel function a theoretical apparent resistivity curve is generated for the layer model. The details of indirect and direct methods of interpretation are given by Koefoed (1979a). There is a variety of software commercially available for 1-D inversion automatic as well as interactive.

The guess model in indirect method affects the quality of inversion; the nearer the initial guess model to the true model, the better and faster is the inversion. Generally, inversion is taken as satisfactory if the field curve matches with computer generated theoretical curve within the acceptable root mean square difference of less than 5% between these curves. The rms difference is only an indicative of physical matching between the field and computed apparent resistivity curves and need not necessarily give the true subsurface hydrogeological model. Several combinations of geoelectrical layered earth models can be obtained within the error limit. It indicates that while a layered-earth model yields a single apparent resistivity curve, an apparent resistivity curve can yield multiple sets of equivalent layered-earth models within the limits of accuracy. This is an inherent non-uniqueness in inversion. Inversion schemes developed so far have attempted to analyze the layer parameter equivalence. The problem still persists. To resolve it to the extent possible certain layer parameters (either resistivity or thickness of layer) in the guess model is fixed with the help of borehole results or known hydrogeological conditions and makes the solution realistic. This is a type of constrained inversion. There are computer source codes available which give an error estimate for all parameters (layer resistivities and thicknesses) as well as a range of values for each parameter and best fit within the range of equivalence for judicious selection of parameters by the interpreter.

Zohdy (1989) developed an innovative method for interpretation of Schlumberger and Wenner sounding curves. It is a fast, automatic, iterative, direct method of interpretation without any need of an initial guess model. The curve is uniformly digitized and the number of layers is taken as equal to the number of digitized points such that the layer boundaries are uniformly spaced on the logarithmic scale for AB/2 or 'a'. That is a digitized point representing AB/2: 10 to 20 m will represent 100 to 200 m in the second log cycle. Hence, the sequence of digitized points represents a succession of layers with increasing thicknesses and there is no scope for including thin layers at depth. The curve is computed with this model and is adjusted with the field curve using factors through iterations for a defined rms difference. The layer model thus derived may represent reasonably satisfying subsurface layering, however, the equivalent models can be further generated by introducing layers satisfying the local geological conditions.

As a general practice, while interpreting the sounding curve, the simple layer model approach is to be adopted which involves a minimum number of layers. Further inclusion of layers should be done if the data demands it. For example, inclusion of thin layers at depth will not affect the sounding curve and it is included in the guess model for inversion and taken for granted that thin layers have been detected. It is not so as the resolving capability of VES does not allow that. It is better to initiate the interpretation through Occam's 1d inversion (Constable *et al.*, 1987) for smoothest model fitting the data which gives an idea about resolvable layers and where no initial guess on the number of layers or bias for a layer parameter is required. The smooth model is initiated with a number of layers equal to measurement points. Vedanti *et al.* (2005) present a modified or generalized Occam's inversion yielding better convergence. The inversion scheme for smoothest model uses constant-thickness layer and resistivity varies. Then it can be used to generate a simple layer model with minimum required rms difference. The simple layer model may be constrained by the geological or borehole information defining the depth to the interfaces. While carrying out inversion it is necessary that the apparent resistivity data input is from smoothed field VES curve, otherwise getting an rms error within 4–5% may be difficult.

6.12.6   Equivalence in layer parameters

It is observed that in a multi-layer geoelectrical sequence, parameters (resistivity and thickness) of intermediate layers can be altered to a certain extent without producing any appreciable change in the shape of resistivity sounding curve obtained over it. That is, the alteration may produce a small variation which is hardly detectable within the accuracy of measurement. A set of sounding curves thus obtained with varied layer parameters are said to be "equivalent" or practically equivalent and the phenomenon is known as "equivalence". If the root mean square difference between two apparent resistivity curves is less than 5%, they cannot be differentiated within the accuracy of observation. It is dominant when intermediate layers are thin compared to their depth of burial. That is, the deeper the intermediate layer the more prominent is the phenomena of equivalence for it. This brings in non-uniqueness in surface resistivity interpretations.

Equivalence in resistivity measurement is controlled by the major direction of current flow vis-à-vis subsurface geoelectrical layering. In a conductive layer surrounded by resistive layers the flow is along the layer and in a resistive layer surrounded by conductive layers it is perpendicular to the layer. Accordingly, Maillet (1947) indicated that in the case where the resistivity of the intermediate layer is less than that for the layer underlying it, i.e., for H or A type curves, equivalent cases could be obtained by changing the intermediate layer parameters in such a way that its longitudinal unit conductance- the ratio of its thickness to resistivity-remains the same. This is known as S equivalence. For an intermediate layer more resistive than the layer underlying it, producing either K or Q type curves, the equivalent layer parameters are obtained by keeping the transverse unit resistance, i.e., the product of thickness and resistivity of the intermediate layer as constant. This is known as T equivalence. The parameters S and T were named by Maillet as Dar Zarrouk parameters. Maillet indicated that condition for equivalence is governed by resistivity of the layer underlying the equivalent layer and the above mentioned conditions are derived for infinitely conductive or resistive substratum. However, the conditions hold good for $\rho_n/\rho_{n-1} > 10$ and $\rho_n/\rho_{n-1} < 0.1$, where ρ_{n-1} and ρ_n are resistivities of n-1th and nth layer underlying it. Also, in a multi-layer sequence, it will be affected by both S as well as T equivalence and by the overlying as well as underlying layers. So, for an intermediate conductive layer longitudinal conductance and for a resistive layer transverse resistance only can be computed accurately.

Using T and S equivalence, Pylaev's nomograms (1948) given in Bhattacharya and Patra (1968) were the first nomogram available to define the range of equivalence for the 4 types of 3-layer sounding curves. Chandra (1983) prepares such nomograms for lesser rms difference of 1% and revealed that the region of equivalence can be narrowed down by reducing the rms. It is shown in figure 6.36. It is observed that the higher the value of rms difference, the wider is the region of equivalence. Since, apparent resistivity curves with 4 to 5% difference in rms between them cannot be resolved, it is the maximum value of rms difference which is taken for determining the limits of the region of equivalence. Zohdy (1974) through construction of Dar Zarrouk (DZ) curves proposed by (Maillet, 1947) analyzes the limits of equivalence beyond the constraints of these nomograms for multi-layer geoelectrical sections. Koefoed (1969) mathematically analyzed equivalence through raised kernel function and indicated the basic constraint in resolving resistivity and thickness independently.

After primary interpretation of field curve is done the intermediate layer parameters obtained could be modified, if needed, for correlation with the lithological sequence under equivalence condition. Sharma and Kaikkonen (1999) opine that K type earth structure is easiest to resolve and A-type is the most difficult. Simms (1991) presented different inversion schemes for analyzing the equivalence. All these present an insight into the problem of equivalence. Presently, the limits of equivalence are obtained through parameter resolution matrix using computer codes to define the best fit model and the range for each layer parameter of the model. Equivalence or non-uniqueness cannot be resolved independently. It can be minimized if one of the layer parameters is known, viz., the depth to lithological interfaces or layer thicknesses through borehole data or seismic surveys. Also, it can be minimized to a great extent by joint inversion of resistivity sounding data with other geophysical data collected at

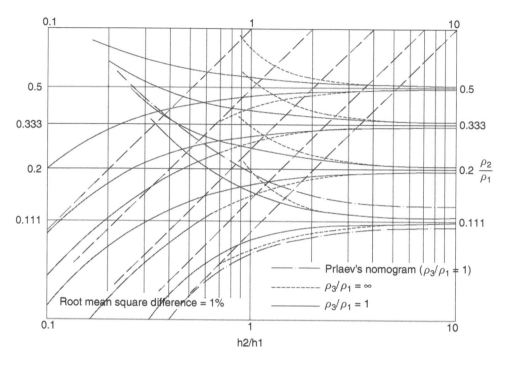

Figure 6.36 Equivalence nomogram for H type with $\rho_3/\rho_1 =$ infinitely high and $\rho_3/\rho_1 = 1$ for 1% rms difference (Source: Chandra, 1983).

that point which may have different sensitivity (Fitterman *et al.*, 1988; Sharma and Kaikkonen, 1999).

6.12.7   Poor resolution or suppression of a geoelectrical layer

Poor resolution of a layer results when it gets suppressed by the effect of the overlying and underlying layers immediately surrounding it. This happens either when the resistivity of the layer is in between the resistivities of the surrounding layers or the layer is surrounded by layers with contrasting resistivities. The effect becomes prominent if the layer to be resolved is thin, i.e., the depth of its occurrence is much higher than its thickness. The problem of suppression of a layer with intermediate resistivity is quite common in A type curves, generally obtained in hard rock. It becomes difficult to distinguish a layer from the ascending last segment of A type VES curves.

If a geoelectrical layer is thin it is not reflected or poorly reflected in the sounding curve even with distinct resistivity contrast. This phenomenon of suppression is more prominent for layers at depth. To explain this an example of 5-layer geoelectrical layer sequence is discussed (Chandra, 1983). The parameters of a 5-layer model are: layer resistivities as $\rho_1 = 500$ ohm.m, $\rho_2 = 100$ ohm.m, $\rho_3 = 5$ ohm.m, $\rho_4 = 200$ ohm.m and $\rho_5 = 20$ ohm.m and layer thicknesses as $h_1 = 20$ m, $h_2 = 40$ m, $h_3 = 15$ m and $h_4 = 50$ m.

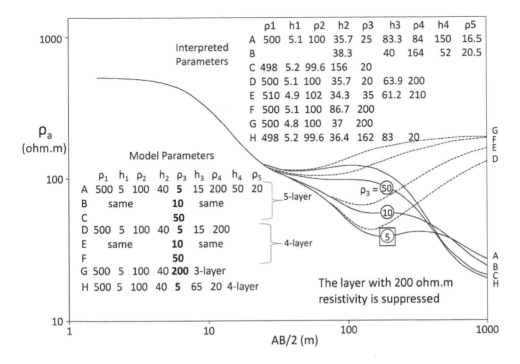

Figure 6.37 Suppression of a resistive layer by conductive surrounding. The 4th layer with 200 ohm.m resistivity and 50 m thickness is suppressed by the overlying and underlying conductive layers (Source: Chandra, 1983).

The aim is how best the parameters of the layer which is actually 50 m thick and associated with 200 ohm.m resistivity can be estimated from the sounding curve computed on the basis of the layer parameters given above. The example represents the sequence of Tertiary sandstones underlain by a resistive basaltic trap layer and then Gondwana sandstones and shales encountered in parts of Andhra Pradesh, India. The 50 m thick layer of 200 ohm.m resistivity represents the resistive basaltic trap embedded in conductive surroundings (Fig. 6.37). Interpretation of the model curve by curve matching and modification by computer based techniques revealed an equivalent curve with parameters as $\rho_1 = 500$ ohm.m, $\rho_2 = 100$ ohm.m, $\rho_3 = 25$ ohm.m, $\rho_4 = 84$ ohm.m and $\rho_5 = 16.5$ ohm.m, and $h_1 = 5.1$ m, $h_2 = 35.7$ m, $h_3 = 83.3$ m and $h_4 = 150$ m. That is, the parameters of the 3rd layer are overestimated while the resistivity of the target 4th layer is underestimated and its thickness is over estimated. It was observed that an increase in resistivity (from 5 to 10 ohm.m) of the 3rd layer or its removal altogether in the original model helped estimate the parameters of the 4th layer within permissible range. That is, the 4th layer which is resistive is getting suppressed because of highly conductive layers overlying and underlying it. In such cases it is essential that borehole information in the form of litholog and geophysical log or information from seismic surveys be considered to constrain the interpretation and assess the limitations. Similarly, the thin conductive layers – the fractured zones at depths get suppressed in

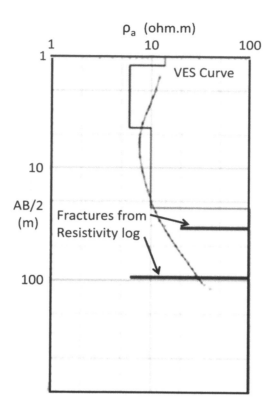

Figure 6.38 Suppression of a conductive thin layer (fractured zone) by resistive surrounding.

the resistive surrounding. Figure 6.38 presents a VES curve from which the fractured zones could not be delineated but the borehole drilled at the VES point encountered the fractures.

6.12.8 Depth-wise transition in resistivity

When there is a continuous fall or rise in resistivity with depth, the intermediate layers are not reflected prominently in a sounding curve, particularly, if the thickness of the intermediate layer is considerably less in comparison to that of the overburden. This phenomenon is referred to as "transition" in electrical behavior (Mallick and Roy, 1968 and Mallick and Jain, 1979). A common occurrence is in granitic terrain where a gradual change occurs from top weathered material to underlying saprolite or from saprolite to underlying fractured granite and compact granite. Since the intensity and density of fractures diminishes with depth, a gradual increase in resistivity with depth is observed within the fractured granite. It gives rise to a transitional increase in resistivity with depth. Chandra and Kumar (1982) conducted field studies in granite terrain in India and identified the transition effect in soundings which could not be resolved

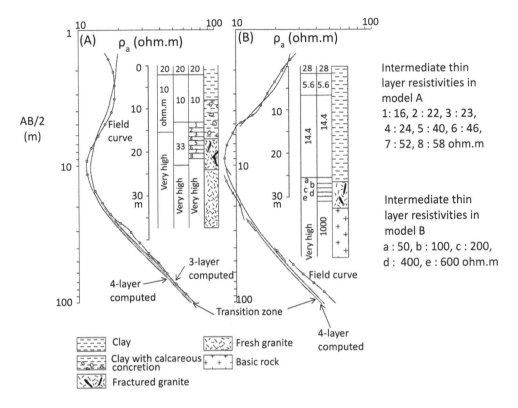

Figure 6.39 Transition in layer resistivity: field VES curve from Bundelkhand granitic terrain, Uttar Pradesh, India (Source: Chandra and Kumar, 1982).

from sounding interpretations by assuming discrete homogeneous three-layered earth (Fig. 6.39).

The gradual increase in resistivity with depth can be seen from the resistivity log from hard rocks (Fig. 6.40). The log records resistivity from 20 m depth. Coincidently at this borehole the weathered zone and saprolite extend to 20 m depth. The resistivity of formations underlying the saprolite upto 40 m depth increases gradually and not in steps The increase or decrease in resistivity over a certain depth range i.e., within the intermediate layer, could be linear or exponential (Fig. 6.41).

Mundry and Zschau (1983) consider a linear gradient in resistivity of the intermediate layer. Banerjee *et al.* (1980) consider an exponential one. While Mallick and Roy (1968) and Jain (1972) consider the intermediate layer with linearly transitional conductivity, Koefoed (1979b) considers linearly transitional resistivity which is more common. A linear increase in resistivity is not equivalent to a linear decrease in conductivity. Chandra and Kumar (1982) through a field study found that the resistive bottom layer in a three-layer case can have a linear increase in resistivity instead of an abrupt high. Further, the linear increase in resistivity can be approximated by a splitting into a number of thin homogeneous layers of equal thickness and increased

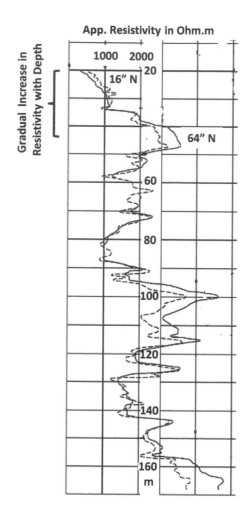

Figure 6.40 Electrical resistivity log of borehole showing gradual increase in formation resistivity with depth upto about 48 m. The weathered zone is upto 14 m followed by saprolite upto 20 m depth.

resistivities in steps. The apparent resistivity curve computed by this multi-layering matches with the field curve (Fig. 6.39), indicating that it is quite difficult to identify transitional increase in resistivity with depth from the VES curve.

6.12.9 Effect of top soil conductivity

The saturated fractured zone in hard rocks should be associated with resistivity 'lows' in a resistivity profile traversing it. However, identifying fractured zones from the 'lows' at times may be misleading because of the presence of conductive overburden, which may also indicate lows. Figure 6.42 illustrates two Wenner profiles with a = 30 m and

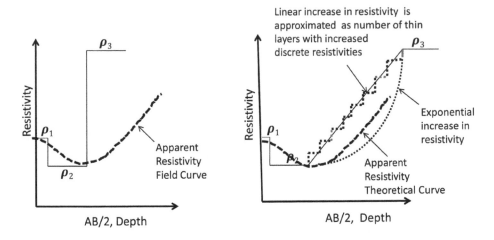

Figure 6.41 Geoelectrical layer models for (a) discrete layer resistivities and (b) transitional increase in intermediate layer resistivity: linear or exponential.

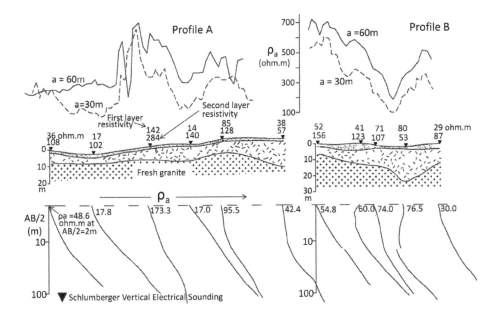

Figure 6.42 Wenner resistivity profiling showing the effect of top soil conductivity (Source: Chandra *et al.*, 1983).

a = 60 m run across the suspected fractured zone in granitic terrain (Chandra *et al.*, 1983). The resistivity 'low' in profile A and B are of the same order. The layer parameters of soundings observed over the low in profile A indicate shallow occurrence of compact rock (absence of fractured zone) and low resistivity (14–17 ohm.m) overburden, while those observed over the low of profile B indicate deeper occurrence of compact rock with overburden resistivity as 80 ohm.m. This could be explained by a simple set of apparent resistivity curves for 3-layer geoelectrical models with varying resistivity of the first layer, observed in hard rocks (Fig. 6.7). For H or A type curves a reduction in top layer resistivity from 400 to 1 ohm.m shifts the curve towards the right and less apparent resistivities are recorded for all the current electrode spacings. However, for descending type curves the effect is reversed. The presence of a conductive top layer shifts the curve upward at higher current electrode separations (Satpathy, 1974). In this case the presence of conductive overburden changes the apparent resistivity value in an opposite manner. After a critical value of current electrode separation, the apparent resistivity values increase if the top resistive layer is replaced by a conductive one, and values decrease if the top conductive layer is replaced by a resistive one.

REFERENCES

Angenheister, G. (ed.) (1982) Physical properties of rocks. In: Landolt-Bornstein, New series, Vol.VI b, Berlin, Heidelberg New York, Springer-Verlag Publication.

Anjorin, M P & Olorunfemi, M.O. (2011) A short note on comparative study of Schlumberger and Half-Schlumberger arrays in vertical electrical sounding in a basement complex terrain of Southwest Nigeria. *Pacific Journal of Science & Technology*, 12 (2), 528–533.

Apparao, A. (1991) Geoelectric profiling. *Geoexploration*, 27 (3–4), 351–389.

Balakrishna, S. & Ramanujachary, K.R. (1985) Resistivity investigations in different geological terrains. *Groundwater News*, 102–118

Ballukraya, P.N., Sakthivadivel, R. & Baratan, R. (1983) Breaks in resistivity sounding curves as indicators of hard rock aquifers. *Nordic Hydrology*, 14 (1), 33–40.

Banerjee, B. & Pal, B.P. (1986) A simple method for determination of depth of investigation characteristics in resistivity prospecting. *Exploration Geophysics*, 17 (2), 93–95.

Banerjee, B., Sengupta, B.J. & Pal, B.P. (1980) Apparent resistivity of a multilayered earth with a layer having exponentially varying conductivity. *Geophysical Prospecting*, 28 (3), 435–452.

Barker, R.D. (1981) The offset system of electrical resistivity sounding and its use with a multicore cable. *Geophysical Prospecting*, 29 (1), 128–143.

Barker, R.D. (1989) Depth of investigation of collinear symmetrical four-electrode arrays. *Geophysics*, 54(8), 1031–1037.

Bernard, J., Leite, O. & Vermeersch, F. (2006) *Multi-electrode resistivity imaging for environmental and mining applications*. IRIS Instruments, Orleans, France, 6 p.

Bernard, J. & Valla, P. (1991) Groundwater exploration in fissured media with electrical and VLF methods. *Geoexploration*, 27 (1–2), 81–91.

Bhattacharya, B.B. & Dutta, I. (1982) Depth of investigation studies for gradient arrays over homogeneous isotropic half-space. *Geophysics*, 47 (8), 1198–1203.

Bhattacharya, B.B. & Sen, M.K. (1981) Depth of investigation of collinear electrode arrays over homogeneous anisotropic half-space in direct current methods. *Geophysics*, 46(5), 768–780.

Bhattacharya, P.K. & Patra, H.P. (1968) *Direct current geoelectrical sounding, principles and interpretation*. Amsterdam, Elsevier Publication.

Central Ground Water Board (1982) *Ground water studies in the Sina and the Man river basins, south Maharashtra – Project findings and recommendations*. CGWB, Govt. of India, Project Tech. Report No. P-1

Chandra, P.C. (1983) Geophysical studies for aquifer delineation in Lower Maner Basin of Andhra Pradesh & the Ganga-Tons inter-stream area of Uttar Pradesh. *Ph.D Thesis, National Geophysical Research Institute, Hyderabad/Banaras Hindu University, Varanasi, India*.

Chandra, P.C. & Kumar, P. (1982) Geological investigation for groundwater in drought affected areas of Banda district, UP. *Journal of Association of Exploration Geophysicists*, 3 (2).

Chandra, P.C., Ramakrishna, A. & Singh, H. (1983) Tracing shear zones by geoelectrical techniques for groundwater in Dindigul area of Madurai district, Tamil Nadu. In: *Proceedings of the Seminar on Assessment, Development & Management of Ground Water Resources*, CGWB, New Delhi, pp. 187–194.

Chandra, P.C., Reddy, P.H.P. & Singh, S.C. (1994) Geophysical studies for groundwater exploration in Kasai and Subarnarekha River basins, *(UNDP Project) Central Ground Water Board, Min. of Water Resources, Govt. of India Tech. Report.*

Chandra, P.C., Srivastava, M.M., Adil, M., Bhowmic, M.K., Pandey, K.S., Haq, S. & Singh, U.B. (2002) Geoelectrical investigations for groundwater in quartzitic sandstones and granites of Sonbhadra district, U.P. In: *Proceedings of the Int. Conf. on Hydrology and Watershed Management, 18–20 December 2002, JNTU Hyderabad.*

Chandra, P.C., Tata, S. & Raju, K.C.B. (1987) Geoelectrical response of cavities in limestones: an experimental field study from Kurnool district, Andhra Pradesh, India. *Geoexploration*, 24 (6), 483–502.

Chandra, S., Rao, V.A., Krishnamurthy, N.S., Dutta, S. & Ahmed, S. (2006) Integrated studies for characterization of lineaments used to locate groundwater potential zones in a hard rock region of Karnataka, India, *Hydrogeology Journal.* 14 (6), 1042–1051.

Chandra, S., Nagaiah, E., Reddy, D.V., Rao, V. A. & Ahmed, S. (2012) Exploring deep potential aquifer in water scarce crystalline rocks. *Journal Earth System Science*, 121 (6), 1455–1468.

Chaturvedi, P.C., Prasad, G.J. & Bandyopadhyay, R.R. (1979) *A report on resistivity surveys in the Deccan Trap.* Sina-Man Project, Central Ground Water Board, Govt. of India, Report.

Coggon, J.H. (1973) A comparison of IP electrode surveys. *Geophysics.* 38 (4), 737–761.

Constable, S.C., Parker, R.L. & Constable, C.G. (1987) Occam's inversion: A practical algorithm for generating smooth models from electromagnetic sounding data. *Geophysics*, 52 (3), 289–300.

Corbett, J.D. (1992) A comparison of dipole-dipole and gradient arrays. *Exploration Geophysics*, 23 (1&2), 75–82.

Dahlin, T. (2000) Short note on electrode charge-up effects in DC resistivity data acquisition using multi-electrode arrays. *Geophysical Prospecting*, 48 (1), 181–187.

Dahlin, T. (2001) The development of DC resistivity imaging techniques, *Computers & Geosciences*, 27 (9), 1019–1029.

Dahlin, T. & Zhou, B. (2004) A numerical comparison of 2D resistivity imaging with 10 electrode arrays. *Geophysical Prospecting*, 52 (5), 379–398.

Daily, W., Ramirez, A., Labrecque, D. & Nitao, J. (1992) Electrical resistivity tomography of vadose zone movement. *Water Resources Research*, 28 (5), 1429–1442.

Edwards, L.S. (1977) A modified pseudosection for resistivity and induced-polarization. *Geophysics*, 42 (5), 1020–1036.

Edwards, R.N. & Howell, E.C. (1976) A field test of the magnetometric resistivity (MMR) method. *Geophysics*, 41 (6A), 1170–1183.

Ehirim, C.N. & Essien, E.E. (2009) Comparative investigation of offset Wenner, Square and Schlumberger arrays in electrical anisotropy studies. *Scientia Africana*, 8 (2), 53–60.

Eloranta, E.H. (1986) Potential field of a stationary electric current using Fredholm's integral equations of the second kind. *Geophysical Prospecting*, 34 (6), 856–872.

Engalenc, M. (1978) Methode d'etude et de recherché de l'eau souterraine des roches cristallines de l'Afrique de l'ouest: Bull. *Comite interafricain d'etudes hydrauliques (C.I.E.H.) serie hydrogeologie*, 41.

Everett, M.E. & Meju, M.A. (2005) Near-surface controlled-source electromagnetic induction: background and recent advances. In: Rubin, Y. & Hubbard, S.S. (eds.)

Hydrogeophysics: *Chapter 6, Water Science and Technology Library, Vol. 50.* The Netherlands, Springer. pp. 157–183.

Evjen, H.M. (1938) Depth factor and resolving power of electrical measurements. *Geophysics*, 3 (2), 78–95.

Falco, P., Negro, F., Szalai, S. & Milnes, E. (2013) Fracture characterisation using geoelectric null-arrays. *Journal of Applied Geophysics*, 93 (1), 33–42.

Fitterman, D.V., Meekes, J.A.C. & Ritsema, I.L. (1988) Equivalence behavior of three electrical sounding methods as applied to hydrogeological problems. Presented at: *The 50th Annual Meeting and Technical Exhibition of the European Association of Exploration Geophysicists, 6–8 June 1988, The Hague, The Netherlands.*

Flathe, H. & Leibold, W. (1976) *The smooth sounding graph – a manual for field work in direct current resistivity sounding.* Federal Institute for Geosciences and Natural Resources, Hannover, Germany.

Furness, P. (1993) Gradient array profiles over thin resistive veins. *Geophysical Prospecting*, 41 (1), 113–130.

Ghosh, D.P. (1971) The application of linear filter theory to the direct interpretation of geoelectrical resistivity sounding measurements. *Geophysical Prospecting*, 19 (2), 192–217.

Griffiths, D.H. & Barker, R.D. (1993) Two-dimensional resistivity imaging and modeling in areas of complex geology. *Journal of Applied Geophysics*, 29 (3–4), 211–226.

Griffiths, D.H. & Turnbull, J. (1985) A multi-electrode array for resistivity surveying. *First Break*, 3 (7), 16-20.

Griffiths, D.H. & Turnbull, J. & Olayinka, A.I. (1990) Two-dimensional resistivity mapping with a computer-controlled array. *First Break*, 8 (4), 121–129.

Habberjam, G.M. (1972) The effect of anisotropy on square array resistivity measurements. *Geophysical Prospecting*, 20 (2), 249–266.

Habberjam, G.M. (1975) Apparent resistivity, anisotropy and strike measurements. *Geophysical Prospecting*, 23 (2), 211–247.

Hagrey, S. A.al (1994) Electric study of fracture anisotropy at Falkenberg, Germany. *Geophysics*, 59 (6), 881–888.

Jain, S.C. (1972) Resistivity sounding on a three-layer transitional model. *Geophysical Prospecting*, 20 (2), 293–292.

Keller, G.V. (1968) Comments on "An inverse slope method of determining absolute resistivity" by Sanker Narayan, P.V. & Ramanujachary, K.R. (Geophysics, Vol. 32, No. 6, pp. 1036–1040). *Geophysics*, 33 (5), 843–845.

Keller, G.V. & Frischknecht, F.C. (1966) *Electrical method in geophysical prospecting.* Oxford, *Pergamon Press.*

Koefoed, O. (1969) An analysis of equivalence in resistivity sounding. *Geophysical Prospecting*, 17 (3), 327–335.

Koefoed, O. (1979a) *Geosounding Principles, 1 – Resistivity sounding measurements.* Amsterdam, Elsevier Scientific Publishing Company.

Koefoed, O. (1979b) Resistivity sounding on an earth model containing transition layers with linear change in resistivity with depth. *Geophysical Prospecting*, 27 (4), 862–868.

Kunetz, G. (1966) *Principles of direct current resistivity prospecting.* Geoexploration Monographs Series 1- No. 1, Braekken, H. & van Nostrand, R. (eds.), Geopublication Associates, Berlin-Nikolasse, Gerbruder Borntraeger.

Lane Jr, J.W., Haeni, F.P., & Watson, W.M. (1995) Use of square-array direct-current resistivity method to detect fractures in crystalline bedrock in New Hampshire. *Ground Water*, 33 (3), 476–485.

Loke, M.H. (2000) Electrical imaging surveys for environmental and engineering studies: a practical guide to 2-D and 3-D surveys. [Online] Available from: www.http://moho.ess.ucla .edu/~pdavis/ESS135_2013/LITERATURE/%20LokeDCREsistivity.pdf.

Loke, M.H. (2001) Tutorial: 2-D and 3-D electrical imaging surveys. [Online] Available from: www.https://pangea.stanford.edu/research/groups/.../DCResistivity_Notes.pdf.

Louis, I.F., Louis, F.I. & Grambas, A. (2002) Exploring for favorable groundwater conditions in hard rock environments by resistivity imaging methods: synthetic simulation approach and case study example. *In: International Conference on Earth Sciences & Electronics (ICESE-2002); Journal of Electrical & Electronics Engineering, Special Issue*, October 2002. pp. 1–15.

Maillet, R. (1947) The fundamental equation of electrical prospecting. *Geophysics*, 12 (4), 529–556.

Mallick, K. & Roy, A. (1968) Resistivity sounding on a two-layer earth with transitional boundary. *Geophysical Prospecting*, 16 (4), 436–446.

Mallick, K. & Jain, S.C. (1979) Resistivity sounding on a layered transitional earth. *Geophysical Prospecting*, 27 (4), 869–875.

Matsui,T., Park, S.G., Park, M.K. & Matsuura, S. (2000) Relationship between electrical resistivity and physical properties of rocks. In: *GeoEng2000 Intenational Symposium on Geotechnical & Geological Engineering,19–24 November 2000, Melbourne, Australia*. International Society for Rock Mechanics.

Middleton, M.F. (1974) On rule of thumb interpretation of resistivity gradient array data. Bulletin Australian Society of Exploration *Geophysicists*, 5 (4), 134–135.

Miele, M., Laymon, D., Gilkeson, R., & Michelotti, R. (1996) *Rectangular Schlumberger resistivity arrays for delineating vadose-zone clay-lined fractures in shallow tuff*. Los Alamos National Laboratory, Report No. LA-UR 96–512.

Mundry, E. & Zschau, H.-J. (1983) Geoelectrical models involving layers with a linear change in resistivity and their use in the investigation of clay deposits. *Geophysical Prospecting*, 31 (5), 810–928.

Muralidharan, D. (1996) A semi-quantitative approach to detect aquifers in hard rocks from apparent resistivity data. *Journal Geological Society of India*, 47 (2), 237–242.

Olayinka, A. & Barker, R. (1990) Borehole siting in crystalline basement areas of Nigeria with a microprocessor-controlled resistivity traversing system. *Ground Water*, 28 (2), 178–183.

Oldenburg, D.W. & Li, Y. (1999) Estimating depth of investigation in DC resistivity and IP surveys. *Geophysics*, 64 (2), 403–416.

Orellana, E. & Mooney, H.M., (1966) Master tables and curves for vertical electrical sounding over layered structures. *Madrid, Interciencia, 150 p., 66 tables*.

Palacky, G.J. (1991) Resistivity characteristics of geologic targets. In: Nabighian, M.S. (ed.) *Electromagnetic Methods in Applied Geophysics-Theory*, Vol.1 Chapter 3. Tulsa, Oklahoma, USA, SEG Publication USA, Society of Exploration Geophysicists, pp. 53–129.

Palacky, G.J. & Kadekaru, K. (1979) Effect of tropical weathering on electrical and electromagnetic measurements. *Geophysical Prospecting*, 44 (1), 69–88.

Palacky, G.J., Ritsema, I.L. & De Jong, S.J. (1981) Electromagnetic prospecting for groundwater in Precambrian terrains in the Republic of Upper Volta. *Geophysical Prospecting*, 29 (6), 932–955.

Pratt, D.A. & Whiteley, R.J. (1974) Computer simulation and evaluation of electrode array responses in resistivity and I P prospecting. *Bull. Aust. Soc. Expl. Geophys*. 5 (2), 65–87.

Pylaev, A.M. (1948) Rykovodstvo po Interpretatsi Vertikal'nykh Electricheskikh Zondirovanii. *Gosgeolizdat, Moskva*.

Rijkswaterstaat (1969) Standard graphs for resistivity prospecting. The *Netherlands, EAEG*.

Roy, A. & Apparao, A. (1971) Depth of investigation in direct current methods. *Geophysics*, 36 (5), 943–959.

Sanker Narayan, P.V. & Ramanujachary, K.R. (1967) An inverse slope method of determining absolute resistivity. *Geophysics*, 32 (6), 1036–1039.

Satpathy, B.N. (1974) A paradox in apparent resistivity measurements over a ground section with conductive substratum. *Geophysics*, 39 (1), 93.

Schulz, R. (1985) Interpretation and depth of Investigation of gradient measurements in direct current geoelectrics. *Geophysical Prospecting*, 33 (8), 1240–1253.

Seaton, W.J. & Burbey, T.J. (2002) Evaluation of two-dimensional resistivity methods in a fractured crystalline-rock terrane. *Journal of Applied Geophysics*, 51(1), 21–41.

Sharma, S.P. & Biswas, A. (2013) A practical solution in delineating thin conducting structures and suppression problem in direct current resistivity sounding. *Journal Earth System Science*, 122 (4), 1065–1080.

Sharma, S.P. & Kaikkonen, P. (1999) Appraisal of equivalence and suppression problems in 1D EM and DC measurements using global optimization and joint inversion. *Geophysical Prospecting*, 47 (2), 219–249.

Shettigara, V.K. & Adams, W.M. (1989) Detection of lateral variations in geological structures using electrical resistivity gradient profiling. *Geophysical Prospecting*, 37(3), 293–310.

Simms, J.E. (1991) Uniqueness and resolution in resistivity interpretation. *Ph.D Thesis, Texas A&M University, USA.*

Stummer, P., Maurer, H. & Green, A.G. (2004) Experimental design: Electrical resistivity data sets that provide optimum subsurface information. *Geophysics*, 69 (1), 120–139.

Sumner, J.S. (1972) A comparison of electrode arrays in IP surveying. *Presented at: The AIME Annual Meeting, 20–24 February 1972, San Francisco, California, USA.*

Sumner, J.S. (1976) *Principles of induced polarization for geophysical exploration.* Amsterdam, Elsevier Scientific Publishing Company.

Stummer, P. (2003) New developments in electrical resistivity imaging. *Ph.D Dissertation (DISS. ETH No. 15034) Swiss Federal Inst. Technology, Zurich.*

Szalai, S., Koppan, A. & Szarka, L. (2007) Effect of positional inaccuracies on multielectrode. In: *Near Surface 2007: Thirteenth European Meeting Environmental & Engineering Geophysics, 3–5 September 2007, Istanabul, Turkey.*

Szalai, S. & Szarka, L. (2008) On the classification of surface geoelectric arrays. *Geophysical Prospecting*, 56 (2), 159–175.

Szalai, S., Novak, A., & Szarka, L. (2009) Depth of investigation and vertical resolution of surface geoelectric arrays. *Journal of Environmental & Engineering Geophysics*, 14 (1), 15–23.

Taylor, R.W. & Fleming, A.H. (1988) Characterizing jointed system by azimuthal resistivity surveys. *Ground Water*, 26 (4), 464–474.

Telford, W.M., Geldart, L. P. & Sheriff, R.E. (1990) *Applied Geophysics.* Cambridge University Press.

Van Nostrand, R.G. & Cook, K.L. (1966) *Interpretation of resistivity data.* U.S. Geological Survey Prof. Paper 499, U.S.G.S.

Van Overmeeren, R.A. & Ritsema, I.L. (1988) Continuous vertical electrical sounding. *First Break*, 6 (10), 313–324.

Vedanti, N., Srivastava, R.P., Sagode, J. & Dimri, V.P. (2005) An efficient 1 d Occam's inversion algorithm using analytically computed first- and second-derivatives for dc resistivity soundings. *Computer & Geosciences*, 31 (3), 319–328.

Ward, S.H. (1990), Resistivity and induced polarization methods. In: Ward, S.H. (ed.) *Geotechnical and Environmental Geophysics, Vol. I: Review and Tutorial.* Tulsa, Oklahoma, USA, Society of Exploration Geophysicists, pp. 147–189.

Watson, K.A. & Barkar, R.D. (1999) Differentiating anisotropy and lateral effects using azimuthal resistivity offset Wenner sounding. *Geophysics*, 64(3), 739–745.

Whiteley, R.J. (1973) Electrode arrays in resistivity and IP prospecting: a review. *Exploration Geophysics*, 4 (1), 1–29.

Yadav, G.S. (1988) Pole-Dipole resistivity sounding technique for shallow investigations in hard rock areas. *Pure & Applied Geophysics*, 127 (1), 63–71.

Zohdy, A.A.R. (1965) The auxiliary point method of electrical sounding interpretation, and its relationship to the Dar Zarrouk parameters. *Geophysics*, 30 (4), 644–659.

Zohdy, A.A.R. (1968) *The effect of current leakage and electrode spacing errors on resistivity measurements*. U.S.Geological Survey Prof. Paper, 600-D, pp. D258–D264, U.S.G.S.

Zohdy A.A.R. (1974) *Use of Dar Zarrouk curves in the interpretation of vertical electrical sounding data*. U.S. Geological Survey, Geological Survey Bulletin 1313-D, U.S.G.S.

Zohdy, A.A.R., Eaton, G.P. & Mabey, D.R. (1974) *Application of surface geophysics to groundwater investigations*. U.S. Geological Survey Techniques of Water Resources Investigations, Book 2, Chapter D1, U.S.G.S.

Zohdy, A.A.R. (1989) A new method for the interpretation of Schlumberger and Wenner sounding curves. *Geophysics*, 54 (2), 245–253.

Chapter 7

The self potential method

7.1 INTRODUCTION

Self Potential (SP) or Spontaneous Potential is a noninvasive, passive method to measure natural electrical potentials developed in the ground. It has been used to locate sulphide ores by measuring the electrochemical potentials and groundwater movement by electrokinetic or streaming potentials. It can provide useful information on the direction of movement of naturally and/or artificially recharged groundwater and that induced by pumping of water well. In hard rock, groundwater flow through preferential drainage paths of near surface fractured zones can be identified on the surface, particularly when flow velocity is increased by a well tapping the fractured zone. Since an electrical potential gradient is created by groundwater flow along the fractured zone, measurement helps establishing its orientation.

Measurement of SP is made for a variety of applications. It is used to detect, map and monitor seepages in earth dams (Ogilvy et al. 1969; Bogoslovsky and Ogilvy, 1970a and b). SP being affected by the hydraulic gradient the time-lapsed or continuous measurements of SP on the ground around a well being pumped and their distance versus time plot can help define the zone of influence (Costar et al., 2008; Straface et al., 2010). Also, SP being affected by geological structures it has been used to delineate fault zones (Richards et al., 2010) associated with water springs (Monteiro Santos et al., 2002). Fitterman (1979) analyses SP anomalies near a vertical resistivity discontinuity and shows the effect of resistivity on anomaly amplitude. Jinadasa and de Silva (2009) indicate a possible correlation between resistivity low and negative SP anomaly with subsurface geological structure. Wishart et al. (2006) point out the usefulness of azimuthal SP measurements in delineating the fractured zone in combination with resistivity measurements. Robert et al. (2011) reveal the usefulness of combined application of SP and resistivity imaging in locating fractured zones. Corwin (1990) presents an excellent tutorial while Revil et al. (2006) presents a summary of the applications of SP in hydrogeological issues and the future perspective.

Though SP is mostly associated with near surface hydrogeological or hydrochemical variations, Jouniaux (2011) reports its possible link to deeper injected groundwater flow in a fault zone at about 1600 m depth. It has been used for assessing the regional groundwater flow characteristics (Satou et al., 2005). SP can be used to monitor the long term environmental variables and also the relatively short term changes caused by artificial recharge of groundwater and infiltration of water through karstic limestones and the vadose zone (Erchul and Slifer, 1989; Suski et al., 2006). Long term monitoring of anomalous SP variations across fractured zone can reveal the stress variations (Gensane et al., 1999). Time-lapsed or continuous SP measurements as well as 2d imaging (Colangelo et al., 2006) for monitoring the subsurface fluid flow and infiltration process are gaining importance. Mapping SP near the radial arms of a collector well can help demarcate the effective part of the radial arms in capturing groundwater (Muralidharan et al., 2005). SP profiles provide information on variations in depth to water table in an area (Birch, 1998; Suski et al., 2006; Jardini et al., 2009) and can be used to estimate hydraulic conductivity also (Straface et al., 2010). Fagerlund and Heinson (2003) carried out SP surveys during pumping tests and indicate the possibility of obtaining hydrogeological parameters. Chandra et al. (1994) make an attempt to identify the direction of groundwater flow into a pumped well. Schiavone and Quarto (1984) conducted SP studies in coastal area to relate the upward flow of water from an aquifer. SP can be used in groundwater pollution studies to estimate the spreading of a pollutant plume (Naudet et al., 2003 and 2004) by measuring the redox potential. It can be used for hydrocarbon pollution of groundwater (Minsley et al., 2007) if its effect changes the SP significantly (Forte and Bentley, 2013). Also, SP measurements can be used for monitoring biogeochemical processes (Snieder et al., 2007).

Measuring SP is one of the oldest and simplest geophysical techniques and has been in practice in mineral prospecting since early 19th century. Due to uncertainty of the source causing the anomaly, noise, non-repeatability and non-availability of quantitative interpretational procedures, the application of SP has not been extensive. With the development of inversion schemes and procedures in quantitative estimations over the past 35–40 years for separating potentials of different origins, the technique gained renewed interest. SP field survey is quite economical and fast and can be used for reconnaissance also. Generally, it is combined with other geophysical surveys like resistivity imaging and electromagnetic imaging to facilitate hydrogeological transformation of the interpretations. The publication by Corwin (1990) presents the most detailed procedure and applications of SP method. The method is briefly described below through selected case studies emphasizing its usefulness in hard rock.

7.2   BASICS

Self Potential is a natural electric potential observed anywhere on the earth. Electrical potential difference is observed between two electrodes inserted at a distance in the ground even without injecting current. This natural potential difference between any two points on the ground surface, not due to any application of external electric current, may range from a fraction of millivolt to a few tens or hundreds of millivolts. The natural potentials can develop between different formations due to variations in mineralization, flow of groundwater, thermal gradient and variation in interstitial fluid conductivity. SP developed in the ground varies with space and time. The two main sources of SP are electrochemical and electrokinetic. The electrochemical one is due to the electrochemical reactions in the near surface formations. The electrokinetic component, also known as streaming potential, is generated by flow of water through porous formation and fractured zones and is of our interest in groundwater investigations in hard rock. The development of streaming potential was first recorded by Quincke in 1859 (Berube, 2004). An electric potential gradient is developed due to movement of ions with flow of water and therefore streaming potential difference on ground can be observed when groundwater flows underneath. The SP developed due to electrokinetic phenomena is generally less than that due to electrochemical one.

At the level of rock-fluid interface, the mechanism of streaming potential is the development of surface charge on a solid surface by chemical interaction with the liquid – the pore water when in contact, including the accumulation of unbalanced surface charge on clay particles (Berube, 2004). The concentration of one type of charge on the solid surface attracts the opposite charge in the liquid in contact and helps develop a concentration of counter-ions at the interface. This rearrangement of charges at the interface is seen as an electric double layer. The concentration of counter-ions reduces with distance from the interface in the liquid phase and forms a diffused layer. With the movement of liquid under hydraulic gradient, the excess liquid phase counter-ions accumulated at the solid-liquid interface get sheared off and move with the liquid within the diffused part of the electric double layer while some remain attached as a fixed layer also known as the Stern layer at the solid surface. The movement of excess charge will cause a variation in counter-ion concentration along the flow. There will be a surplus of ions downstream and deficit at the upstream end. This will cause a potential difference – the streaming potential and the electric current to flow. Thus, the streaming potential is developed due to the drag of excess electrical charge by the flow (Ishido and Mizutani, 1981). The streaming potential is proportional to the potential difference between the fixed layer and the moving liquid. The potential distribution in the electric double layer is shown in figure 7.1 taken from Fagerlund and Heinson (2003) and Kim et al. (2004). The potential at the shear plane within the electric double layer from where counter-ions in the liquid start moving with the liquid is known as 'zeta-potential'. The Helmholz-Smoluchowski equation relating streaming potential to zeta potential is

$$\Delta V = (\varepsilon \zeta / \mu \sigma) \Delta P$$

or the Electrokinetic Coefficient

$$C_S = \Delta V / \Delta P = \varepsilon \zeta / \mu \sigma$$

where ΔV is streaming potential, ε is dielectric constant of the fluid, ζ is zeta potential, μ is viscosity of the fluid, σ is conductivity of the fluid and ΔP is pressure head difference. The value of ζ ranges from a few tens of mV to more than $100\,mV$ (Lorne et al., 1999). The value of Cs as per laboratory studies, ranges approximately from -10^{-6} V/Pa to -10^{-7} V/Pa (or -10^{-2} V/m to -10^{-3} V/m) for electrical conductivity of water ranging from 0.01 to 0.1 S/m and pH around 7 (Jouniaux et al., 2009).

The development of the zeta potential is proportional to the surface charge accumulation on the solid-liquid interface and hence controlled by pH value of the liquid (Ishido and Mizutani, 1981). The higher the pH value the lesser is the surface charge accumulation and negative zeta-potential. For low pH zeta potential is positive. This indicates that in groundwater flow assessment through SP measurements, the chemical quality of groundwater plays a significant role. The SP decreases with increase in salinity of pore fluid (Erchul and Slifer, 1989). In liquids with higher ionic strength due to chemical composition the diffused part of the double layer will get compressed and

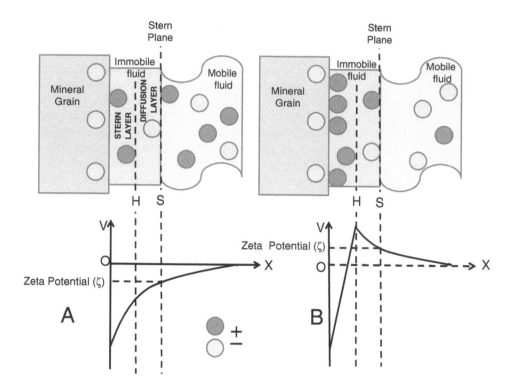

Figure 7.1 The electrical double-layer at the rock-water interface according to the Stern model. The electric potential (V) is a function of distance (x) from the pore wall. The hydrodynamic slipping plane (S) separates the mobile and immobile phases of the fluid. The potential at this plane is called the zeta-potential (ζ). Depending on amount of specific adsorption in the Stern layer between the pore wall and plane H, ζ can be negative (A) or positive (B) For a negative ζ, more positive than negative ions are transported with the fluid (Source: Fagerlund and Heinson, 2003; Kim *et al.*, 2004).

the zeta potential is reduced (Berube, 2004) and hence the streaming potential is also reduced (Vichabian and Morgan, 2002). Besides, it depends on hydraulic conductivity, pressure and temperature (Kim *et al.*, 2005).

7.3 INSTRUMENT

A sensitive high-impedance digital voltmeter is basically required for SP measurements. Since SP survey is generally carried out along with a resistivity survey, a resistivity meter having potential measuring facility can be used. The amplitude of anomalies due to streaming potential being quite low, the instrument should be sensitive to measure potentials in millivolt range with a resolution preferably 0.1 mV. Instrument with micro-processor based stacking facility helps rejection of noise. Besides, the accessories required are electrodes and winches with insulated cable of required length. A simple metal stake electrode is not used as the electrochemical reaction due to the metal electrode (electronic conductor) in contact with moisture (ionic conductor) in the ground will allow accumulation of electrical charges at the interface and the potential developed on the electrode will obscure the natural SP measurement accurately. To minimize this effect non-polarizable electrodes are used. The most commonly used non-polarizable electrode is a copper electrode immersed in a porous pot containing concentrate copper sulfate solution. Similar non-polarizable electrodes can be prepared by other metal electrodes like lead, zinc, silver immersed in their salt solutions. Porous pots are either porous plugs of wood or unglazed porcelain permitting ionic diffusion from the salt solution and making electrical contact with the soil. In case, copper electrodes in porous pots containing copper sulfate solution type non-polarizable electrodes are used, they are kept connected in a tub containing copper sulfate solution for 8 to 12 hours in advance to minimize any potential difference between the electrodes. Before doing this, the copper rods are cleaned thoroughly and the coating if any on the rods is removed. To maintain concentration of the solution an excess amount of copper sulfate crystals or powder is put into it. The concentration of copper sulfate solution in the porous pot is maintained throughout the survey to avoid any potential difference arising due to variations in concentration. The solution in the porous pot is replaced by a fresh solution whenever the connected electrode pair in the solution tub shows polarization potential. When there is no survey, the porous pots are emptied and cleaned by fresh water. The potential of these non-polarizing electrodes can change with moisture content of the soil in which it is placed as well as the surrounding temperature. Corwin (1990) indicates a potential variation of the order of $+0.3$ to $+1.0$ mV per percentage change in soil moisture content for copper rod in copper sulfate solution type electrode and further mentions that there could be a potential difference of the order of 70 mV if the electrodes are kept in dry and saturated clay soil and a change of $+0.5$ to $+1.0$ mV per degree Centigrade of temperature change of solution in the porous pot. Erchul and Slifer (1989) used copper-clad steel rods for long term monitoring of large anomalies and systematic change of SP with minimum electrode polarization effect. According to them these bi-metallic electrodes are cheaper, easy to handle and durable. However, they preferred the use of porous pots for one-time survey over large area where greater accuracy is desired in measuring smaller anomalies.

7.4 FIELD PROCEDURES

The SP values related to groundwater flow being generally within tens of millivolts, care has to be taken for precise field measurements to identify anomalies and perturbations. The profile layout, spacing and station interval depend on the objective. For this, Corwin (1990) suggests to consider also the anticipated wavelength of anomaly. Profile lines are laid across the expected anomaly trend and the profile interval is kept larger than the station interval. In general, an SP survey is carried out, along 20 to 50 m spaced parallel long profile lines or along 8 to 12 radial lines covering 360°, originating from a borehole or a point of interest (Fig. 7.2a and b). The measurement stations can be kept at 2 to 10 m interval. Since in most of the cases the trend of anomaly is not known before hand, it is better to conduct measurements in a grid pattern. The survey can be conducted along two concentric circumferential profiles with 24 or 32 electrodes as suggested by Erchul and Slifer (1989) and Kim *et al.* (2004).

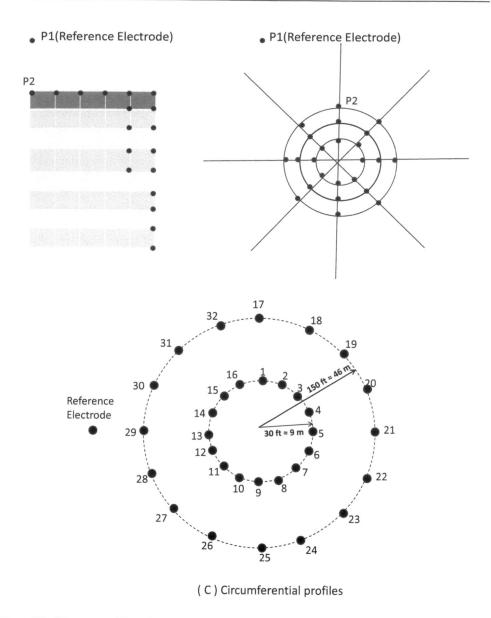

(C) Circumferential profiles

Figure 7.2 SP survey in (A) grid pattern and along (B) radial profiles and (C) concentric circumferential
profiles (Erchul and Slifer,1989), the latter two are used while the survey is conducted
around a well or sinkhole.

The concentric circles may have diameters of 15 and 45 m with 8 electrodes on inner
circle and 16 electrodes on outer circle or 10 and 45 m with 16 electrodes on each
circle (Fig. 7.2c). The circumferential profiling was used by Erchul and Slifer (1989)
for assessing groundwater movement in karstic terrain. The radial and circumferential

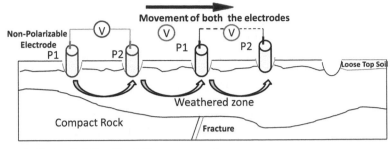

A. Potential Gradient Measurement

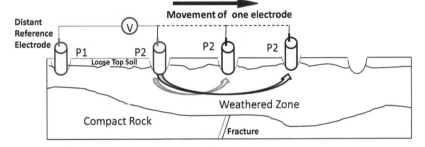

B. Total Potential Measurement

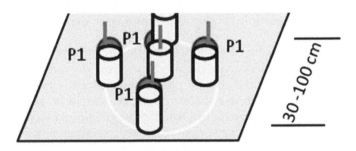

C. Electrode Group

Figure 7.3 Electrode arrays for Self Potential measurements; (a) potential gradient measurement, (b) total potential measurement and (c) potential measurement by a group of electrodes.

profilings are preferred when surveys are conducted around a well so that measurements can be taken systematically in a regular fashion before, during or after pumping the well.

The measurements can be made by two commonly used electrode configurations viz., total field measurement and gradient measurement. The gradient measurement (Fig. 7.3a) is also known as "leap-frog" or dipole or fixed-spaced electrode configuration. In total field measurement (Fig. 7.3b) one of the potential electrodes is kept fixed as 'base' or 'reference electrode' in a geologically suitable and topographically compatible remote area, i.e., it should not be on a local geological structure. Also there should not be any geological structure like a dyke or quartz reef or a large extensive rock exposure in between the base electrode and the area to be surveyed. It should neither be in a deep ditch nor on highly elevated ground. There should not be a flowing water body and cultural noise near the base electrode giving rise to much variation in its potential and drift. To reduce the effect of contact potential and soil moisture variation at a station Sill and Johng (1979) suggest averaging the measurement through multiple electrodes planted in a regular fashion around the station within a metre or so, as shown in figure 7.3c. It improves data quality and gives an average value of potential and the variations. However, the standard deviation of the readings should be reasonably low.

As a practice, the distance between base electrode and survey area is kept at least equal to one profile length. That is, if a profile length of 100 m is to be covered the base electrode is to be at least 100 m away from the measuring station nearest to the base. However, it depends on the availability of a noise free site, objective and easy access to base for repeat measurements. For very shallow investigations, the base electrode can be kept even within the survey area away from an anomalous zone if measurements are stable. The other measuring electrode is moved along the profile lines with connected cable and referenced to the base electrode. The measuring instrument may be kept stationary or moved along the profile. The electrodes are connected to the instrument in such a way that the base or reference electrode has negative polarity and the moving electrode has positive polarity. In the case where the reference electrode position is to be shifted, new reference electrode is tied with the previous one and a few measurements are overlapped. In this arrangement a large cable length on a portable winch is required.

The inhomogeneities visible on the ground surface near the measurement stations, like metallic bodies or rock pieces are removed and blank or water filled cavities are filled in properly. It is better if electrode pits are located away from such inhomogeneities. In the case of flowing surface water, the pit should be away from it as it causes drift in SP values. The presence of such inhomogeneities are recorded while taking measurements. These inhomogeneities create large spurious potential differences and thus introduce noise in SP data. Electrode pits are dug about 20 to 30 cms deep. All the pits should be almost of the same depth and it is observed that the electrode is placed in the undisturbed soil. In the case of dry soil, pits are sufficiently watered to minimize contact resistance, but it is done well in advance of measurements so that extraneous potentials developed due to percolation or movement of free water, get stabilized. Also, the inhomogeneities in soil moisture at the electrode locations are to be checked. Pits are not watered during measurements and in case it has to be done electrodes are placed in the pits and measurements made only after water gets fully stabilized otherwise drift in SP values will be observed. The electrodes should be firmly placed into the pit in a single attempt as SP values change with repeatedly adjusted placing of the electrode into the pit. The measurements are completed in a minimum time possible to avoid drift due to polarization and temperature variations.

In measurements with a gradient configuration, both the electrodes are simultaneously moved at a constant separation along profile lines and electrodes are connected to resistivity meter or voltmeter in such a way that the polarity of the electrodes is not changed, for example, the positive terminal of resistivity meter is always connected to leading electrode and negative terminal is connected to trailing electrode unless a check on measurement is warranted. Monteiro Santos *et al.* (2002) suggest alternating the lead and rear electrodes to remove the effect of electrode polarization. The distance between the electrodes is kept small. Particularly for measuring through gradient configurations a smaller station interval is preferred. Conveniently, the distance between the electrodes is kept equal to one station interval so that while traversing along the profile the leading electrode is moved successively to the next station and accordingly the trailing electrode occupies the last station left by the leading electrode. The gradient of SP values thus obtained is for the mid points between the stations. As compared to gradient measurement, the total field measurement is preferred, as it gives a large potential difference and the error associated with electrode polarization as well as spurious local effects is less. In gradient configuration electrode polarization can be minimized by interchanging the electrodes (not polarity) after a couple of measurements. In cases where the profile line is quite long a gradient array is preferred over total field measurements as it requires very small cable length.

It is necessary to have repeat measurement at the reference electrode or base station. Short interval or continuous measurement at the base station, if possible, is preferred. In general base station measurement is repeated at certain time intervals. For short profile lengths Birch (1993) proposes multiple round-trip loop measurements that yield a set of observations for each station which enables corrections for drift due to telluric currents, transients and fast variations. Besides, to estimate the variations in streaming potential due to pumping of a well and to decipher the direction of groundwater movement to the well induced by pumping the measurements are repeated at different stages, viz., before and during pumping and immediately and at late times after pump shut-down (Chandra *et al.*, 1994). As pointed out by Revil *et al.* (2006) repeat measurements are also essential to differentiate the static electrochemical potentials from streaming potentials. The precaution in measurements at a pumped well is that water coming out of the well should not flow near the electrode pits. Channeling of pumped water through specific arrangement is essential to get the correct SP measurements. SP being affected by topographic and near-surface formation resistivity variations it is useful to determine topographic heights and conduct a few resistivity soundings, profilings or resistivity tomography along the SP profile lines. Corry *et al.* (1983) and Corwin (1990) give the details of field procedures and precautions. In a long term monitoring programme the variations in temperature and moisture content of the surface soil can cause a systematic shift in SP values as shown by Erchul and Slifer (1989). These effects can be filtered by studying the shift and averaging.

7.5 PROCESSING OF DATA

The SP data are mainly subjected to 'electrode polarization' and 'drift'. Corwin (1990) defines 'polarization' as the potential difference between a pair of electrodes in the absence of an external electric field at a given time and 'drift' as the variation in the value of polarization potential with time caused by changes in temperature, soil moisture, soil type and concentration of the solution. Since drift is observed with time lapse, it is mostly required for long profile lines. Here the 'time' is related to the duration of field survey on a particular day. In total field measurements all the data are necessarily to be reduced to a common point as if all the data were collected at one time and for that drift corrections are incorporated. For this, the potential difference between the electrode pair at the beginning of the survey and after completion of the survey, by repeat base reading are recorded and the difference in the initial and final electrode potential differences are linearly distributed among the measurements. Corrected data can be plotted as profiles or as iso-potential contour maps. The data obtained through gradient configuration can also provide iso-potential maps, for this the measurements are referred to a base station electrode as in total field measurements and the potential difference obtained in each gradient measurement is successively added along the profile to convert them into total field measurement at the successive stations. For example, let the gradient SP values between stations 1–2, 2–3, 3–4 and 4–5 be 10, −5, 20 and 30 mV, then the total field values at stations 2, 3, 4 and 5 are 10, 5, 25 and 55 mV respectively.

7.6 INTERPRETATION

Interpretation of SP ranges from qualitative inspection of SP profiles, its gradient and contour maps, generally sufficient for groundwater flow studies to quantitative interpretation with the help of modeling and inversion. Quantitative interpretation of SP data is difficult and becomes ambiguous at times, as a number of factors, noise and other natural potentials (due to lithological variations) distort the streaming potential anomaly. Also, the order of magnitude of noise may be same as that of the anomaly (±5 to 10 milliVolts). Wanfang *et al.* (1999) attempt separating the SP response from karstic features from that due to topographic effect and noise. The main difficulty in interpreting streaming potential is complexity in groundwater flow, the potential it generates and the subsurface resistivity distribution. Since the flow of current and hence equipotential line is also dependent of subsurface resistivity distribution, the variations in resistivity will modify the SP anomalies. Information on subsurface resistivity distribution is essential in interpreting SP anomalies. A resistive zone will produce positive SP anomaly and a conductive zone a negative anomaly (Vichabian and Morgan, 2002). The SP gradient is large where near surface layer resistivity is high (Hashimoto and Tanaka, 1995).That is, streaming potentials are reduced with increasing clay content. Qualitative interpretation is done by studying the amplitude and wavelength of anomaly and matching with patterns for known source geometries. Smooth anomalies with long wavelength indicate a deeper source. Conversely, short wavelength and higher amplitude indicate a shallow source. As a convention, SP values generally increase, i.e., become more positive in the direction of groundwater flow. That is, more negative upstream and more positive downstream. This is because mostly the solid surface is negatively charged and positive ions from the liquid get attracted towards the solid surface and with the movement of liquid positive ions move downstream. Likewise, flow towards a well due to pumping increases (more positive) the SP values near it. The increase in SP value will be proportional to the intensity of groundwater flow and the hydraulic gradient and the SP contours are, in general, perpendicular to the flow direction. SP is affected by topography. In hilly areas the higher the topographic elevation the lower is the SP. This could be due to the downward flow of groundwater i.e. variation in water table height (Hashimoto and Tanaka, 1995; Ernstson and Scherer, 1986). The equipotentials are nearly parallel to the interface between the palaeo-channel and surrounding sediments. A negative self-potential anomaly is associated with the presence of palaeo-channel because of the horizontal flow of ground water (Bol'eve *et al.*, 2007). The association of negative SP anomaly with inferred low resistivity zones supports the presence of yielding fractures in hard rock (Robert *et al.*, 2011). It is supported by the observations made by Jinadas and de Silva (2009).

In recent years quantitative interpretation has been developed through numerical modeling that needs prior information on subsurface variations of hydraulic head, hydraulic conductivity and electrical resistivity in the modeled area (Berube, 2004). The SP anomalies can be modeled, e.g., vertical lithological contacts or structural discontinuities give steep, asymmetrical anomaly with its amplitude depending on resistivity ratio (Fitterman, 1979; Corwin and Hoover, 1979). For a better understanding, SP profile data are compared with geoelectrical and geological cross-sections. The inversion of self-potential signals has significant applications. It helps to understand the pattern of ground water flow in the subsurface, identify preferential flow path and also variations in permeability (Minsley, 2007; Jardani *et al.*, 2007; Bol'eve *et al.*, 2007).

Successful application of SP in understanding groundwater flow through fractures in hard rock needs careful consideration as the SP development in fractured rock differs from that developed in porous formations (Erchul and Slifer, 1989). The development of SP is related to hydraulic head, however it will be smaller in a fractured zone than in a porous formation at comparable hydraulic head. According to Erchul and Slifer SP values for larger fracture openings decrease or are positive while for smaller fractures SP values are more negative. They further add that filling of a fracture opening by sand increases the SP while that by clay reduces the SP.

7.7 CASE STUDIES ON EFFECT OF WELL PUMPING ON SP

7.7.1 Changes in SP after 24 hrs pumping

SP measurements were made at a high yielding well in metasediments in the eastern flank of Singhbhum Shear zone in Jharkhand, India. The schists, quartzites and phyllites exposed in the area are highly folded with steeply dipping flanks. A well (EW) was drilled up to 137.97 m depth on a NE-SW trending lineament shown in figures 5.10 and 13.10. It tapped fractured zones in the depth ranges of 40.91–41.91 m, 65.77–66.77 m and 129-130 m. The cumulative yield of the well from these fractured zones was 11.49 lps. SP survey was conducted on 6 parallel profiles oriented in a N 39° E direction, covering an area of 210 m × 250 m around the pumped well (EW). The survey was conducted before pumping and after 24 hrs of pumping the well (Chandra *et al.*, 1994). The development of SP was different from that expected in porous homogeneous formations. Being a hard rock area with varied heterogeneities, the development of SP before pumping and changes induced by pumping the well reflect the combined effect of electrokinetic actions, magnitude of flow in different fractures, lithological variations and the structures. It is brought out in the contour maps shown in Figure 7.4a and b. It is observed that before pumping SP values at most of the points

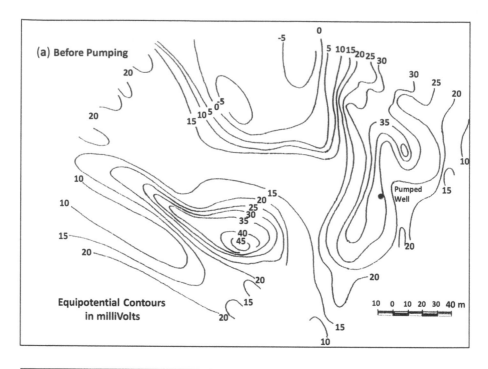

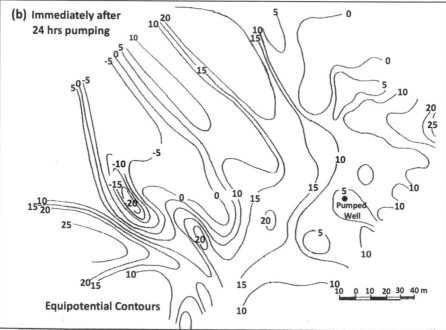

Figure 7.4 Variations in SP values due to groundwater flow after 24 hrs pumping of a well in metasediments (Chandra et al., 1994).

except in northern part were positive. The contours have two different near-orthogonal trends. In the eastern part surrounding the pumped well the trend is N-S to NNE-SSW and in the western part it is NW-SE to WNW-ESE. While the eastern trend conforms to the lineament and aeromagnetic linear (Fig. 5.10), the western trend conforms to the strike of a syncline. After 24 hrs pumping the well, the observations show a general reduction in SP values with a maximum in the west central part. The SP value of the order of 40 mV in the western part is reduced to 20 mV and further west a negative SP (−20 mV) centre is developed, while towards the north the negative SP (−15 mV) centre becomes positive after pumping. In the eastern part the positive SP values of 35 to 40 mV around the well as well as towards northeast reduce to 5 to 10 mV after pumping. The SP values in the southwestern part are not much affected by pumping. It reflects varied reaction to pumping as well as the minimum area influenced. As far as resistivity variation is concerned, there is a general decrease in resistivity towards northwest revealed by gradient resistivity profiling and it is reflected by negative SP value towards northwest. The variations in SP in the western part can be related to subsurface resistivity variations and structural controls on groundwater movement by the steeply dipping flanks of highly folded schists, quartzites and phyllites.

7.7.2 Changes in SP after 1 hr pumping

An SP survey was carried out around a well located between a basic dyke and a quartz reef in a dyke infested granitic terrain of Odisha, India (Fig. 7.5). The well pierced through the weathered zone up to 15.24 m depth followed by saprolite up to 20 m and a fractured zone in the depth range of 23–24 m. The cumulative discharge from the saprolite and the fractured zone was 1.25 lps during the compressor test. The SP measurements were made before pumping and immediately after 1 hr of pumping the well. The negative SP values of the order of −20 mV around the well became positive after pumping. There was not much change in the SP values towards north. The results indicate the effect of low discharge pumping over a smaller area only.

7.7.3 Groundwater flow through cavernous limestone

An SP survey was carried out at a well in cavernous limestone in parts of Andhra Pradesh, India before pumping and during pumping. The measurements were made along 3 parallel profiles viz., W30, E20 and E50 encompassing the well (Chandra et al., 1987) with base station at SW3 on profile W30 (Fig. 7.6). The SP profiles and equipotential map are shown. The well tapped a cavity 13 m below ground with a discharge of 15.8 lps and had a fast rate of recovery. The magnitude of streaming potential is generally low, of the order of a few millivolts. The without pumping SP values on profile W30 are, in general, negative and are comparatively positive at stations NE1 and NE2 which are nearer to the well located between profiles W30 and E20 (Fig. 7.6b). SP values measured during pumping, increased at all the stations and became positive except at the same stations, NE1 and NE2. The maximum increase is observed at station SW1 on profile W30. The contrast in SP values at SW1 with and without pumping of the well is anomalous. However, it indicates the presence of a water filled cavity near station SW1 which reacted at a maximum to the pumping of the well. The SP values along profile E50 varied over a larger range. The maximum

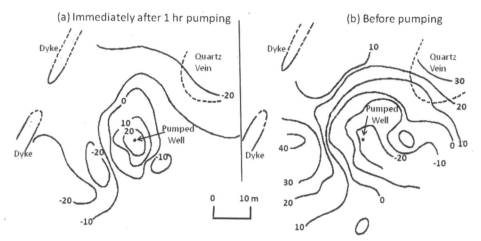

SP measurement at 2m station interval along radial profiles from the well

Figure 7.5 Variations in SP due to groundwater flow induced by 1 hr pumping of a well in granite. The measurements were made along radial profiles from the well (Chandra *et al.*, 1994).

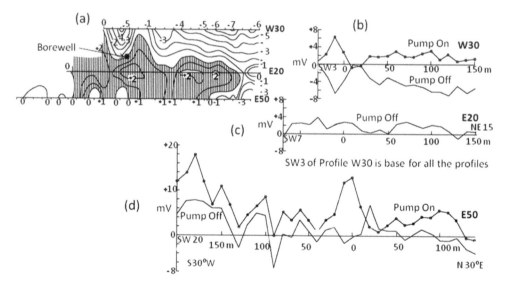

Figure 7.6 SP measurements over cavernous limestones (Chandra *et al.*, 1987).

variation is seen towards southwestern part. The equipotential map (Fig. 7.6a) based on SP values without pumping reveals a prominent negative SP zone towards the N and NE and a positive zone towards the E and SE of the well. Streaming potentials changed in the general direction of groundwater flow to the well. The gradient of anomaly can be related to the magnitude of flow.

In hydrogeological investigations SP anomalies are of low amplitude. Ambiguity in interpretation persists because of the complex character of the anomaly. In long profile lines, SP anomalies are affected by telluric current variations which may be of much higher order and anomalies are also affected by magnetic storms (Corry *et al.*, 1983; Corwin, 1990). Surveys yield erroneous anomalies on sloping ground. SP is affected by topography. In hilly areas the higher the topographic elevation the lower is the SP. This could be due to the downward flow of groundwater i.e. variation in the water table height (Hashimoto and Tanaka, 1995). The presence of near surface inhomogeneities, conductive overburden, variations in soil moisture, electrochemical effects, overhead power lines and corroded pipe lines would obscure the anomalies due to the streaming potential. Also measurements are affected by the location of the reference electrode and watering of electrodes during measurement. Besides, the SP is affected by heterogeneities and the anisotropy induced by geological structures. The SP varies with conductive/resistive bodies in the subsurface; the gradient is large where the near surface layer resistivity is high. The SP value varies with soil temperature. The 'zeta potential reduces with reducing temperature. With increasing temperature more electric charges flow. With so many variables and uncertainties, the SP survey needs to be evaluated for its effective applicability in delineating the flow path in hard rocks and relating it to flow rate and fracture characterization.

REFERENCES

Berube, A.P. (2004) Investigating the streaming potential phenomenon using electric measurements and numerical modeling with special reference to seepage monitoring in embankment dams. *Ph.D. Thesis, Dept. Civil and Environ. Eng., Div. Appl. Geop, Lulea University of Technology, Sweden.*

Birch, F.S. (1993) Testing Fournier's method for finding water table from self-potential. *Ground Water,* 31 (1), 50–56.

Birch, F.S. (1998) Imaging the water table by filtering self-potential profiles. *Ground Water,* 36 (5), 779–782.

Bogoslovsky, V.A. & Ogilvy, A.A. (1970a) Natural Potential anomalies as a quantitative index of the rate of seepage from water reservoirs. *Geophysical Prospecting,* 18 (2), 261–268.

Bogoslovsky, V.A. & Ogilvy, A.A. (1970b) Application of geophysical methods for studying the technical status of earth dams. *Geophysical Prospecting,* 18 (S1), 758–773.

Bol'eve, A., Revil, A., Janod, F., Mattiuzzo, J.L. & Jardani, A. (2007) A new formulation to compute self-potential signals associated with ground water flow. *Hydrology & Earth System Sciences, Discuss,* 4 (3) 1429–1463.

Chandra, P.C., Tata, S. & Raju, K.C.B. (1987) Geoelectrical response of cavities in limestones: an experimental field study from Kurnool district, Andhra Pradesh, India. *Geoexploration,* 24 (6), 483–502.

Chandra, P.C., Reddy, P.H.P. & Singh, S.C. (1994) *Geophysical studies for groundwater exploration in Kasai and Subarnarekha River basins* (UNDP Project) Central Ground Water Board, Min. of Water Resources, Govt. of India Tech. Report.

Colangelo, G., Lapenna, V., Perrone, A., Piscitelli, S. & Telesca, L. (2006) 2D Self-Potential tomographies for studying groundwater flows in the Varco d'Izzo landslide (Basilicata, southern Italy). *Engineering Geology,* 88 (3–4), 274–286.

Corry, C.E., DeMoully, G.T. & Gerety, M.T. (1983) *Field procedure manual for self-potential surveys.* Publication Zonge Engineering & Research Organization, Arizona, USA.

Corwin, R.F. (1990) The self-potential method for environmental and engineering applications. In: Ward, S.H. (ed.) *Geotechnical and Environmental Geophysics, Vol. I: Review and Tutorial*. Tulsa, Oklahoma, USA, Society of Exploration Geophysicists, pp. 127–145.

Corwin, R.F. & Hoover, D.B. (1979) The self-potential method in geothermal exploration. *Geophysics*, 44 (2), 226–245.

Costar, A., Wilson, T., Heinson, G., Lovel, A. & Smit, Z. (2008) Remote monitoring of ground water flow in fractured rock using electrokinetic methods. Presented at: The *2nd International Salinity Forum, Adelaide, Australia.*

Erchul, R.A. & Slifer, D.W. (1989) *Geotechnical applications of the self potential (SP) method*, Report 2, The use of self potential to detect ground-water flow in karst. USAEWES Tech. Report REMR-GT-6.

Ernstson, K. & Scherer, H.U. (1986) Self-potential variations with time and their relation to hydrogeologic and meteorological parameters. *Geophysics*, 51 (1), 1967–1977.

Fagerlund, F. & Heinson, G. (2003) Detecting subsurface groundwater flow in fractured rock using self-potential (SP) methods *Environmental Geology* 43 (7) 782–794.

Fitterman, D.V. (1979) Calculations of self-potential anomalies near vertical contacts *Geophysics*, 44 (2), 195–205.

Forté, S. & Bentley, L.R. (2013) Effect of hydrocarbon contamination on streaming potential. *Near-Surface Geophysics, European Association of Geoscientists & Engineers*, 11 (1), 75–83.

Gensane, O., Konyukhov, B., Le Moul, J.-L. & Morat, P. (1999) SP coseismic signals observed on an electrodes array in an underground quarry *Geophysical Research Letters*, 26 (23) 3529–3532.

Hashimoto, T. & Tanaka, Y. (1995) A large self-potential anomaly on Unzen volcano, Shimabera peninsula, Kyushu island, Japan. *Geophysical Research Letters*, 22 (3), 191–194.

Ishido, T. & Mizutani, H. (1981) Experimental and theoretical basis of electrokinetic phenomena in rock-water systems and its applications to geophysics *Journal of Geophysical Research*, 86 (B3), 1763–1775.

Jardani, A., Revil, A., Bole've, A., Crespy, A., Dupont, J.-P., Barrash, W. & Malama, B. (2007) Tomography of the Darcy velocity from self-potential measurements *Geophysical Research Letters* 34 (24), doi:10.1029/2007GL031907.

Jardani, A., Revil, A., Barrash, W., Crespy, A., Rizzo, E., Straface, S., Cardiff, M., Malama, B., Millers, C. & Johnson, T. (2009) Reconstruction of the Water Table from Self-Potential Data: A Bayesian Approach *Ground Water*, 47(2), 213–227.

Jinadasa, S.U.P. & de Silva, R.P. (2009) Resistivity imaging and self-potential applications in groundwater investigations in hard crystalline rocks. *Journal of the National Science Foundation, Sri Lanka*, 37(1), 23–32.

Jouniaux, L. (2011) Electrokinetic techniques for the determination of hydraulic conductivity. In: L Elango, I. (ed.) *Hydraulic conductivity – issues, Determination and Applications* Chapter 16. Croatia, InTech Available from: http://www.intechopen.com/books/hydraulic-conductivity-issues-determination-and-applications/electrokinetic-techniques-for-the-determination-of-hydraulic-conductivity.

Jouniaux, L., Maineult, A., Naudet, V., Pessel, M. & Sailhac, P. (2009) Review of self-potential methods in hydrogeophysics. *C.R. Geoscience*, 341 (10–11), 928–936. doi:10.1016/j.crte.2009.08.008.

Kim, S., Heinson, G. & Joseph, J. (2004) Electrokinetic groundwater exploration: a new geophysical technique. In: Roach, I.C. (ed.) *Regolith 2004:* Proceedings of the CRC LEME Regional Regolith Symposia, November 2004, Adelaide, Perth and Canberra. Bentley, CRC LEME. pp. 181–185.

Kim, S., Heinson, G. & Joseph, J. (2005) Laboratory measurements of electrokinetic potential from fluid flow in porous media. In: Roach, I.C. (ed.) *Regolith 2005 – Ten years of*

CRC LEME: Proceedings of the CRC LEME Regional Regolith Symposia, November 2005, Canberra and Adelaide. Bentley, CRC LEME. pp. 176–178.

Lorne, B., Perrier, F. & Avouac, J.P. (1999) Streaming potential measurements: 1. Properties of the electrical double layer from crushed rock samples. *Journal of Geophysical Research,* 104 (B8), 17857–17877.

Minsley, B.J. (2007) *Modeling and inversion of self potential data.* Ph.D Thesis, Department of Earth, Atmospheric & Planetary Sciences, Massachusetts Institute of Technology, USA.

Minsley, B.J., Sogade, J. & Morgan, F.D. (2007) Three-dimensional self-potential inversion for subsurface DNAPL contaminant detection at the Savannah River Site, South Carolina. *Water Resources Research,* 43 (4), W04429, doi:10.1029/2005WR003996.

Monteiro Santos, F.A., Almeida, E.P., Castro, R., Nolasco, R. & Mendes-Victor, L. (2002) A hydrogeological investigation using EM34 and SP surveys. *Earth, Planets & Space,* 54, 655–662.

Muralidharan, D., Rangarajan, R., Murthy, J.V.S. & Rao, Y.P. (2005) Mapping of hydrodynamic changes around radial arms of collector well by streaming potential survey. *Current Science, Scientific Correspondence,* 88 (12), 1901–1904.

Naudet, V., Revil, A. & Bottero, J.-Y. (2003) Relationship between self-potential (SP) signals and redox conditions in contaminated groundwater. *Geophysical Research Letters,* 30 (21), 2091, doi: 10.1029/2003 GL018096.

Naudet, V., Revil, A., Rizzo, E., Bottero, J.-Y. & Begassat, P. (2004) Groundwater redox conditions and conductivity in a contaminant plume from geoelctrical investigations. *Hydrology & Earth System Sciences,* 8 (1), 8–22.

Ogilvy, A.A., Ayed, M.A. & Bogoslovsky, V.A. (1969) Geophysical studies of water leakages from reservoirs. *Geophysical Prospecting,* 17 (1), 36–62.

Revil, A., Titov, K., Doussan, C. & Lapenna, V. (2006) Applications of the self potential method to hydrological problems. In: Vereecken, H., Binley, A., Cassiani, G. Revil, A. & Titov, K. (eds.) *Applied Hydrogeophysics, Chapter 9, (IV-Earth and Environmental Sciences, NATO Science Series, 71), Proceedings of the NATO Advanced Research Workshop on Soils and Groundwater Contamination: Improved Risk Assessment, 25–29 July 2004, St. Petersburg, Russia.* Dordrecht, The Netherlands, Springer. pp. 255–292.

Richards, K., Revil, A., Jardani, A., Henderson, F., Batzle, M. & Haas, A. (2010) Pattern of shallow ground water flow at Mount Princeton Hot Springs, Colorado, using geoelectrical methods. *Journal of Volcanology and Geothermal Research,* 198 (1–2), 217–232.

Robert, T., Dassargues, A., Brouyere, S., Kaufmann, O., Hallet, V. & Nguyen, F. (2011) Assessing the contribution of electrical resistivity tomography (ERT) and self-potential (SP) methods for a water well drilling program in fractured/karstified limestones. *Journal of Applied Geophysics,* 75 (1), 42–53.

Satou, S., Shimada, J. & Goto, T. (2005) Use of self-potential (SP) method to understand the regional groundwater flow system. *Eos Trans. American Geophysical Union,* 86(52) *Fall Meet. Suppl.* Abstract H 23 E–1482.

Schiavone, D. & Quarto, R. (1984) Self-potential prospecting in the study of water movements. *Geoexploration,* 22 (1), 47–58.

Sill, W.R. & Johng, D.S. (1979) *Self potential survey, Roosevelt hot springs, Utah.* Topical Report, Contract No. DE-AC07-78ET28392, Dept. of Geology and Geophysics, University of Utah, USA.

Snieder, R., Hubbard, S., Haney, M., Bawden, G., Hatchell, P., Revil, A. & DOE Geophysical Monitoring Working Group, (2007) Advanced noninvasive geophysical monitoring techniques. *Annual Review of Earth and Planetary Sciences,* 35, 653–683.

Straface, S., Rizzo, E. & Chidichimo, F. (2010) Estimation of hydraulic conductivity and water table map in a large-scale laboratory model by means of the self-potential method. *Journal of Geophysical Research,* 115 (B6), B06105, doi:10.1029/2009JB007053.

Suski, B., Revil, A., Titov, K., Konosavsky, P., Voltz, M., Dages C. & Huttel, O. (2006) Monitoring of an infiltration experiment using the self-potential method. *Water Resources Research,* 42 (8) W08418, doi:10.1029/2005WR004840.

Vichabian, Y. & Morgan, F.D. (2002) Self potentials in cave detection. *The Leading Edge,* 21 (9), 866–871.

Wanfang, Z., Beck, B.F. & Stephenson, J.B. (1999) Investigation of groundwater flow in karst areas using component separation of natural potential measurements. *Environmental Geology,* 37 (1–2), 19–25.

Wishart, D.N., Slater, L.D. & Gates, A.E. (2006) Self potential improves characterization of hydraulically-active fractures from azimuthal geoelectrical measurements. *Geophysical Research Letters* 33 (17), L17314, doi:10.1029/2006GL027092.

Chapter 8

The mise-a-la-masse method

8.1 INTRODUCTION

In hard rock borehole sites are generally located on the basis of favourable surface geophysical anomalies. Once a moderate to high yielding well is drilled tapping a saturated shallow fractured zone, the location of another drilling site may be required in the same area as accurately as possible to tap the same fractured zone for a similar yield. This is necessary because wells drilled a few metres apart often show quite different yields due to rapid variations in fracture occurrences within a short distance and the uncertainty in success and yield prevails. Wells cannot be located at random even if some fracture yielding wells exist in the area. It requires demarcation of the lateral hydraulic continuity of such fractured zones. To make the operation economical, avoiding well failures, a technique of mapping is used by electrically charging the fractured zone encountered in the borehole and tracing its galvanic continuity through the potential developed on the ground surface. It is well-accepted in mineral prospecting as 'hole-to-surface' technique popularly known as mise-a-la-masse (MAM) which actually means 'excitation at the mass' (Parasnis, 1967). It is also known as 'charged body potential' technique (Jamtlid *et al.*, 1984). The essential requirement for applying the technique is availability of a borehole that encounters a prominent saturated yielding shallow fractured zone.

The technique was developed in the beginning of twentieth century by C. Schlumberger (Osiensky and Donaldson, 1994) to assess the dimension of an ore body either encountered in a borehole or outcropping at the surface. It is quite effective in mineral prospecting where contrast in the conductivity of the mineralized zone and the host rock is quite high (Bhattacharya *et al.*, 2001). It has been widely used to demarcate the extensions of ore bodies encountered in boreholes (Parasnis, 1967; Beasley and Ward, 1985). Though the contrast in conductivity of a water yielding saturated fractured zone with the surrounding host is not as high as metallic ore deposits, still the technique has been used successfully for lateral demarcation of shallow fractured zones. The objective is to map the developed potential distribution on the ground surface by energizing the low resistivity fractured zone putting a current electrode in it. Its geometry is ascertained from the equipotential lines. The technique can also be used for subsurface characterization by establishing the continuity of fractures amongst the boreholes drilled in an area. Related to radioactive waste disposal Jamtlid *et al.* (1984) interpreted the orientation of deep fractured zone up to a lateral distance of 150 m having contrasting resistivity with host rock by conducting measurements on surface as well as in boreholes. Chandra *et al.* (1994) attempted delineation of saturated fractured zone encountered at a depth of 31 m. Chandra *et al.* (1987) used the technique to locate water filled cavities in limestone. Paananen and Kuivamaki (1992) used the technique to decipher migration of radionuclides associated with flowing groundwater through fractured zones in crystalline rocks by identifying their orientation and continuity. The technique has been used to delineate the extension and orientation of electrically conductive contaminant plume (Osiensky and Donaldson, 1994; Nimmer and Osiensky, 2003; Osiensky *et al.*, 1999) and monitor its movement. Though such studies in hard rock will be tedious because of the thin groundwater flow path through fractured zones, attempts can be made. Supriyanto *et al.* (2005) used the technique to locate geothermal fields associated with highly conductive hydrothermal fluid flowing through permeable faults, fractured zones and contact plains between formations. Bevc and Morrison (1991) utilized the technique in a modified form to monitor the movement of saline water injected into a shallow aquifer. There are several such applications of the technique and only a few are mentioned here.

8.2   BASICS

In this technique one electrode of a current electrode pair is placed in the conductive saturated fractured zone which is accessible through the borehole intersecting it and the other return current electrode is placed on the ground surface at infinite distance, i.e., practically at a large distance from the borehole. The basic objective is to increase the current density in the fractured zone. As shown by Parasnis (1967), for a homogeneous, isotropic medium the equipotential surfaces for a buried point current source are concentric spheres around the current source (Fig. 8.1) but, in the case where the current electrode is positioned against a conductive fractured zone in a borehole and the current gets focused along the elongated fractured zone the equipotential surfaces are elongated. The fractured zone, if highly conductive, acts as a long current electrode and has almost the same potential throughout with very little potential drop across it (Ketola, 1972; Beasley and Ward, 1985). Thus, the potential on the ground surface due to energization of the fractured zone manifests its orientation as the potential distribution will be affected by the preferential flow of current along the relatively conductive saturated fractured zone and the equipotential lines get elongated. The trend and distortion in the equipotential lines drawn on the ground surface will show the trend of the current lines and their convergence along the fractured zone. Though the equipotential line will be affected by the anisotropic behavior of the country rock, the equipotential maximum contour pattern will coincide with the strike of the conductive fractured zone. In mineral exploration the technique has been used for deciphering the geometry of highly conductive ore bodies even at depths beyond 100 m (Ketola, 1972), however in the case of groundwater exploration where conductivity contrast is not high, it can be used only for shallow saturated fractures underlying a moderately resistive weathered zone.

8.3   INSTRUMENT

Any resistivity meter, preferably with multi-selection constant current input can be used to measure the potential. A microprocessor based stacking facility for rejection of

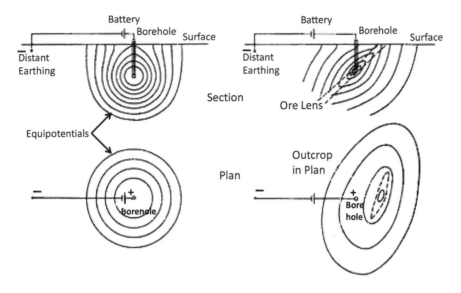

Figure 8.1 Mise-a-la-masse method (Parasnis, 1967).

noise is preferred. A simple electrode can be lowered in the borehole through the cable. Since the current input can vary, the potential measured at each station is normalized by the current input concerned.

8.4 FIELD PROCEDURES

To conduct a survey by this technique a borehole, which has encountered a water bearing fractured zone and where the level of groundwater is above the fractured zone, is the essential prerequisite. The other requirement is that there should be other geophysical data like electrical resistivity soundings and magnetic, resistivity image or electromagnetic profiles available around the site for supplementing the interpretations. The field operation is easy, fast and cost effective. One of the current electrodes is lowered in the borehole and positioned against the fractured zone whose lateral extensions are to be mapped. If required the exact depth of the fractured zone can be obtained from geophysical logging of the borehole. The other current electrode is put on the ground at a large distance from the borehole; minimum 10 times the potential measuring profile line lengths is considered as an infinite distance. Potentials are measured along a number of parallel profile lines laid on the ground surface surrounding the borehole in which the fractured zone is being energized or along radial profile lines from the borehole. The well or borehole in which the electrode is lowered is not pumped during the survey. The profiles are laid in such a manner that the area where orientation of the 'borehole-encountered- fracture' is to be demarcated is effectively covered. A parallel profile line spacing can be 10 to 50 m, while the radial ones diverging out from the borehole can be in regularly spaced in 8 to 12 directions.

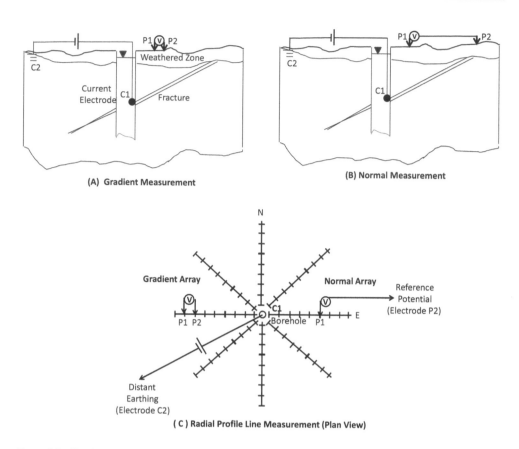

Figure 8.2 Gradient and normal arrangements of potential electrodes for radial profile mise-a-la-masse survey around a well.

Potentials measured along these profile lines can be at station intervals of 2 to 10 m. It can vary as per requirement. For measuring the potential on the ground surface two types of electrode configurations are used, viz., normal and gradient (Fig. 8.2). In normal configuration, one of the potential electrodes is fixed at a large distance on the opposite side of the distant current electrode. The other potential electrode is moved from one station to another along the profile lines. It is also known as 'pole-pole' configuration. In gradient configuration, the potential difference is measured by a potential dipole. The length of the dipole is kept as small as possible to measure the potential gradient. It is always kept as a multiple of the station interval for easy field operation. The dipole length could be increased if measured potential difference is too low vis-à-vis the sensitivity of instrument deployed. The dipole length is however not varied from measurement to measurement but is decided and fixed for the total set of measurements. This is also known as 'pole-dipole' configuration. Generally, normal configuration is preferred as in gradient configuration a very small potential difference is obtained and the error is added for both the electrodes. Also, the speed of coverage is better in normal configuration.

Proper location of "infinite" or distant current and potential electrodes with local geological considerations is important. The location of infinite current or potential electrode affects the measured value of the potential, however the relative amplitude of the potentials and trend remain almost the same (Mwenifumbo, 1980). The potential maps prepared for different locations of infinite potential electrodes are shown in figure 8.3 (Chandra *et al.*, 1987). The current input against the fractured zone should be sufficient for recording a considerable potential on the surface. The pits or holes for planting the potential electrodes are made in advance and watered to avoid spurious potentials. Also the non-polarizing electrodes are used for potential measurements. Mise-a-la-masse measurements can also be made within the borehole to establish the continuity between fractured zones encountered in two nearby boreholes. In this technique while a fractured zone in a borehole is energized by one of the current electrodes lowered in it, one of the potential electrodes is moved in another nearby borehole. It has proved to be quite useful in establishing the continuity of ore intersections in boreholes (Mwenifumbo, 1997).

8.5 PROCESSING OF DATA

If a constant current source is not used, potential data are normalized to a fixed current input say 1A for comparison of the amplitudes of potentials measured. The potential data can also be converted to apparent resistivity data. The data can be presented as profile or contour maps of iso-potential or iso-resistivity.

8.6 INTERPRETATION

The mise-a-la-masse anomalies can be interpreted qualitatively as well as quantitatively. The details of quantitative interpretation can be obtained in Mwenifumbo (1980), Eloranta (1985, 1986) and Wang *et al.* (1991). However, the general practice has been to study the trend of equipotential lines and make a qualitative interpretation. The development of a potential on the surface depends on the thickness and width of the fractured zone, its depth of burial, extent, dip, resistivity contrast with the surroundings and resistivity of the overburden. The elongation of equipotential lines along the axis of a conductor is the characteristic feature to identify the orientation of a fractured zone and it was established through a tank model experiment by Braekken (1963). Mise-a-la-masse anomalies can be either in the form of apparent resistivity or potential. Eloranta (1985) indicated that recording of mise-a-la-masse anomaly as potential is most reliable for interpretation. Mwenifumbo (1980) conducted laboratory scale electrolytic tank modeling as well as finite element formulation for quantitative interpretations of conductive targets buried in a uniform half-space. The general shape of potential anomaly above a conductive point source target in a resistive homogeneous half space, according to Mwenifumbo, is a sharp peak in 'normal' (pole-pole) measurement and a sharp negative to positive cross-over in 'gradient' measurement. The potential and potential gradient profiles across the target broaden as well as amplitude reducing with increasing depth of burial of the target. That is, the shallower the target the sharper is the anomaly. A deep saturated fractured zone may not show up on

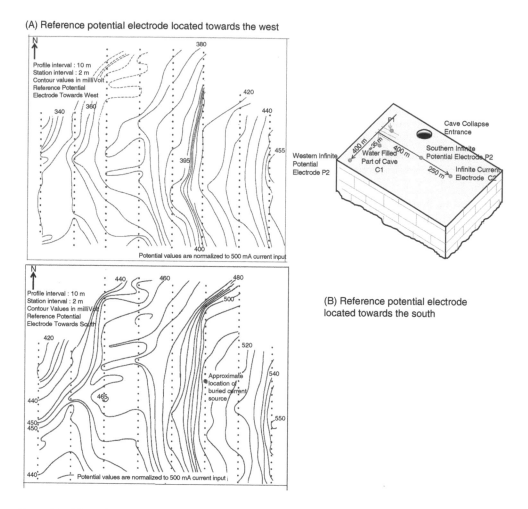

Figure 8.3 Mise-a-la-masse survey with two different locations of reference potential electrode (Chandra *et al.*, 1987).

a potential distribution map. Also the normal as well as gradient anomalies broaden for a buried line source. Asymmetry in potential distribution is introduced if the fractured zone is inclined. The dip of the fractured zone is better indicated in gradient measurement. The amplitude of sharp negative peak on the up-dip side is greater and sharper than the amplitude of broadened positive peak on the down-dip side (Eloranta, 1985). Whereas, in normal measurements up-dip direction is indicated by a steeper slope while the downdip side is indicated by a gentler slope. The presence of a conductive overburden reduces the magnitude of anomaly, while a variation in its thickness would complicate the anomaly. Near surface inhomogeneities would also affect the anomaly. A fractured zone is better defined if it has higher conductivity contrast with

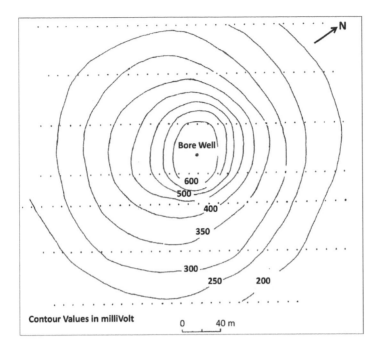

Figure 8.4 Mise-a-la-masse survey to identify the orientation of cavity in limestone encountered at 13 m depth (Chandra *et al.*, 1987).

the surroundings. In contour maps elongated contours indicate the trend of the fractured zone. The axis of the elongated contour with maximum value should be above the fractured zone. High ellipticity of the equipotential lines indicates the presence of a fractured zone along its major axis. Mwenfumbo (1997) indicated that it is not necessary to observe maximum potential at a point nearest to the source due to current channeling along a more conductive path. In hard rock, elongation of equipotential lines or elliptic equipotential lines on the surface can be obtained due to anisotropy also, e.g., in schists, and can be mistaken for the orientation of the conductive fractured zone. Asten (1974) studied the effect of host rock anisotropy on mise-a-la-masse measurements and showed the displacement of the centre of surface equipotential ellipse. In such cases, it is better to conduct radial resistivity sounding or square array sounding along with mise-a-la-masse and determine the effect of anisotropy on it.

The location of an active current electrode in the borehole affects the anomaly pattern. For a dipping fractured zone, if the current electrode is located in the borehole towards its upper part, the anomaly will be relatively sharp compared to that if located towards its bottom part. In this technique the resolution between two fractured zones located nearby is poor. As such, resolution is better in gradient measurements than in potential difference measurements. Deeper zones or zones with little contrast in resistivity with the surroundings will not be picked-up. Any other fractured zone in vicinity of the charged fractured zone would affect potential values. Also, near-surface

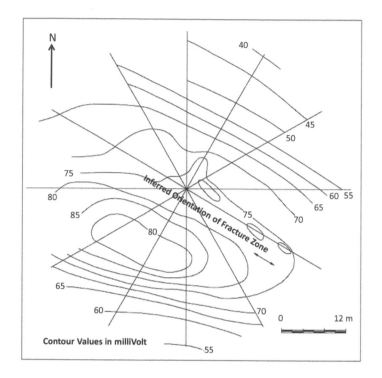

Figure 8.5 Mise-a-la-masse survey for identifying fractured zone encountered in the borehole in granites at 31 m depth (Chandra *et al.*, 1994).

inhomogeneities affect the potential values. While interpreting, both resistivity and potential data should be studied jointly. The technique can be fruitfully used to ascertain lateral continuity of a saturated fractured zone or cavity. It can be used to assess the impact of "hydro-fracturing" conducted in a borehole. Its adoption as a regular practice would be very useful in mapping potential water bearing fractured zones in hard rock. A few case studies are discussed below.

8.7 CASE STUDIES

A mise-a-la-masse survey was carried out over a part of a cave in limestones in water scarce Kurnool district of Andhra Pradesh, India (Chandra *et al.*, 1987). The objective was to identify on the ground surface the exact location of the water filled part of the cave at around 35 m depth, so that water in it can be tapped through a well. The placing of the current electrode was through the cave entrance, negotiating a very difficult narrow path, of more than 700 m of underground cave, with a few segments by crawling. Figure 8.3 shows the mise-a-la-masse potential contour maps of the electrically charged water filled limestone cavity of approximately 2 m × 2 m cross-sectional area. The surveys were carried out with the reference potential electrode at

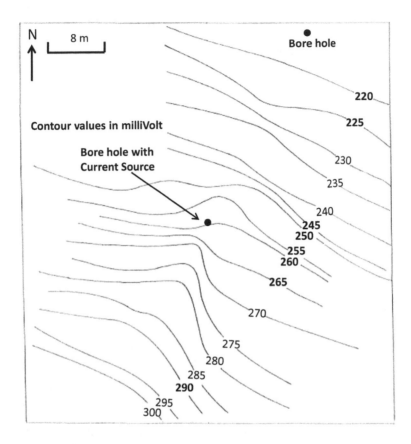

Figure 8.6 Mise-a-la-masse measurements experimented for a fractured zone encountered at 137.5 m depth in metasediments of Jharkhand, India. No significant anomaly was picked up as the charged fractured zone was quite deep.

two locations. Fig 8.3b shows the location of the cavity approximated prior to the surveys. The location of the highest value equipotential line is closest to the actual location. The cave orientation at the point of measurement is almost N-S.

Results of the mise-a-la-masse potential measurement carried out placing the current electrode in a cavity encountered in a borehole at 13 m depth is shown in figure 8.4. In this case the 'infinite' current and potential electrodes were placed respectively at 750 m and 350 m north of the borehole. The potential values were normalized to 500 mA current. Qualitatively it indicates N to NW trend and deepening of cavity towards the west of the borehole. The delineation of lateral extension of a shallow fractured zone in metasediments by mise-a-la-masse has been successful. A fractured zone encountered at 31 m depth in an exploratory borehole in metasediments in Jharkhand, India was traced (Fig. 8.5). Another borehole was drilled at 15 m distance on the deciphered SE extension of the fractured zone. It encountered the fracture at the same depth. The higher potential value (90 mV) closed contour indicates southwest shallowing of the fractured zone encountered in the borehole. The mise-a-la-masse measurement for a deeper fractured zone is not effective as shown in figure 8.6. The high yielding (11.5 lps) fractured zone at 137.5 m encountered in a borehole in metasediments was charged. The discharges of other fractured zones at 105 and 117 m depth were insignificant. The potential measurements on the surface indicated only the trend of the formation.

REFERENCES

Asten, M.W. (1974) The influence of electrical anisotropy on Mise-a-la-masse surveys. *Geophysical Prospecting*, 22 (2), 238–245.

Beasley, C.W. & Ward, S.H. (1985) *Three-dimensional Mise-a-la-masse modeling applied to mapping fracture zones*. Utah University, Dept. of Geology & Geophysics, Utah University, Research Institute, Earth Science Laboratory, Salt Lake City, Utah, USA Report DOE/SAN/12196-3 & ESL-143; ON: DE86002007.

Bevc, D. & Morrison, H.F. (1991) Borehole-to-surface electrical resistivity monitoring of a salt water injection experiment. *Geophysics*, 56 (6), 769–777.

Bhattacharya, B.B., Gupta, D., Banerjee, B. & Shalivahan (2001) Mise-a-la-masse survey for an auriferous sulfide deposit: case history. *Geophysics*, 66 (1), 70–77.

Braekken, H. (1963) A model experiment in charged potential studies. *Geoexploration*, 1 (2), 48–56.

Chandra, P.C., Tata S & Raju, K.C.B. (1987) Geoelectrical response of cavities in limestones: an experimental field study from Kurnool district, Andhra Pradesh, India *Geoexploration*, 24 (6), 483–502.

Chandra, P.C., Reddy, P.H.P. & Singh, S.C. (1994) *Geophysical studies for groundwater exploration in Kasai and Subarnarekha River basins* (UNDP Project) Central Ground Water Board, Min. of Water Resources, Govt. of India Tech. Report.

Eloranta, E.H. (1985) A comparison between Mise-a-la-masse anomalies obtained by pole-pole and pole-dipole electrode configurations. *Geoexploration*, 23 (4), 471–481.

Eloranta, E.H. (1986) The behavior of Mise-a-la-masse anomalies near a vertical contact. *Geoexploration*, 24 (1), 1–14.

Jamtlid, A., Magnusson, K.-A., Olsson, O. & Stenberg, L. (1984) Electrical borehole measurements for the mapping of fracture zones in crystalline rock. *Geoexploration*, 22 (3–4), 203–216.

Ketola, M. (1972) Some points of view concerning Mise-a-la-masse measurements. *Geoexploration*, 10 (1), 1–21.

Mwenifumbo, C.J. (1980) Interpretation of Mise-a-la-masse data for vein type bodies. *Ph.D. thesis, University of Western Ontario, London, Ontario, Canada.*

Mwenifumbo, C.J. (1997) Electrical methods for ore body delineation, In: Gubins, A.G. (ed.) *Proceedings of Exploration 97: Fourth Decennial International Conference on Mineral Exploration, 14–18 September, 1997, Toronto, Canada*, GEO F/X Div. of AG Inf. Sys. Ltd. Canada, pp. 667–676.

Nimmer, R.E. & Osiensky, J.L. (2003) Charged body potential monitoring of an electrolyte plume emanating from a dripping source. *Journal of Environmental Science & Health-Part A Toxic/Hazard Substances & Environmental Engineering (Abstract)*, 38 (5), 737–752.

Osiensky, J.L. & Donaldson, P.R. (1994) A modified Mise-a-la-masse method for contaminant plume delineation. *Ground Water*, 32 (3), 448–457.

Osiensky, J.L., Williams, R.E., Ralston, D.R., Johnson, G.S. & Mink, L.L. (1999) Simulation of electrical potential differences near a contaminant plume excited by a point source of current. *Mine Water and the Environment, International Mine Water Association,* 18 (1), 29–44.

Paananen, M. & Kuivamaki, A. (1992) *Structural interpretation for mapping of groundwater flow paths at Palmottu U-Th-Mineralization.* The Palmottu Analogue Project, Geological Survey of Finland, pp. 33–40.

Parasnis, D.S. (1967) Three-dimensional electric Mise-a-la-masse survey of an irregular lead-zinc-copper deposit in central Sweden. Presented at: *The 29th meeting of EAEG, Stockholm, Sweden.* pp. 407–437.

Supriyanto, S., Daud, Y., Sudarman, S. & Ushijima, K. (2005) Use of a Mise-a-la-masse Survey to determine new production targets in Sibayak Field, Indonesia. In: *Proceedings of the World Geothermal Congress, 2005, Antalya, Turkey.* International Geothermal Association, pp. 1–7.

Wang, T., Stodt, J.A., Stierman, D.J. & Murdoch, L.C. (1991) Mapping hydraulic fractures using a borehole-to-surface electrical resistivity method. *Geoexploration,* 28 (3–4), 349–369.

Chapter 9

The frequency domain electromagnetic method

9.1 INTRODUCTION

Application of the electromagnetic (EM) method in mineral exploration has been quite extensive and successful. In mineral exploration targets are mostly metallic ore bodies which are highly conductive, generally occurring in a very low conductivity (high resistivity) host rock. In hard rock groundwater exploration a major difference is in the range of conductivities of the targets and in conductivity contrasts with the host. The freshwater bearing weathered and fractured zone targets have much less conductivity as well as less contrast with the host medium. According to Kaikkonen (1979) for a good conductor, generally the ratio of conductivity of the conductor to that of the host medium should be more than 100. As such, fractured zones are not good conductors. The EM method is, however, in use for groundwater exploration in hard rock over several decades and has been successful in many areas. It has been used to delineate saturated weathered and fractured zones and select sites for water well drilling. The EM method is applied in frequency domain (FEM) as well as Time domain or Transient (TEM) with controlled source. It includes the very low frequency (VLF) method also which uses higher frequencies and a remote source. The FEM and VLF profiling are quite common for reconnaissance and to identify groundwater priority areas in hard rock. The applications of EM methods have been on the ground, in boreholes as well as airborne.

There are several case studies to illustrate the usefulness of the FEM method in groundwater investigations in hard rock. Some of them are Palacky *et al.* (1981) and Palacky (1991) using FEM, VLF and electrical resistivity (ER) profiling to delineate groundwater zones in Precambrian crystallines of Burkina Faso; Chandra *et al.* (1994) employing FEM, VLF and gradient resistivity profiling (GRP) to delineate fractured zones in Singhbhum granites of India, Meju *et al.* (2001) using FEM and TEM to delineate fractured aquifers and locating well sites in granitic terrain of northeast Brazil and Vogelsang (1987) using FEM to locate karstic features and faults. Koefoed and Biewinga (1976) and Patra (1976) use FEM soundings to address the groundwater problems in arid areas of South Tunisia and in India respectively. McNeill (1990 and 1991) presents various applications of EM methods in groundwater. Ong *et al.* (2010) use FEM and ER in delineating a saltwater plume surrounded by fresh groundwater. EM methods can be used to address the environmental problems (Fitterman and Labson, 2005) associated with subsurface hydrogeological conditions, solid waste site characterization, groundwater contaminations, sea water intrusion and the groundwater salinity interface where they produce considerable contrast in electrical conductivity with the surrounding. Even the detection of groundwater contamination by organic compounds has been attempted as it creates a difference in the dielectric properties. Using FEM Jin *et al.* (2008) attempt an assessment of diesel contamination in groundwater. EM methods are also used in geotechnical investigations. The controlled source FEM method is described in detail by Parasnis (1986) and Frischknecht *et al.* (1991).

Application of resistivity sounding and profiling – the ER method in groundwater exploration is quite popular. However, it has some procedural and technical constraints such as, injection of current into ground through electrodes, difficulty in current penetration when the surface layer is extremely resistive, e.g., exposed rocks, dry sands, desert sand dunes, permafrost or ice etc. and the effect of shallow near-surface inhomogeneities on measurements. In the FEM method also subsurface resistivity or conductivity variations are measured but the constraint of current penetration by electrodes is overcome as it is induced in the earth magnetically. Since an ER survey (VES) requires measurements stretched over a large area, it is more vulnerable to lateral changes in resistivity and also non-availability of large open stretches of land required to spread the current electrodes, about 3 to 5 times the desired depth of exploration, prevents its application in places. The ratio of surface spread to depth of exploration in FEM is less compared to VES and hence FEM measurements take less surface spread. The depth of investigation can also be increased by reducing the frequency of EM field. The FEM survey is faster in data acquisition than the ER survey. Another important aspect is that unlike the ER method, the resolution in EM depends on absolute conductivity of the target and not on conductivity contrast with the surrounding (Auken *et al.*, 2006). The FEM method is effective in detecting conductive layers and the presence of a highly resistive overburden does not hinder detection of deeper conductive targets. But in the presence of thick conductive overburden, like clay or a saline water saturated layer the deeper moderately conductive freshwater saturated fractured zones may not get detected. Besides, the effect of anthropogenic features like power lines, rails, buried cables, fencing etc., is relatively pronounced in FEM. The FEM method is generally used for profiling along with ER profiling or imaging to decipher the lateral variations in subsurface conductivity. FEM profiling with several frequencies can be conducted to generate a series of close-spaced soundings and assess depth-wise conductivity variations along a profile. A variety of ground FEM equipment covering a wide range of frequencies is available. The FEM equipment is expensive compared to conventional ER equipment.

9.2 BASICS

The FEM is a continuous-source method. It uses a wide range of frequency from a few hundreds of Hertz to a few kilo Hertz. The frequency of source (transmitter) current can be discretely varied. The primary EM field produced by a time-varying electric current in the transmitter coil spreads all around and penetrates into the ground. The primary magnetic field acts as that of a magnetic dipole (at distances much larger than the coil radius). When the primary magnetic field impinges upon a subsurface

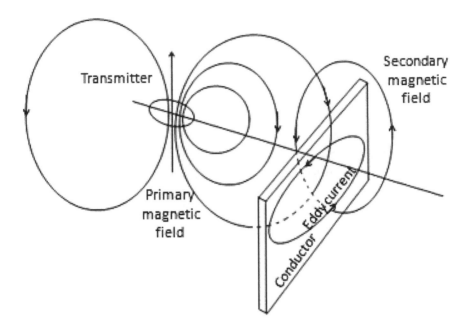

Figure 9.1 The eddy current induced by primary magnetic field in a conductor and the secondary magnetic field.

conductor (a body of electrical conductivity higher than the surrounding and of limited extensions) voltage is induced in the conductor which produces a small induced current in it known as an eddy current. The eddy current flows in closed loops in planes perpendicular to the inducing magnetic field. The propagation of magnetic field and eddy current in the subsurface are influenced by transmitter frequency, geometry of primary field, electrical conductivity distributions and geometry of conductive target. The eddy current density decreases exponentially with depth which is known as the skin effect. The secondary electromagnetic field (in the opposite sense to the primary at the conductor) produced by eddy current conveys the characteristics of the subsurface target. It is picked up by a sensitive receiver coil placed at a distance (Fig. 9.1). Since FEM is a continuous-source method, the primary field also reaches the receiver coil and measurements are made in presence of primary field. The receiver coil senses the primary as well as the secondary field which is much smaller than the primary field and the resultant of the primary and secondary fields is measured.

The primary and secondary fields are at the same frequency but the phase of the secondary field will be different from that of the primary field. The phase relation between the primary and secondary fields can be understood from induction by a time-varying primary magnetic field of frequency $v \ (= \omega/2\pi)$ in a single turn circular coil in free space as explained by Parasnis (1986). A coil with resistance Z and self-inductance L is placed in the primary field. The induced emf in the coil due to the primary magnetic field is proportional to the time rate-of-change of this field and lags behind in phase with the primary field by $\pi/2$. The induced current in the coil and

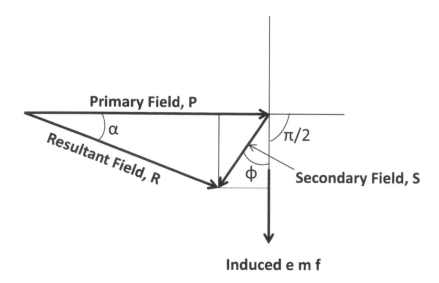

Figure 9.2 Phase relation between the primary, secondary and resultant fields for a single-turn loop (Source: Parasnis, 1986).

the resulting secondary magnetic field produced by it further lag behind the induced emf by an angle ϕ, i.e., the secondary magnetic field lags $\pi/2 + \phi$ in phase behind the primary field. While the lag of $\pi/2$ is associated with induction, ϕ is associated with electrical properties – resistance and self-inductance of the conductor. The relation is $\phi = \tan^{-1}(\omega L/Z)$ where ω is angular frequency, i.e., the magnitude of ϕ will increase with increase in conductivity of the conductor. The resultant of the primary and secondary fields differs from the primary field in intensity, phase and direction and is elliptically polarized. The magnitude of the resultant field reveals the presence and characteristics of the conductor. The vector relation between primary and secondary fields to get the resultant field and phase relation are shown in figure 9.2. It shows that at a given frequency for a good conductor $(Z \rightarrow 0)$ the maximum value of ϕ could be $\pi/2$, i.e., the secondary field will be 180° out of phase with the primary field and for a poor conductor $(Z \rightarrow \infty)$ the value of ϕ will be zero and the secondary field lags behind the primary by $\pi/2$ only. The component of the secondary field in phase with the primary field is known as the 'inphase' or 'real' component and that lagging $\pi/2$ from the primary field is the 'out of phase', quadrature or 'imaginary' component. For a good conductor the 'inphase' component will be large and the 'out of phase' component will be small or the ratio of 'real' to 'imaginary' will be more than one. There are several ways to measure the secondary field. It can be expressed as a percentage of the primary field at the receiver. Another is to measure the inphase and out of phase components of the secondary field. For this, a phase reference is required which is obtained by connecting the receiver to the transmitter through a cable.

The magnitude of the secondary field at the receiver is a complex function of transmitter-receiver spacing, frequency of primary field and subsurface conductivity

(McNeill, 1980). A term 'skin depth' is involved which is a measure of the depth where the current density falls to 1/e of its value at the surface. Skin depth δ is the depth travelled by a propagating EM field in the half space where its amplitude is attenuated by a factor e or 37% of amplitude at the surface. At 2δ depth the amplitude is decreased to $(1/e)^2$ or 13.5%. Skin depth (δ) is expressed as

$$\delta = \sqrt{\frac{2}{\omega\mu_o\sigma}}$$

where, $\omega = 2\pi f$, f is frequency, μ_o is the magnetic permeability of free space and σ is the conductivity of a homogeneous half space with the assumption that magnetic permeability and electrical conductivity are independent of frequency. This shows that skin depth is controlled by the frequency used for measurement, varying inversely with square root of frequency as well as subsurface conductivity. The 'skin depth' is not 'depth of investigation' which includes other factors like instrument sensitivity and accuracy and noise level. In ideal conditions the depth of investigation may be much greater than skin depth, while in noisy conditions it may be much less than the skin depth (Spies, 1989).

The dimensionless ratio of transmitter (T) – receiver(R) spacing (a) to skin depth (δ): (a/δ is known as the 'induction number' (B). As shown by McNeill (1980), for a value of $B = [(\omega\mu_o\sigma/2)^{1/2}\ a]$ much less than 1 by keeping T-R spacing much less than the skin depth, i.e., at a low induction number, the ratio of the secondary magnetic field to the primary magnetic field is linearly proportional to conductivity and simplifies as

$$\left(\frac{H_s}{H_p}\right) = \frac{i\omega\mu_o\sigma a^2}{4}$$

With this approximation, the magnitude of the secondary magnetic field becomes proportional to conductivity. The apparent conductivity can be directly measured from this ratio as

$$\sigma_a \cong \frac{4}{i\omega\mu_o a^2}\left(\frac{H_s}{H_p}\right)_{out\ of\ phase\ component}$$

indicating that the out of phase component can be converted to conductivity. McNeill further indicates that for $a/\delta \ll 1$ the value of $\omega \ll \frac{2}{\mu_o\sigma a^2}$. So for a fixed T-R spacing and operating frequency, the approximation fails for higher conductivities and is the limitation of measurement at low induction number. Readers may consult McNeill (1980) for details of terrain conductivity measurement at low induction number and for the FEM method Chapter 3 by Frischknecht et al. in the SEG publication (1991) on 'Electromagnetic Methods in Applied Geophysics- Applications Part A and Part B' edited by M N Nabighian.

9.3   INSTRUMENT

A variety of multi-frequency EM instruments are available. The moving T-R FEM system comprises transmitter, induction coil receiver, console and cable. The transmitter is energized through a battery. The number of operating frequencies and their magnitude vary from instrument to instrument and the manufacturer. The instruments presently available provide a good number of selective discrete frequencies over a wide range of 3 log cycles. They can be sequentially operated at several frequencies and at a number of transmitter-receiver separations, say 50, 100 and 200 m and up to a maximum of 400 m. Cable is used to send the phase reference signal from transmitter to receiver. More recently, fiber optic cable has been put to use for accurate signals (Spies and Frischknecht, 1991). The instruments are developed to resolve the small secondary field accurately. The primary field is compensated to measure the secondary field response. The refinement in low-noise electronic amplification and filtering in receivers and analog to digital conversion improve the performance (Auken *et al.*, 2006). FEM instruments known as 'terrain conductivity meters' (McNeill, 1980), with short T-R spacing are available for near-surface investigations for a fast coverage at low induction number. These are single frequency, fixed T-R coil separation instruments operated with horizontal or vertical coplanar configurations at very short station intervals. The overall efficiency of a FEM system depends on how accurately it measures the small secondary field in the presence of a primary field under varied ground conditions.

9.4   FIELD PROCEDURES

The FEM surveys are mostly conducted in profiling mode to investigate lateral variations in conductivity along a traverse. There are several ways to conduct FEM profiling. The earlier methods were compensator or Sundberg and Turam where a large source of primary field – the transmitter (T) was kept fixed and the receiver (R) was sequentially placed at selected stations along profile lines (Parasnis, 1986). Measurement with a fixed source being quite cumbersome, the convenient method of synchronized T-R movement is used, though the depth of investigation is relatively less compared to a fixed large source (Frischknecht *et al.*, 1991). The moving T-R method is advantageous to use in rugged topography. It is quite convenient for reconnaissance surveys and can be conducted by two persons. This is also known as the Slingram method or loop-loop method. It is described here as it has been widely used in groundwater investigations. In this method the transmitter and receiver are small loops and no large energizing is required. A number of T-R configurations can be used by varying the position and orientation of the transmitter and receiver loops. The configuration can be selected according to the mode of survey i.e., profiling or sounding. The most commonly used ones are horizontal coplanar loops, vertical coplanar loops and vertical coaxial loops (Fig. 9.3). These configurations give maximum coupling of primary field between T and R loops. Since the anomalies obtained in the horizontal loop or HLEM configuration profiling is comparatively large (Frischknecht *et al.*, 1991), in hard rock groundwater exploration dealing with relatively less conductive targets compared to

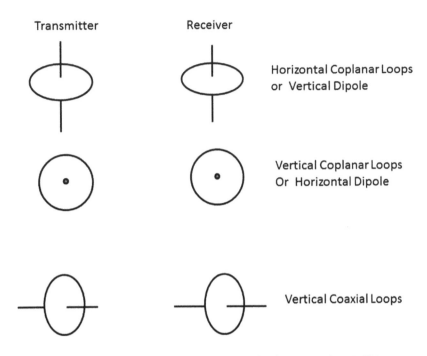

Figure 9.3 The transmitter-receiver configurations for frequency domain EM surveys.

mineral exploration, HLEM configuration is preferred. Some of the references are Palacky *et al.* (1981), Chandra *et al.* (1994) and Meju *et al.* (2001).

In the Slingram method in-phase or real and out of phase or imaginary components of the secondary field are measured and the anomaly values are presented as a percentage of primary magnetic field. The anomaly obtained depends on traverse direction with respect to the trend of conductor. The traverses or profile lines are laid across the probable strike of the conductive fractured zone and profile lengths are kept much larger than the expected lateral extents of anomaly attributable to a fractured zone so that the background is well manifested. Generally, the transmitter and receiver placed at a selected spacing and connected by a cable are moved along traverses in unison and in-line to successive equidistant stations (Fig. 9.4). T-R positions can be interchanged, i.e., either the transmitter or the receiver leads, but whatever configuration is adopted it is maintained throughout the survey in that area. Conventionally, the mid-point between transmitter and receiver position is considered as the measurement plotting point. It is also possible to move along a traverse keeping fixed spaced T-R perpendicular to the traverse. This is known as broadside profiling (Fig. 9.4). According to Frischknecht *et al.* (1991), broadside profiling gives a narrower anomaly resulting in better resolution as compared to in-line profiling.

The EM measurements are sensitive to T-R orientations and spacing and therefore they are checked for accuracy at each consecutive measurement, especially in hard rock areas where topography is mostly uneven. Selection of T-R spacing and station

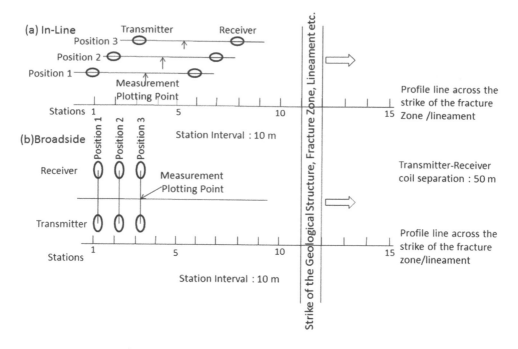

Figure 9.4 In-line and Broadside field layout for HLEM profiling.

interval is coupled with the expected dimensions of the target and therefore their proper selection is quite important for success and should be such that the anomaly is not missed. With an increase in T-R separation the anomaly of long conductors increases but large T-R separation is not suitable for detecting small conductors, while very small T-R separation yields only near-surface local conductivity variations acting as 'geological noise' (Parasnis, 1986). There is no standard rule for T-R spacing and station interval for measurements. For a thin steeply dipping conductor, T-R spacing for HLEM profiling is kept twice the expected depth to its upper edge (Frischknecht *et al.*, 1991) and station interval is generally kept at one-half the T-R spacing but not more than 25 m. The traverse can be repeated with larger T-R spacing if required, which can go up to 300 or 400 m depending on the objective, depth to be explored and availability of T-R connecting cable. However, the depth of penetration includes other factors like frequency used and formation conductivity. The frequencies are selected considering the skin depth and amplitude of the response. The reconnaissance HLEM survey can be conducted with a single moderate to high frequency, 50 or 100 m T-R spacing at a reasonable station interval of 25 m. For better definition and lateral resolution of the anomaly due to the underlying fractured zone, particularly in areas with conductive weathered zone overburden, detailed profiling is conducted with minimum 2 frequencies; one within 100–300 Hz and another at 2000–5000 Hz for better response and at an appropriate station interval which could be as small as 10 m, if required. Ketola (1968) suggests a station interval of at least one-tenth the T-R spacing.

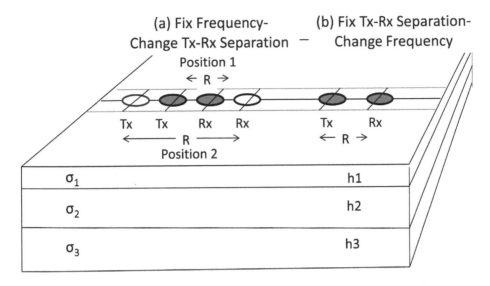

Figure 9.5 Field layout for HLEM sounding: (a) geometric sounding (fix frequency change Tx-Rx separation) and (b) parametric sounding (fix Tx-Rx separation change frequency).

Higher station density makes the survey time-consuming and expensive. A possible compromise could be to follow-up with high density multi-frequency data generation only over the suspected zone. To cover an area systematically parallel traverses are laid with reasonable profile spacing that can be maximum upto T-R spacing used.

Besides profiling, sounding can also be performed by FEM to assess depth-wise electrical conductivity variations. For this, the equipment should have a multi-frequency arrangement and provision of cables to have different T-R spacings. The soundings can be conducted in two ways, either by increasing the frequency in discrete steps keeping the T-R spacing fixed or by increasing the T-R spacing in discrete steps keeping frequency fixed (Fig. 9.5). The former is known as parametric sounding or frequency sounding whereas the latter is known as geometric sounding (Spies and Frischknecht, 1991). In geometric sounding the maximum spacing should be comparable to the depth to be investigated and the minimum spacing should be such that response from the near-surface layer is picked up. Frequency sounding utilizes the dependence of the skin effect on frequency. The low frequency eddy current diffuses to a greater depth as compared to high frequency eddy currents and therefore by making measurements at several frequencies depth wise variations in conductivity can be assessed. Therefore, if the subsurface can be assumed as horizontally layered, FEM profiling with multi-frequency or multi T-R spacing measurements at closely spaced stations will yield a series of close-spaced 1-d soundings that can generate a 2-d subsurface geoelectrical cross section which is contextually translated into a hydrogeological transect. Either it is multi-frequency sounding or multi-spacing sounding the number of frequencies and spacing should be sufficient (maximum 10 observations per decade of log cycle) to get the depths to expected interfaces. The multi-frequency EM system should have at least 5 well-separated frequencies for reliable conductivity estimates (Palacky, 1981 and Sinha, 1983) and as a practice measurements should be made at all the available frequencies. As mentioned by Howard (1983) referred in Spies and Frischknecht (1991) to resolve depth to an interface the number of measurements should be much greater than the ratio of depth of interface to skin depth. Korhonen et al. (2009) use a system in which the transmitter generates a primary magnetic field at 82 discrete frequencies in the range of 2 to 20 kHz and receiver measures the radial, tangential and vertical components of the secondary magnetic field at different T-R spacings. The ratio of vertical to radial components computed at each frequency is converted into apparent resistivity vs depth curve and interpreted as 1d model. As a practice, before attempting field measurements for sounding, forward modeling should be conducted for the expected hydrogeological model to visualize the response and select frequencies.

Prior to HLEM field operations, a reconnaissance survey may be required to select the orientation of profile lay-out and station interval. The orientation of the profile line is kept across the geological strike of the target or structure for a good electromagnetic coupling. Since the strike of fractured zones, basic intrusives and quartz reefs may not be visible on the ground surface in places, the profile is placed across the geological strike or photo-lineaments inferred from satellite imagery. The transmitter and receiver are moved along the profiles as per defined T-R configuration and spacing. It requires two persons, one with transmitter and the other with receiver. The straight profile line laying and distance pegging is done by the surveying crew and GPS may be used. Since accuracy in T-R spacing is desired, adequate time should be given for surveying. The maximum error of 20 cm in 100 m T-R separation is permissible (Palacky, 1991). After checking the specified T-R spacing, transmitter and receiver are connected by appropriate cables and alignment of transmitter and receiver is checked. The operator at the receiver directs the transmitter operator to start the transmitter at a specified frequency, in the case where the transmitter has options for frequency selections. The measurements are made through the attached console. Once the measurements are found satisfactory after stacking, they are recorded either in the console memory or in a note book available with the receiver operator. On completion of a set of readings, the next frequency is selected only after the transmitter is put in the 'off' position; frequency is not to be changed in transmitter the 'on' position. After the measurements with all specified frequencies are over at a station, the entire system is shifted to the next pre-defined station. In case T-R spacing is to be changed, it is not changed at every station but the entire profile is repeated by other spacing. Every day data should be plotted after the field surveys so that anomaly profiles can be preliminarily interpreted for the objective like fractured zone and thick weathered zone etc. and the one which has not yielded satisfactory anomalies can be repeated for different frequencies and T-R spacings or additional measurements can be made on extended profiles.

Errors and noise in EM measurements are described by Sinha (1980) and Spies and Frischknecht (1991). The noise can be grouped as instrumental, geometrical, geological, electromagnetic and cultural. The geometrical errors are due to inaccuracy in T-R spacing and loop misalignment i.e., orientation of transmitter and receiver, probability of which increases in uneven or rugged topography in hard rock areas. It can be minimized by proper and cautious surveying. The profile line should be laid across the strike direction, compromising with accessibility and topography. According to Spies and Frischknecht (1991) error in spacing affects the estimation of resistivity more than the layer thickness. Further, they point out that measurements by horizontal coplanar loop are affected more by misalignment than vertical coplanar loop configuration and inphase quantities are much more sensitive to geometrical errors than out of phase. The T-R spacing error affects inphase measurements and deviation in T-R coplanar orientation affects out of phase measurements (Jansen, 1991). Sinha (1980) indicates that sloping ground and T-R altitude difference affect the response and tilt of the T axis towards R or away from R are more pronounced at low frequencies.

The profiling is conducted to investigate lateral variations and vertical variations are not considered, while sounding is conducted to investigate vertical or depth-wise variations and lateral variations are not considered or act as noise. For example, if there is no lateral conductivity variation there will not be any anomaly in profiling. In sounding, later variations in conductivity in the form of near-surface inhomogeneities, dip and edges of layers, conductivity discontinuities in the form of fractured zones, faults etc., affect the measurements. In general, this can be considered as 'geological noise'. Mostly the 'geological noise' could be the objective of a profiling. The localized or small-scale near-surface conductivity inhomogeneities affect the moving T-R measurements less compared to those located near potential electrodes in DC resistivity method. Before taking up sounding or profiling possible 'geological noise' may be considered and accordingly corrections are incorporated in interpretations. Electromagnetic noise such as geomagnetic signals including atmospheric lightning discharge, power-line radiation and wind causing vibration of the sensor in the Earth's magnetic field are rejected by filters (Spies and Frischknecht, 1991). Cultural noise or manmade noise such as iron fencing, electric power-lines, buried metallic pipes affect the measurements and are to be avoided.

9.5 PROCESSING OF DATA

The HLEM profiling data can be interpreted qualitatively as well as quantitatively or semi-quantitatively. For qualitative interpretations, conventionally the inphase and out of phase data sets are plotted together for different frequencies and T-R spacings separately. Alternatively, inphase and out of phase data can be plotted separately for all the frequencies used. It helps visual interpretation of promising anomalies. The parallel profile inphase and out of phase data can be presented as stacked profiles on the profile lines located on an appropriate large scale map. It helps define the trend of the conductor. The profile data may contain noise as spikes. These spikes are removed in such a way that the shape of the anomaly is preserved and a smooth anomaly is obtained.

9.6 INTERPRETATION

Basically it is assumed that conductor is in resistive host medium. A conductor produces detectable anomaly if it has high conductivity contrast with the host medium, its thickness is comparable to T-R coil separation and occurs at a moderate depth. However for drawing inferences it is necessary that various factors influencing the response, e.g., its strike, lateral extents, dip, conductivity of host medium, effect of overburden conductivity and geometry are studied and limitations are known. In this regard an exhaustive

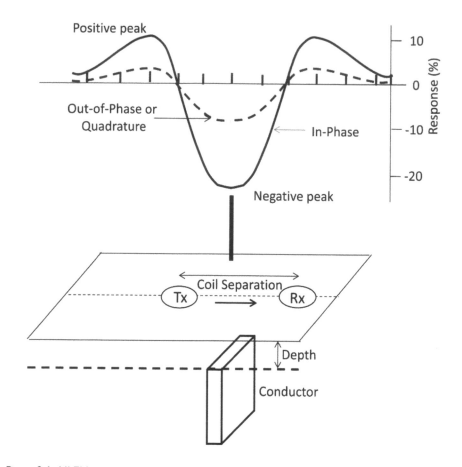

Figure 9.6 HLEM response over a thin plate-like vertical conductor in a resistive host medium.

description is presented by Frischknecht *et al.* (1991). The laboratory model studies and numerical computation of standard anomaly curves for different target geometries are quite useful in comparing theoretical and observed anomalies, interpreting them and getting an insight into the possible physical model and the noise. Based on modeling studies sets of model anomaly curves prepared by several researchers are available. Some of the references are Hedstrom and Parasnis (1958), Poddar and Bhattacharya (1966), Nair *et al.* (1968), Ketola (1968), Eadie (1979), Gupta *et al.* (1980) and Scintrex (1986). Besides, inversion schemes are available for interpretation of HLEM data (Pirttijarvi *et al.*, 2002 and Sasaki and Meju, 2006).

A simple geometric model generally considered is of a vertical or steeply dipping shallow conductor with infinite (large) strike length and depth extent and a small thickness. The components of secondary field response over a vertical conductor at depth in a resistive host medium with commonly used HLEM or Slingram T-R configuration, moving along a line passing through the mid-point and orthogonal to the geological strike of the conductor are shown in figure 9.6. The response is generally

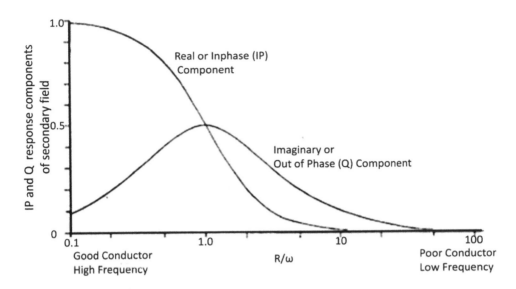

Figure 9.7 Variation of the real and imaginary components with resistivity and frequency (single turn loop, exact curves), where R is resistance and ω is angular frequency of a sinusoidal field; (Source: Parasnis, 1986).

presented as a percentage of the primary field. The anomaly over a vertical sheet conductor is trough-like and symmetrical comprising minimum (negative peak) exactly over the conductor flanked by two maxima (positive peaks). The in-phase and out of phase responses depend on the conductivity of the conductor and host medium, its dimension compared to T-R coil separation, its location with respect to the coil, its dip, depth, frequency of transmitting primary EM field and the presence of conductive overburden. The conductivity characteristics of an isolated conductor in a resistive host, without any conductive overburden which complicates the anomaly, can be qualitatively assessed through a simple analysis of relative amplitudes of inphase and out of phase components of the secondary field at two frequencies. It can be visualized from figure 9.7 showing the plots of real (inphase) and imaginary (out of phase) components in arbitrary units against the parameter resistance/frequency (Parasnis, 1966 and 1986). As a rule of thumb with an increase in frequency for a good conductor inphase response increases and out of phase response decreases and for a poor conductor inphase response decreases and out of phase response increases or the out of phase response reduces while in-phase response increases with increase in conductivity. For a moderate conductor, both the inphase and out of phase responses are comparable and large. The effect of frequency change on inphase response is more than that on out of phase component. The presence of conductive overburden however, brings in variations from these observations.

The anomaly becomes asymmetrical for a dipping conductor (Fig. 9.8). The effect of dip is described in detail by Nair *et al.* (1968). A higher amplitude of positive shoulder is observed on the down dip side compared to that on the up dip side. For a

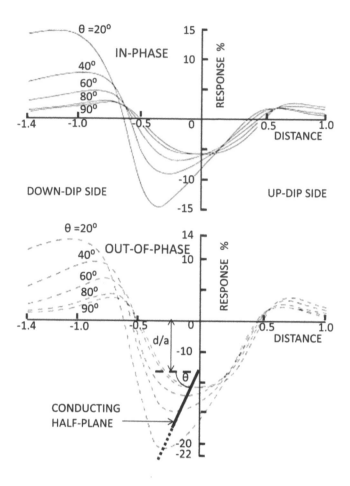

Figure 9.8 Asymmetrical anomaly over dipping 'poor' or relatively more resistive conductor; small in-phase and out-of-phase responses out of which out-of-phase component predominates; here d/a = 0.2, where d is depth to the the top of the conductor and a is T-R spacing and dimensionless response parameter λ/a = 30 (poor conductor), where λ/a = 10^5/σaμ νt, σ: conductivity, μ: magnetic permeability, ν: frequency of the transmitter and t: thickness of the sheet conductor (Adopted from Nair *et al.*, 1968).

good conductor while inphase negative peak gets shifted towards down dip, the out of phase anomaly gets shifted towards up dip and in-phase response is larger than out of phase response. In the case of a poor conductor, both the anomalies are small, the minima get shifted towards down dip side and out of phase response is more prominent than in-phase (Fig. 9.8). An example is shown in figure 9.12. The amplitude of the positive shoulders and the minimum depends on depth extent of the conductor. With a fixed depth to the top, the amplitude increases with increasing depth extent and also the shift of the negative peak is significant for shallow conductors with low dip (Nair *et al.*, 1968). However, the response decreases with increasing depth to the top of the conductor. In general, under normal conditions, the depth of investigation is 3/4th of

T-R spacing. Since the resolution decreases with depth, increasing the T-R spacing for achieving greater depth of investigation does not work well.

In reality subsurface structures in hard rock are quite complex. The saturated conductive fractured zones are generally overlain by a weathered zone of widely varying thickness leading to a substantial variation in the relatively conductive saturated bottom part of the weathered zone which may or may not be in contact with the fractured zones locally. The fractured zone conductors have limited dimensions. At places, the resistivities of the saturated weathered zone, the fractured zone and the host medium are comparable. If the host medium including the weathered zone is extremely resistive, it will not affect the response, and the 'free-space' anomaly models are valid for comparison. Otherwise the conductive weathered zone or the host medium plays a significant role in modifying the response and causing ambiguities in interpretation. With a conductive host medium in contact with a conductor, in addition to an eddy current in the conductor, the channeling of another type of regionally induced current known as the 'galvanic current' flows through the host medium and passes through the boundary of the target and influences the response, increasing its amplitude, which becomes more prominent at higher frequencies and with the length of conductor (Lajoie and West, 1976; West and Macnae, 1991). When contrast in the conductivity of the target and host is less, the galvanic current dominates over the eddy current (Frischknecht et al., 1991). That is, for a poor conductor the anomaly enhancement is greater. Lajoie and West (1976) through a modeling study further indicate that current channeling helps identify targets at greater depth, but it produces strong unwanted response; the dipping conductor appears more vertical than actual and depth estimation may be erroneous.

In presence of a conductive weathered zone, the response suffers phase rotation (change in phase angle relating inphase and out of phase) and attenuation. If it is in contact with a moderately conductive fractured zone, the galvanic current induced in the conductive weathered zone may enhance the response and also complicate the interpretation of anomalies from the underlying conductor-the fractured zone. The presence of conductive weathered zone/ overburden of almost uniform character over a wide area may not complicate the interpretation as much as is done by its patchy occurrence or local thickening with lateral non-uniformity in conductivity and thickness. Such inhomogeneities may produce similar response as that from a deeper conductor and also affect the response. Lowrie and West (1965) conducted a model experiment to study the effect of conductive overburden considering it as a thin infinite horizontal sheet. According to them due to the presence of a conductive overburden the out of phase anomaly is inverted and apparently the conductor appears as a better conductor and deeper. Villegas-Garcia and West (1983) study the horizontal loop coplanar (HLEM) response for three types of non-uniform conductive overburden viz., ramp, ridge and valley through analog modeling and indicate the necessity of conducting surveys at several frequencies and transmitter-receiver coil separations to distinguish the variations in response of the overburden from that of a conductor in the underlying resistive host medium. This also helps in differentiating weak anomalies due to deep buried good conductors from the weak anomalies due to shallow poor conductors. Out of these, the valley shaped overburden or localized thickening of the weathered zone is quite common in hard rocks and produces a misleading anomaly similar to that from a vertical plate conductor underlying a uniform overburden. In the case

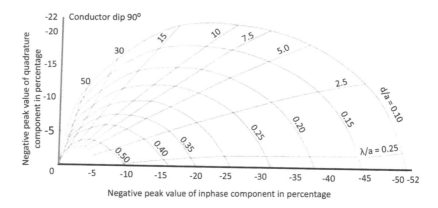

Figure 9.9 Phase diagram of the negative peak responses of horizontal loop EM profiling over conducting half-planes dipping at 90°, d/a is the ratio of conductor depth to T-R spacing, λ/a is the dimensionless response parameter of the half-plane (Source: Nair *et al.*, 1968).

of a valley structure the current channeling phenomenon becomes significant and the anomaly is comparatively large. The out of phase component gets reversed at higher frequency. According to Villegas-Garcia and West (1983) the valley structure is characterized by a prominent shoulder anomaly and the anomaly cross-over separation is controlled by the ratio of valley width to T-R coil separation. These criteria help differentiate the anomalies from valley structure and deep conductor. The ramp shaped overburden produces an asymmetrical anomaly and may be mistaken for a dipping conductor.

The HLEM data collected at several frequencies at a particular T-R separation can be interpreted manually to get the depth and conductance (conductivity/thickness product) of a conductor using phase or phasor or Argand diagrams. Phase diagrams (Fig. 9.9 taken from Nair *et al.*, 1968) are plots of negative peak values of in-phase vs out of phase response in percent for different values of d/a and λ/a. $λ/a = 10^5/σa\mu\nu t$ is a dimensionless response parameter like $α = σa\mu\omega t$, where d: depth to conductor, a: T-R spacing, t: conductor thickness, μ: magnetic permeability, ν; frequency, ω: angular frequency and σ: conductivity. To use phase diagrams, the negative peak values of in-phase and out of phase data for each frequency are plotted on a transparent sheet on the same scale as that of the theoretical phase diagram and superimposed on it to get a match and deduce the parameters. Similarly, for a stratified region layer conductivities and thicknesses are obtained from the matching phase diagram and induction number. The detailed procedure is given in Eadie (1979). To improve the estimation of shallow as well as deep layer conductivities and thicknesses Korhonen *et al.* (2009) and Schamper *et al.* (2012) showed the applicability of measuring radial, tangential and vertical components of the secondary magnetic field at a number of frequencies and T-R spacing and laterally constrained 1 d joint inversion of the radial and vertical components.

To assess the effect of conductive overburden and host medium the field in phase-out of phase peak negative values measured with respect to background response for the frequency used can be plotted on a free-air phase diagram (Fig. 9.9). In the

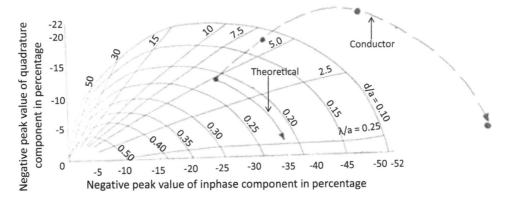

Figure 9.10 Plot of field measured inphase-quadrature (out of phase) negative peak values at 222 Hz, 444 Hz, 1777 Hz and 3555 Hz on the phase diagram by Nair *et al.* (1968, figure 9.9). The theoretical (free-air or highly resistive host medium) frequency migration of anomaly amplitude is compared to that observed for the suspected vertical conductor in conductive host medium with conductive overburden (Source: Lajoie and West, 1977).

case where frequency migration follows the free-air anomaly phase diagram it can be assumed that the effect of the conductive overburden and/or conductive host medium is least. Otherwise, the plot of field inphase-out of phase peak negative values for different frequencies would deviate substantially, indicating phase rotation of the conductor anomaly due to conductive overburden as shown in figure 9.10 (Lajoie and West, 1977).

9.7 DELINEATION OF SATURATED FRACTURED ZONES

The FEM profiling can be usefully applied in delineating the saturated fractured zones in compact rock underlying the weathered zone. By FEM measurements it is difficult to delineate an individual thin fracture at depth but thick fractured zones can be delineated. As discussed in Chapter 2 the fracturing is intense where weathering is deep and vice versa. The fractured zone could be vertical as well as steeply dipping. Horizontal sheet fracturing is also possible but they are generally shallow. The fractures are generally interconnected but could be localized also. They could be connected to the weathered zone regionally as well as locally. The resistivity of a saturated fractured zone is less than that of compact host rock but much higher than resitivities generally observed with metal ore deposits. Also, the contrast with the host rock is less. The resistivity of the fractured zone could be of the order of 200 to 300 ohm.m, whereas that for compact rock ranges beyond 1000 ohm.m and for the overlying weathered zone generally ranges within 100 ohm.m. The resistivity of the weathered zone can vary over a wide range depending on the degree of weathering, clay content, saturation and quality of water. At places fractures may be filled with conductive altered-clay material. The fractured zones are generally considered as weak conductors. Therefore, the low frequency inphase response which is, in general, low, becomes almost negligible over fractured zones and surveys with higher frequencies are essentially required for obtaining a considerable response. Palacky *et al.* (1981) conduct higher frequency (3555 Hz) horizontal loop coplanar (HLEM) surveys for delineating fractured zones for groundwater well siting in the Precambrians of the Republic of Upper Volta (now named Burkina Faso). Results with coil separation of 50 m were found satisfactory in delineating the fractured zones. This has also been observed with HLEM profiling in hard rock in India (Chandra *et al.*, 1994) where surveys were conducted using a maximum frequency of 3037.5 Hz. Olsson *et al.* (1984) conducted high frequency FEM for delineating fracture-free zones related to radioactive nuclear waste disposal. Frischknecht *et al.* (1991) also suggest the use of higher frequencies to get a detectable anomaly. They further suggest for large T-R spacing, if the saturated fracture is deep or it has a large strike and depth extent. But mostly, strike length or depth extent is not known and therefore it is necessary that profiling with higher frequencies and at least two T-R spacings, say 50 and 100 m be conducted so that response with better resolution from fractured zones of smaller dimensions is also recorded.

For detecting fractured zones HLEM configuration is suitable. The amplitude of anomaly depends on the width of the fractured zone; it is large over a thick fractured zone. The individual fractures which are quite thin may not get detected. Also, it would be difficult to differentiate a clay filled fractured zone from water filled ones. Delineation of the fractured zone becomes tricky with the presence of an overlying clay rich conductive weathered zone as variation in its resistivity and thickness also produces a significant anomaly as discussed above and distinguishing the fractured zone anomaly is difficult. Ehinola *et al.* (2006) based on field observations and drilling results from hard rocks in Nigeria attempt to qualitatively categorize the FEM anomalies for selecting well site location. According to them, on the FEM profile the predominance of high density positive inphase and out of phase data points indicates minor fracture under a thick overburden while that of high density negative inphase and out of phase data points indicates major fractures under a thin overburden. The negative anomalies with frequent intersection of inphase and out of phase curves reveal the presence of intense and interconnected fractures. The near-parallel separate inphase and out of phase curves indicate thick overburden. A combination of magnetic, electrical resistivity and FEM profiling can reduce the ambiguities. Some of the case studies from India (Chandra *et al.*, 1994) are discussed below.

9.7.1   Case study from metasediments

The case study site 1 from Jharkhand, India is in highly folded, steeply dipping schists, quartzites and phyllites. The HLEM profiling was conducted with frequencies of 112. 5 Hz and 3037.5 Hz along two profiles P2 and P3 about 450 m apart in the SE-NW direction (Fig. 9.11 a and 9.11 b). The profiles P2 and P3 were laid across a coincident aeromagnetic and photo linear for a distance of 190 and 800 m respectively. The T-R coil spacing was 100 m. The 50 m coil spacing did not yield any appreciable anomaly. The measurements were made at 10 and 25 m station intervals on profiles P2 and P3 respectively. On P2 the low frequency (112.5 Hz) anomaly at 100 m T-R spacing is not prominent except the development of a mild negative in-phase (IP) anomaly. The development of a secondary field is prominent at higher frequencies. The out of

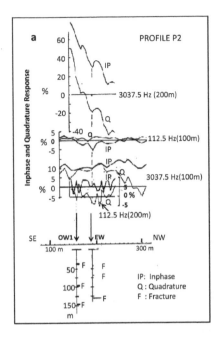
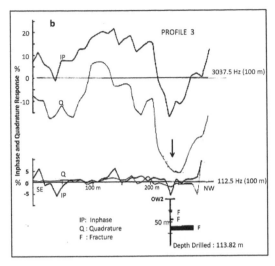

Figure 9.11   HLEM surveys at site I in metasediments in Jharkhand State of India: inphase and quadrature (out of phase) anomalies at 112.5 Hz and 3037.5 Hz on profiles P2 and P3. The fractured zones encountered in the boreholes: exploratory well (EW) and observation wells (OW1) and (OW2) are shown (Source: Chandra *et al.*, 1994).

phase (Q) response developed only at higher frequency (3037.5 Hz) towards the NW of the profile P2. The anomaly developed with T-R spacing at 200 m, in which the low frequency anomaly shifted towards the NW and the high frequency anomaly showed a step like development. Since this is also associated with step like reducing gradient resistivity towards NW, the HLEM anomaly is considered as an indicative of either thickenning or reduced resistivity of weathered zone towards the NW. Two boreholes, EW and OW1 were drilled 35 m apart up to the depths of 137.97 and 171.49 m respectively on profile P2. While EW was located on the IP anomaly developed at 112.5 Hz with T-R spacing 100 m, the OW1 was located on the IP and Q anomaly developed at 112.5 Hz with T-R spacing 200 m.

The borehole EW encountered a weathered zone up to 13 m, followed by a saprolite zone up to 25 m depth, whereas in OW1 these were encountered up to 14 m and 20 m depth respectively. The vertical electrical sounding indicates a resistivity between 64 and 198 ohm.m for the weathered zone. In EW fractured zones associated with low resistivity and thickness about a metre were encountered at 40.91–41.91 (F1: 1 m), 65.77–66.77 (F2: 1 m) and 129-130 m (F3: 1 m) depth which were confirmed through electrical resistivity logging (Fig. 12.5a and b). The resistivity of the fractured zones could not be ascertained precisely from the logging data. The yield from the individual fractured zones F1, F2 and F3 were 0.75, 1.58 and 9.16 litres per second (lps). In OW1 the fractured zones were encountered at 34.28–35.28 (F1: 1 m), 94–97 (F2: 3 m) and 146–147 m (F3: 1 m) depth with respective yields 0.75, 3.5 and 3.16 lps. The weathered and saprolite zones were dry and not very thick. Also the resistivity of the underlying compact rock is more than 1000 ohm.m, as observed from the VES and resistivity log and hence the secondary field response can be considered as due to the relatively conductive fractured zone only.

On profile P3 the weathered zone is relatively thick, about 20 m, while the saprolite thickness is only 8 m. The resistivity of the weathered and saprolite zones ranges from 22 to 70 ohm.m The low frequency (112.5 Hz) HLEM anomalies are not prominent. The anomalies with high frequency (3037.5 Hz) are prominent and the base level shift of IP anomaly and increase in response towards NW indicate the effect of modetrately resistive (22-70 ohm.m) overburden and its thickening towards NW (Fig. 9.11). However, the qualitative nature of the conductor anomaly is preserved and is indicative of the presence of fractured zones towards the NW end of P3. The borehole drilled upto about 113 m depth on the NW anomaly encountered fractured zones in the depth ranges of 27.5–30.0 (2.5 m), 42.62–45.42 (2.8 m) and 57.86–68.1 (10.24 m). These fractured zones are quite thick as compared to those encountered in P2. The yield of the shallowest fracture was 0.75 lps and that of the deepest fracture was maximum, 21.25 lps during drilling.

9.7.2   Case studies from granitic terrain

HLEM profiling was carried out over 500 m across an aeromagnetic and photo linear at site 2 in granitic terrain of Jharkhand State, India (Fig. 9.12). The profiling was conducted with frequencies of 112.5, 337.5, 1012.5 and 3037.5 Hz and T-R spacing of 100 m at an station interval of 25 m. An anomalous development of inphase and out of phase response with 112.5 Hz towards the NW end of the profile and reversals of out of phase response in the central part are observed. These reversals are due to the presence of a conductive overburden. Three sets of trough shaped anomalies develop at higher frequencies. Out of these for the central anomaly, inphase as well as out of phase components develop at 337.5 and 1012.5 Hz frequencies and for the other two only out of phase anomaly develops. It was inferred that the target producing the central anomaly is relatively more conductive than those associated with the other anomalies. At 3037.5 Hz the anomalies get enhanced. The asymmetrical wide central anomaly indicates the presence of more than one dipping conductors underlying the overburden. The gradient resistivity profiling confirms the anomalies. The central anomaly was drilled (EW1) upto a depth of 198.76 m. It encountered 22 m thick weathered zone, followed by 3 m thick saprolite. The resistivity of the weathered zone and saprolite estimated from VES is in the range of 148 to 212 ohm.m. The fractures were encountered at 48.36–49.36 (F1:1 m) and 110.32–110.53 m (F2:1.21 m) depths. The yield from the saprolite was 2.5 lps and that from fractures F1 and F2 were, 2.0 and 3.0 lps respectively. A bore hole EW2 was drilled up to a depth of 300.76 m on NE flank of another HLEM anomaly in the area (Fig. 9.13). It is on a gradient resistivity high. The borehole was dry though it encountered weathered zone and saprolite up to about 25 m depth. The borehole EW3 drilled up to a depth of 200.82 m on this HLEM anomaly (Fig. 9.13) encountered 7 m thick weathered zone underlain by 10 m thick

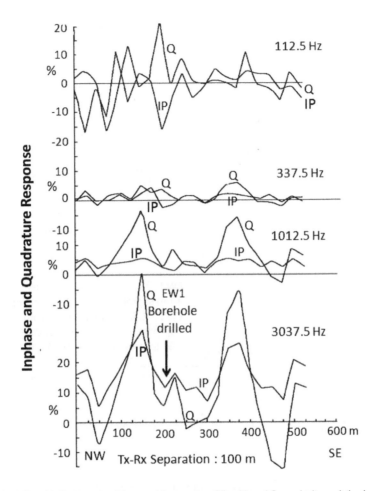

Figure 9.12 HLEM profiling at site 2 in granitic terrain of Jharkhand State, India and the location of the drilled borehole EW1 (Source: Chandra *et al.*, 1994).

saprolite. The fracture was encountered at 44.8 m depth. The yields from the saprolite and the fractured zone were 1.41 1nd 1.91 lps respectively. The causative conductor is inferred as the thickening of the conductive overburden associated with an underlying saturated fractured zone dipping due NE.

The HLEM profiling was carried out at site 3 in the granitic terrain of Jharkhand State, India with four frequencies, viz., 112.5, 337.5, 1012.5 and 3037.5 HZ and at T-R coil spacing 100 m (Fig. 9.14). Two anomalies were developed and enhanced with higher frequencies. The anomalies are associated with resistivity lows on gradient resistivity profiling. Out of these the NE anomaly is associated with less resistivity value. The borehole (c) drilled on NE anomaly encountered a weathered zone up to 14 m and saprolite up to 23 m depth. As inferred from VES these are associated with

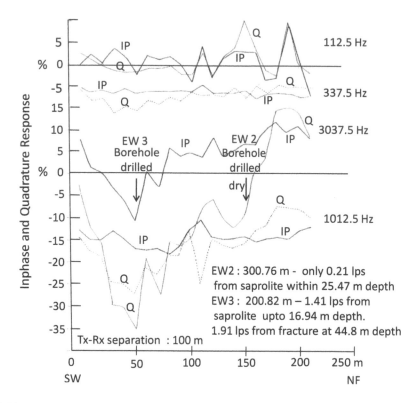

Figure 9.13 HLEM profiling at site 2 and locations of drilled boreholes EW2 and EW3 (Source: Chandra *et al.*, 1994).

resistivity values ranging from 36 to 243 ohm.m. The yielding fractured zone was encountered at 33.32–35.47 m depth. The yield from saprolite was 1.8 and that from fracture was 3.77 lps. The borehole (a) drilled on the SW anomaly encountered almost the same thicknesses of weathered and saprolite zones but a deep fracture at a depth of 176.94–178.94 m was encountered. These results indicate that the deeper fracture is not reflected in the anomaly. The borehole (b) drilled in the central part between a and c was dry. The resistivity logs of boreholes b and c are shown in figure 9.15.

The HLEM survey was conducted at site 4 in granitic terrain using 3 frequencies viz., 337.5, 1012.5 and 3037.5 Hz and 100 m T-R spacing (Fig. 9.16). The broad anomaly is asymmetrical indicating a dipping fractured zone. Since both the inphase and out of phase anomalies get shifted in the down dip side it indicates a poor conductor. The bore hole drilled on the anomaly encountered a 15 m thick weathered zone, 5 m thick saprolite and a fractured zone at a depth of 23–24 m only. The weathered and fractured zones are associated with resistivities within 30 and 90 ohm.m. The yield from the saprolite was 0.5 lps and that from the fractured zone was 0.75 lps only. Here the contact of quartz vein with granite forms the conductor.

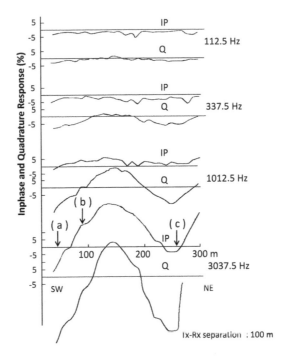

Figure 9.14 HLEM survey at site 3 and locations of boreholes drilled a, b and c in granitic terrain of Jharkhand State, India (Source: Chandra et al., 1994).

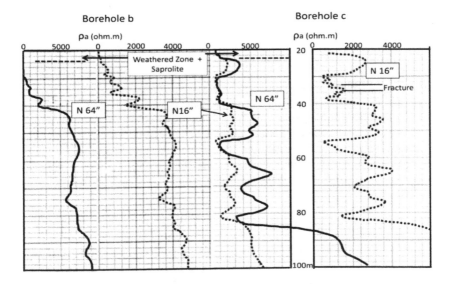

Figure 9.15 The N 16″ and N 64″ resistivity logs of boreholes b and c at site 3 in granitic terrain of Jharkhand State, India (Chandra et al., 1994).

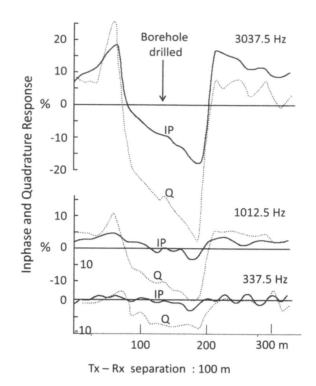

Figure 9.16 HLEM survey at site 4 in granitic terrain of Jharkhand state, India (Source: Chandra *et al.*, 1994).

REFERENCES

Auken, E., Pellerin, L., Christensen, N.B. & Sorensen, K. (2006) A survey of current trends in near-surface electrical and electromagnetic methods. *Geophysics*, 71 (5), G249–G260.

Chandra, P.C., Reddy, P.H.P. & Singh, S.C. (1994) *Geophysical studies for groundwater exploration in Kasai and Subarnarekha River basins* (UNDP Project) Central Ground Water Board, Min. of Water Resources, Govt. of India Tech. Report.

Eadie, T. (1979) *Stratified earth interpretation using standard horizontal loop electromagnetic data*. Research in Applied Geophysics, No.9, Geophysics Laboratory, Univ. of Toronto, Canada.

Ehinola, O.A., Opoola, A.O. & Adesokan, H.A. (2006) Empirical analysis of electromagnetic profiles for groundwater prospecting in rural areas of Ibadan, southwestern Nigeria. *Hydrogeology Journal*, 14 (4), 613–624.

Fitterman, D. & Labson, V.F. (2005) Electromagnetic induction methods for environmental problems. In: Butler, D.K. (ed.) *Near-Surface Geophysics, Chapter 10, Series: Investigations in Geophysics: No. 33*, Tulsa, Oklahoma, U.S.A., Society of Exploration Geophysicists. pp. 301–351.

Frischknecht, F.C., Labson, V.F., Spies, B.R. & Anderson, W.L. (1991) Profiling methods using small sources. In: Nabighian, M.N. (ed.) *Electromagnetic Methods in Applied Geophysics*,

Vol. 2: Applications, Part A, Chapter 3, Tulsa, Oklahoma, U.S.A., Society of Exploration Geophysicists. pp. 105–269.

Gupta, O.P., Joshi, M.S. & Negi, J.G. (1980) Scale model electromagnetic response to inline and broadside systems at skew traverses of a dipping half plane embedded in a conducting host rock. *Geophysical Prospecting*, 28 (1), 119–134.

Hedstrom, E.H. & Parasnis, D.S. (1958) Some model experiments relating to electromagnetic prospecting with special reference to airborne work. *Geophysical Prospecting*, 6 (4), 322–341.

Howard Jr., A.Q. (1983) On resolution in geophysical probing using electromagnetic methods. *IEEE Transactions on Geoscience and Remote Sensing*, GE-21 (1), 102–108.

Jansen, J. (1991) Synthetic resistivity soundings for groundwater investigations. In: *Proceedings of the Fifth National Outdoor Action Conference on Aquifer Restoration, Monitoring & Geophysical Methods, NWWA*, Dublin, Ohio, U.S.A.

Jin, S., Fallgren, P.H., Cooper, J.S., Morris, J.M. & Urynowicz, M.A. (2008) Assessment of diesel contamination in groundwater using electromagnetic induction geophysical techniques. *Journal of Environmental Science and Health*, 43 (Part A), 584–588.

Kaikkonen, P. (1979) Numerical VLF modeling. *Geophysical Prospecting*, 27 (4), 815–834.

Ketola, M. (1968) *The interpretation of Slingram (horizontal loop) anomalies by small-scale model measurements*. Geological Survey of Finland, Report of Investigations No. 2.

Koefoed, O. & Biewinga, D.T. (1976) The application of electromagnetic frequency sounding to groundwater problems. *Geoexploration*, 14 (3–4), 229–241.

Korhonen, K., Ruskeeniemi, T., Paananen, M. & Lehtimaki, J. (2009) Frequency domain electromagnetic soundings of Canadian deep permafrost. *Geophysica*, 45 (1–2), 77–92.

Lajoie, J.J. & West, G.F. (1976) The electromagnetic response of a conductive inhomogeneity in a layered earth. *Geophysics*, 41 (6A), 1133–1156.

Lajoie, J.J. & West, G.F. (1977) Two selected field examples of EM anomalies in a conductive environment. *Geophysics*, 42 (3), 655–660.

Lowrie, W. & West, G.F. (1965) The effect of a conducting overburden on electromagnetic prospecting measurements. *Geophysics*, 30 (4), 624–632.

McNeill, J.D. (1980) *Electromagnetic terrain conductivity measurements at low induction numbers*. Technical note TN-6, Geonics Ltd., Toronto, Canada.

McNeill, J.D. (1990) Use of electromagnetic methods for groundwater studies. In: Ward, S.H. (ed.) *Geotechnical and Environmental Geophysics, Vol. I: Review and Tutorial*. Tulsa, Oklahoma, USA, Society of Exploration Geophysicists, pp. 191–218.

McNeill, J.D. (1991) Advances in electromagnetic methods for groundwater studies. *Geoexploration*, 27 (1–2), 65–80.

Meju, M.A., Fontes, S.L., Ulugergerli, E.U., La Terra, E.F., Germano, C.R. & Carvalho, R.M. (2001) A joint TEM-HLEM geophysical approach to borehole siting in deeply weathered granitic terrains. *Ground Water*, 39 (4), 554–567.

Nair, M.R., Biswas, S.K. & Mazumdar, K. (1968) Experimental studies on the electromagnetic response of tilted conducting half-planes to a horizontal-loop prospecting system. *Geoexploration*, 6 (4), 207–244.

Olsson, O., Duran, O., Jamtlid, A. & Stenberg, L. (1984) Geophysical investigations in Sweden for the characterization of a site for radioactive waste disposal – an overview. *Geoexploration*, 22 (3–4), 187–201.

Ong, J.T., Lane, J.W., Zlotnik, V.A., Halihan, T. & White, E.A. (2010) Combined use of frequency-domain electromagnetic and electrical resistivity surveys to delineate near-lake groundwater flow in the semi-arid Nebraska Sand Hills, USA. *Hydrogeology Journal*, 18 (6), 1539–1545.

Palacky, G.J. (1981) The airborne electromagnetic method as a tool of geological mapping. *Geophysical Prospecting*, 29 (1), 69–88.

Palacky, G.J., Ritsema, I.L. & De Jong, S.J. (1981) Electromagnetic prospecting for groundwater in Precambrian terrains in the Republic of Upper Volta. *Geophysical Prospecting*, 29 (6), 932–955.

Palacky, G.J. (1991) Application of the multifrequency horizontal-loop EM method in overburden investigations. *Geophysical Prospecting*, 39 (8), 1061–1082.

Parasnis, D.S. (1966) Electromagnetic prospecting-C.W. techniques. *Geoexploration*, 4 (4), 177–208.

Parasnis, D.S. (1986) *Principles of Applied Geophysics*. Fourth Edition, London, Chapman and Hall Ltd.

Patra, H.P. (1976) Electromagnetic depth sounding for ground water with particular reference to central frequency sounding: principles, interpretation and applications. *Geoexploration*, 14 (3–4), 254–258

Pirttijarvi, M., Pietila, R., Hattula A. & Hjelt, S.-E. (2002) Modelling and inversion of electromagnetic data using an approximate plate model. *Geophysical Prospecting*, 50 (5), 425–440

Poddar, M. & Bhattacharya, P.K. (1966) On the electromagnetic response of conductors in the inductive method of prospecting (model studies). *Geophysical Prospecting*, 14 (4), 470–486.

Sasaki, Y. & Meju, M.A. (2006) A multidimensional horizontal-loop controlled-source electromagnetic inversion method and its use to characterize heterogeneity in aquiferous fractured crystalline rocks. *Geophysical Journal International*, 166 (1), 59–66.

Schamper C., Rejiba, F. & Guérin, R. (2012) 1D single-site and laterally constrained inversion of multifrequency and multicomponent ground-based electromagnetic induction data-application to the investigation of a near-surface clayey overburden. *Geophysics*, 77 (4), WB19-WB35.

Scintrex, (1986) *IGS-2/EM-4 and SE-88 moving source Genie interpretation manual*. Scintrex Ltd., Ontario, Canada.

Sinha, A.K. (1980) A study of topographic and misorientation effects in multifrequency electromagnetic soundings. *Geoexploration*, 18 (2), 111–133.

Sinha, A.K. (1983) Airborne resistivity mapping using a multifrequency electromagnetic system. *Geophysical Prospecting*, 31 (4), 627–648.

Spies, B.R. (1989) Depth of investigation in electromagnetic sounding methods. *Geophysics*, 54 (7), 872–888.

Spies, B.R. & Frischknecht, F.C. (1991) Electromagnetic sounding. In: Nabighian, M.N. (ed.) *Electromagnetic Methods in Applied Geophysics, Vol. 2: Applications, Part A*, Chapter 5, Tulsa, Oklahoma, U.S.A., Society of Exploration Geophysicists. pp. 285–422.

Villegas-Garcia, C.J. & West, G.F. (1983) Recognition of electromagnetic overburden anomalies with horizontal loop electromagnetic survey data. *Geophysics*, 48 (1), 42–51.

Vogelsang, D. (1987) Examples of electromagnetic prospecting for karst and fault systems. *Geophysical Prospecting*, 35 (5), 604–617.

West, G.F. & Macnae, J.C. (1991) Physics of the electromagnetic induction exploration method. In: Nabighian, M.N. (ed.) *Electromagnetic Methods in Applied Geophysics, Vol. 2: Applications, Part A*, Chapter 1, Tulsa, Oklahoma, U.S.A., Society of Exploration Geophysicists. pp. 5–44.

Chapter 10

The very low frequency electromagnetic method

10.1 INTRODUCTION

The Very Low Frequency (VLF) EM method measures the response of the earth to EM waves transmitted from a distant source. The receiver located at a large distance from transmitter is in the control of the operator. It can be considered as a passive method (Paterson and Ronka, 1971). VLF surveys can be easily conducted in areas where the surface layer is highly resistive and an electrical resistivity survey is difficult due to poor conduction of electrolytic current. It was developed in the 1960s using established military radio communication stations (Paal, 1965; Djeddi *et al.*, 1998). A brief history of its early development is given by Paterson and Ronka (1971). The method was developed for time-efficient, cost-effective shallow subsurface mapping of conductive targets, primarily the ore bodies (Paal, 1968; Paterson and Ronka, 1971; Bayrak, 2002; Shivaji and Gnaneshwar, 1999; Okonkwo and Ezeh, 2012). It can be used for near surface resistivity mapping (Tabbagh *et al.*, 1991), environmental studies (Jeng *et al.*, 2004), groundwater prospecting at shallow levels in hard rock (Palacky *et al.*, 1981; Bernard and Valla, 1991; Chandra *et al.*, 1994; Gomes and Chambel, 2000; Castelo Branco *et al.*, 2004; Sharma and Baranwal, 2005), detecting coal mining-induced fractures (Hutchinson and Barta, 2002), groundwater contamination studies (Stewart and Bretnall, 1986; Benson *et al.*, 1997) and landfill investigation studies (Roberts *et al.*, 1989). McNew and Arav (1995) use VLF measurements to delineate fresh water-salt water interfaces and fresh water aquifer lenses. It has been used to identify steeply dipping structures such as faults and fractured zones (Phillips and Richards, 1965; Gurer *et al.*, 2009). As such, for groundwater exploration in hard rock, the method is mostly used for fast, economic, reconnaissance survey along with magnetic to identify anomalous zones. It is followed by detailed surveys through other geophysical methods to delineate and confirm the location of saturated fractured zones.

The term VLF for 'Very Low Frequency' represents 3 to 30 kHz bandwidth on the electromagnetic spectrum and frequency band generally used for VLF surveys is 15–25 kHz. It is quite a high frequency range for EM surveys. The VLF transmitter is a high tower-fixed transmitting station continuously emitting electromagnetic waves at an assigned frequency. A good number of VLF transmitters are located all over the world transmitting EM waves at different frequencies. Some of the high powered (more than 400 kW) VLF transmitters are NAA, NLK and NPM in the USA operating respectively at 24.0, 24.8 and 21.4 kHz and NWC in Australia presently operating at

19.8 kHz. In central Europe a good number of transmitters can be received (Fischer *et al.*, 1983). The detailed list and worldwide VLF transmitter-wise coverage are available in Wright (1988) and McNeill and Labson (1991). One or several transmitters at different frequencies in this band can be used at a time; however it is basically a single frequency EM method. The selection of distant transmitters depends on the general strike direction of geological structures in the survey area. Like FEM in VLF also the secondary field generated by the conductive target is measured through the receiver coil.

10.2 BASICS

The VLF transmitting tower is taken as a vertical line of current – the vertical electric dipole generates electromagnetic waves at an assigned frequency in the VLF band width. The alternating magnetic field generated is ground parallel i.e., horizontal concentric circles about the vertical line of current (Fig. 10.1). The electric field is vertical

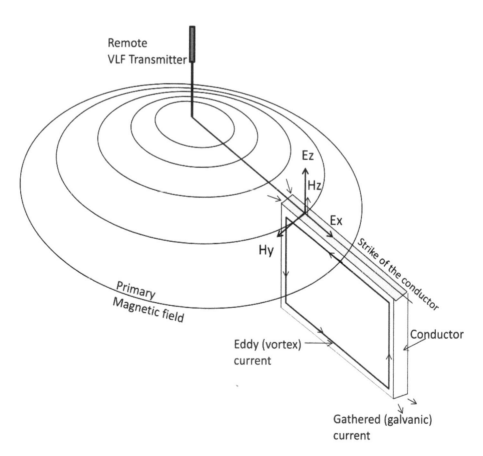

Figure 10.1 VLF EM induction response for a vertical conductor with strike direction in-line with the remote VLF transmitter (after Gurer *et al.*, 2009).

and perpendicular to the magnetic field. The magnetic and electric fields oscillate and the wall of orthogonal magnetic and electric field – the VLF wave front propagates radially outward for thousands of kilometres. Both the fields are perpendicular to the direction of propagation of the wave. At large distances, the radiated field is uniform and considered as an electromagnetic plane wave. Impinging on the ground at a distant point of observation the plane wave with horizontal magnetic component (Hy) and vertical electric component (Ez), interacts with the earth and generates a horizontal electric component (Ex) and a vertical magnetic component (Hz). As indicated by Tabbagh *et al.* (1991) the vertical component of the magnetic field can be observed at any conductivity discontinuity not parallel to the horizontal magnetic field (Hy). In the VLF survey magnetic field components Hy and Hz and electric field component Ex are of interest and are measured.

In the ground, the horizontal primary magnetic field generates an electric field causing an induced current to flow in the conductor which generates a secondary magnetic field. With penetration in the ground the magnetic field attenuates with a phase shift and changes in direction. At the surface the total magnetic field which is the sum of the primary and secondary magnetic fields is measured. The secondary field superimposes the primary field with a phase lag. The effective depth of penetration depends on the resistivity of the ground and the frequency of the electromagnetic field. This is defined through 'skin depth' (δ), the depth at which the amplitude of the wave falls to (1/e) $\sim$0.368 or approximately one thirds of its value at the ground surface. The skin depth is given as

$$\delta = \sqrt{\left(\frac{2}{\sigma\mu\omega}\right)} \approx 503.8\sqrt{\left(\frac{1}{\sigma f}\right)}$$

where σ is conductivity of the subsurface, μ is magnetic permeability and ω ($2\pi f$) is the angular frequency. The depth of penetration decreases with increasing frequency and conductivity. With transmitter frequency in the range of 15 to 30 kHz and conductivity 1×10^{-4} to 3.3×10^{-5} S/cm (resistivity 100 to $\approx$300 ohm.m) for the weathered and fractured zone, the depth of penetration (δ) ranges between 30 and 70 m. Since the wave while returning to surface at the point of measurement further attenuates, the rule of thumb for depth penetration in conductive terrain is half the skin depth (Paterson and Ronka, 1971; Wright, 1988). The variation in maximum depth of penetration with conductivity of the host and/or overburden is shown in figure 10.2 (Wright, 1988). The conductivity controls the attenuation and hence restricts the use of VLF in terrain with conductive overburden. Fraser (1969) restricts the use of VLF up to a minimum resistivity of 200 ohm.m for the overburden.

The plane wave (primary field) penetrating the subsurface of finite conductivity can generate two types of current. In the case where a vertical plate like conductor is present with conductivity much higher than the surrounding host medium, the primary magnetic field crossing the conductor induces a close-looped eddy current within it also known as the 'vortex' current (Wright, 1988, McNeill and Labson, 1991). The vortex current tends to flow in a close loop along the conductor boundary and remains confined within it. Secondly, because of the primary electric field a 'galvanic' or 'gathered' current also flows in the surrounding host medium with finite conductivity. It gathers

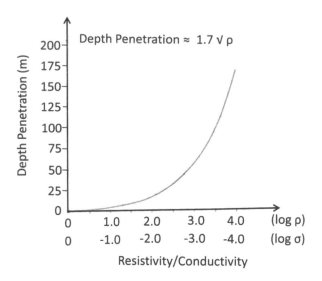

Figure 10.2   Maximum depth of penetration for varied conductivity of host and/or overburden (Source: Wright, 1988).

near the conductor and flows through it and diverges from resistive bodies. It can be said that the vortex current is localized and the galvanic current is regional. The magnetic field component due to the galvanic current decreases linearly with distance from the conductor, while that due to the vortex current decreases with the square of the distance (McNeill and Labson, 1991). The relative strength of these currents depends on the shape of the conductor, its depth extent, orientation, conductivity contrast with the host medium and frequency used. The vortex current component predominates in the case where the conductor is highly conductive as compared to the host. In the general range of conductivity, McNeill (1985) indicates the possibility of a dominating electric field and hence galvanic current at the target and negligible contribution from vortex current. For moderate conductivity contrasts observed with groundwater saturation, the galvanic current component predominates. Gurer *et al.* (2009) indicate that except for very high resistive formations, the vortex current can be ignored and mostly the measured response is due to the galvanic current. The complex combination of these induced currents generates a secondary magnetic field of arbitrary orientation and phase lag with the primary magnetic field. It opposes and creates a perturbation in the primary field and hence causes the anomaly.

At low frequency and/or conductivity the induced current and its secondary magnetic field differs in phase from the primary field by 90° while with the increase in conductivity and frequency the phase lag increases to a maximum of 180°. At the point of measurement the direction of the primary magnetic field is horizontal and the secondary magnetic field makes an angle to the primary field as shown in figure 10.3 taken from Sharma (1997). The total VLF field is the vector sum of primary and secondary fields. A component of the secondary field is 'in-phase' and the other 'out of phase' with the primary field. Since, the secondary magnetic field due to the presence

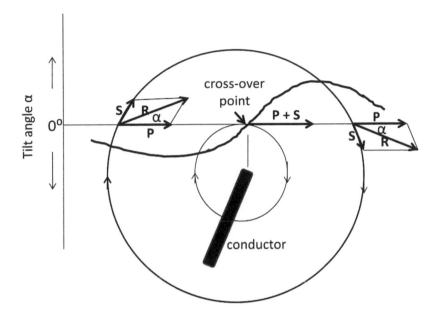

Figure 10.3 The resultant field measured in a VLF profile over a conductor. The tilt angle α from horizontal of the resultant (R) of primary (P) and secondary (S) fields changes sign over the conductor. The location of the conductor is determined from the cross-over point (after Sharma, 1997).

of a subsurface conductive body differs in phase, direction and amplitude from the primary magnetic field, the resultant of these two fields is not obtained directly from vector addition but traces an ellipse in a plane during each time cycle at VLF frequency and the field is called elliptically polarized.

The two fundamental parameters measured from the polarization ellipse are tilt angle α of its major axis from the horizontal and the ratio of its minor axis to the major axis – the ellipticity ε. The theoretical development and related derivations are given in Smith and Ward (1974), Wright (1988), McNeill and Labson (1991) and Gurer *et al.* (2008).

Gurer *et al.* (2008) opine that when the secondary magnetic field components are small compared to the primary magnetic field, real and imaginary components of the secondary magnetic field can be related to tilt angle and ellipticity. The tangent of the tilt angle is approximated with the ratio of the real (inphase) part of the vertical component of the secondary magnetic field H_z to the horizontal total magnetic field H_y (primary field and smaller secondary field). Similarly the ellipticity can be approximated with the imaginary (out of phase or quadrature) part of this ratio. The real and imaginary components are expressed as a percentage of the total primary field. These two parameters indicate the presence of a conductor.

$$Re(H_z/H_y) = \tan \alpha \quad \text{and} \quad Im(H_z/H_y) = \varepsilon$$

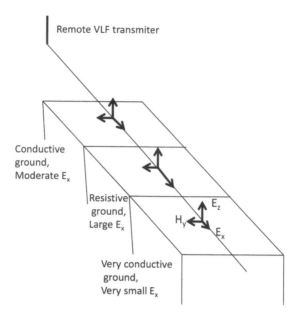

Figure 10.4 Variation in horizontal component of the induced electric field (E_x) with variation in ground resistivity (Source: McNeill and Labson, 1991).

In addition, the variations in the horizontal electric field being associated with conductivity variations (Fig. 10.4, McNeill and Labson, 1991), by measuring the induced horizontal electrical field, the resistivity of the subsurface can be determined. The ratio of the horizontal component of the induced electric field (E_x) to the primary magnetic field is expressed as apparent resistivity $\rho_a = \frac{1}{\omega\mu_o} \left| \frac{E_x}{H_y} \right|^2$ known as Cagniard's equation and the phase angle ϕ between the horizontal electric field and the magnetic field is defined as $\phi = \tan^{-1} \left[\frac{\text{Im}(E_x/H_y)}{\text{Re}(E_x/H_y)} \right]$. The ω and μ_o are angular frequency and magnetic permeability of free space respectively.

10.3 INSTRUMENT

The equipment for conducting a ground VLF survey was first introduced in 1964 by Geonics Ltd. and subsequently a number of firms started manufacturing it (Paterson & Ronka, 1971). The equipment is a light weight, easy to handle receiver. It measures the real (in-phase) and imaginary (out of phase or quadrature) parts of the vertical secondary magnetic field normalized to the horizontal primary magnetic field. Receivers have the facility to measure inphase, quadrature, total field strength, dip angle, primary field direction, apparent resistivity, phase angle etc. Also receivers have the facility to measure all the three components of the magnetic field and do not need any defined orientation with respect to the transmitter (Bernard and Valla, 1991). The receiver can be

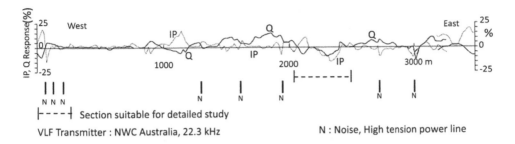

Figure 10.5 A reconnaissance VLF profile from hard rocks, India; the VLF transmitter, NWC Australia at 22.3 kHz was used. The 'noise' due to overhead power transmission line and the section suitable for detailed study is shown (Source: Chandra *et al.*, 1994).

used to measure the fields due to more than one VLF transmitters simultaneously and it need not be oriented in a specific direction. The receiver has provision for resistivity measurement by capacitive electrodes that do not require pegging into the ground.

10.4 FIELD PROCEDURES

The VLF is used as a profiling technique for reconnaissance survey over a long profile as shown in figure 10.5. The field operation is quite fast. The distant transmitter which produces a strong and clear transmission in the survey area is selected. If the instrument being used has the option for selecting 2 or 3 transmitters and from which a strong field is obtained, they are to be selected. The selection of transmitter is such that the strike direction or orientation of suspected fractured zone, lineament or geological formation is in line with the direction of transmitter from the observation point and the primary magnetic field is almost perpendicular to the strike of the target. This will produce a maximum induction response. However, in practice it may be difficult to select a VLF transmitter in-line with the geological strike. Also, it is difficult to ascertain the strike direction of concealed structures. The maximum permissible deviation in strike direction of the target from transmitter orientation could be ±45°.

The profile lines are laid perpendicular to the direction of the transmitting station i.e., tangential to the primary magnetic field. While moving along profile lines, the measurements are made at equally spaced stations, possibly at more than one frequency using different transmitters, if available. The measurement station interval may be 10 m or even less and parallel profiles can be placed at 25 to 100 m separation. By increasing the station interval, shallow, thin features may not be mapped. The profile separation and station interval depend on expected geometry and depth of the target and the nature of the survey – reconnaissance or detailed. Selection of station interval becomes quite significant if anomalies show high rates of curvature as there should be 4 to 5 points on the anomaly. The length of profile is generally kept large, and may be more than a kilometre, if possible, to study the regional trends and estimate the depth extent of the target through filtering discussed later. Generally, the surveys for groundwater exploration are carried out near habitation mostly on flat topography and therefore topographic relief is not considered. The profiles are preferably to be laid on flat topography and it is better to avoid laying the survey profile across topographic relief, such as across elevated hilly terrain or valley depressions. For anomalies on high relief profiles the corrections discussed later are to be applied.

The operational procedure varies with the type of instrument used. If required, orientation of receiver with respect to transmitter is adjusted as given in the manual of the instrument selected. For some instruments, there is no need of keeping the receiver at a specific orientation with respect to the transmitter. In some of the instruments data are direct, digitally displayed, while in others they are recorded by obtaining a minimum intensity of sound signal adjusting the instrument in various positions/orientations. The position of the transmitting station with respect to the movement of operator, i.e., to his right or to his left is to be noted for interpretation of the "cross-overs" of inphase and quadrature components and filtering. Accuracy of data depends on signal-to-noise ratio and selection of transmitter with reference to the orientation of target. The noise spikes on a VLF profile (Fig. 10.5) can be identified and removed. In field operation, repeatability of readings is ensured. With instruments for which a minimum sound is observed, accuracy may vary with operator. By VLF-R measurements resistivity of near surface layer is obtained to get lateral variation in resistivity along profiles. The resistivity measurement is done using two sensors placed on the ground about 5 to 10 m apart, acting as short electric dipoles, at each station along the profile. The sensors are generally capacitive type having no contact with the ground.

Conventionally the transmitter is selected along the geological strike direction and it will facilitate E-polarization mode measurements while a transmitter located perpendicular to the strike direction will facilitate H-polarization mode measurements. In E-polarization mode the wave propagates parallel to the strike of target and the horizontal electric field in the ground is parallel to the target strike and primary magnetic field cuts the target conductor orthogonally and there will be a vortex current within it. In H-polarization mode the wave propagates in a direction perpendicular to the strike of the target and the magnetic field is parallel to the target length and the electric field across the conductor (McNeill and Labson, 1991). Gurer *et al.* (2009) explain that in E-polarization mode the response is due to the gathered (galvanic) current and minor vortex current and in H-polarization mode response is entirely due to the gathered current and hence the VLF-EM anomaly is strong in E-polarization and VLF-R (resistivity) anomaly is strong in H-polarization. In regions where a number of VLF transmitters are available measurements can be made by selecting transmitters along the general strike and dip directions of the subsurface structures to be investigated as it may not be possible to accurately identify the strike direction of the target – the fractured zone. The choice of transmitter is quite often dependent on the signal strength. Where the option is possible the transmitter location controls the survey line direction.

10.5   PROCESSING OF DATA

In-phase and quadrature components are graphically plotted with appropriate scale for the profiles and noise removed. For plotting of data Wright (1988) suggests the scale of horizontal axis for distance to be 2 or 3 times the vertical axis for inphase

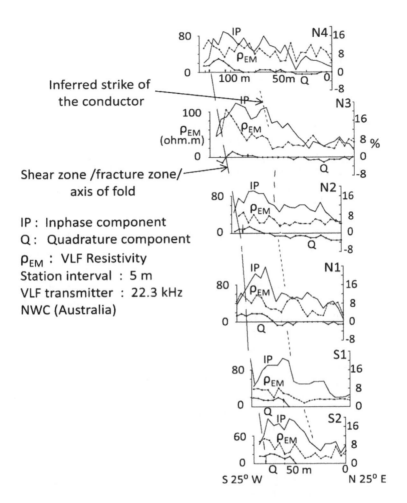

Figure 10.6 VLF parallel-stacked profiles from quartzites and phyllites having a general strike in N 65°W. The profile interval is 10 m (Source: Chandra et al., 1994).

and quadrature values. The data can also be presented as stacked profiles plotted on a map at appropriate locations as shown in figure 10.6. Contour maps can be prepared for in-phase and quadrature components and apparent resistivity with data on parallel profiles. Besides, maps can be prepared for the filtered data using different filtering techniques described latter.

10.6 INTERPRETATION

The method has directional advantages as well as limitations. The fractured zones oriented approximately "in-line" with transmitter are picked up better. A VLF profile apparently presents a plethora of anomalies. Experience plays a big role in removing the noise and interpreting the desired anomalies. The method being generally used for reconnaissance surveys in groundwater exploration the anomalies are mostly interpreted qualitatively. The 'cross-over' point on an anomaly, generally considered as the location of the conductive target – the saturated fractured zone, is required in locating water well drilling sites in hard rocks. The lateral extents of a fractured zone can be demarcated by joining the appropriate cross-over points on stacked parallel profiles. The VLF response characterizing the geometry and conductivity of the target is affected by a number of variables, viz., geological strike of the target with respect to the line joining transmitter and receiver, strike-length, conductance (product of conductivity and thickness) of the target, depth and dip of the target, presence of any thick conductive overburden, conductivity of host rock, near surface inhomogeneities, frequency of the VLF transmitter and topographic relief along the measurement profile. Also, the response is susceptible to cultural noise and may get masked by it. These are considered in quantitative interpretation and drawing inferences.

The high frequency VLF field attenuates quickly with depth and therefore it is effective in detecting fractured zones which are shallow. The use of higher frequencies renders better resolving power in detecting thin closely spaced conductive fractured zones, but higher frequencies make it highly susceptible to variations in overburden characteristics. Saydam (1981) reveals factors controlling the magnitude of response from a vertical thin conductor. They are a) resistivity of host medium, b) presence of conductive overburden, c) depth to the top of conductor and d) conductance (thickness-conductivity product) of the conductor. The magnitude of inphase and quadrature response increases with host rock resistivity and the rate of increase is more for inphase compared to quadrature. The response attenuates fast in conductive surroundings and also the skin-depth reduces. The maximum depth of penetration being half the skin-depth the target detection is affected in conductive medium. For example, at 20k Hz frequency with 25 ohm.m resistivity the depth of investigation would be hardly within 10 to 15 m. In a resistive medium, on the basis of in-phase and quadrature components, it is possible to discriminate between poor and good conductors as the ratio of inphase to quadrature response is suggestive of the conductivity of the target. According to Kaikkonen (1979) a conductor is defined as 'good' when the ratio of conductivity of the conductor to that of the host medium is more than 100. Kaikkonen (1979, 1980) based on numerical modeling indicates that in-phase and quadrature components have the same polarity for a poor conductor. The shape of the anomaly over a conductor (vertical dyke or fractured zone) in a resistive host medium without any conductive overburden is shown in figure 10.7 (Wright, 1988).

As compared to FEM, the presence of conductive overburden affects the VLF response severely. It attenuates the response faster and the technique becomes ineffective in detecting the underlying fractured zones. The effect of conductive overburden depends on its conductivity contrast with the host. With conductive overburden inphase response is attenuated while quadrature response may be attenuated or enhanced which depends on the overburden conductivity contrast and depth of conductor (Saydam, 1981). According to Wright (1988) for a good conductor the conductive overburden enhances the amplitude of quadrature anomaly and distorts its shape and the quadrature response may get totally reversed (Fig. 10.8). The effect of overburden depends on its thickness also (Vozoff, 1971) and anomalies will be mainly due to variations in conductivity and thickness of the overburden – the weathered zone in hard

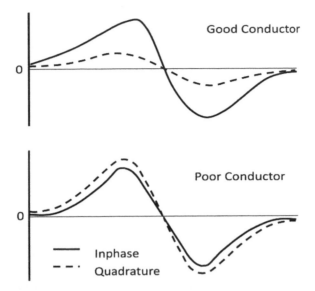

Figure 10.7 The VLF response of a conductor in highly resistive terrain. The ratio of inphase to quadrature response is indicative of the conductivity of conductor (Source: Wright, 1988).

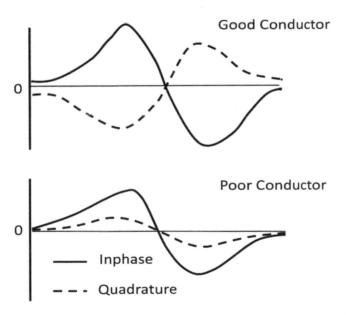

Figure 10.8 The general VLF response of a conductor in resistive terrain with conductive overburden. The conductive overburden may cause reversal of polarity of the quadrature component (Source: Wright, 1988).

rock. According to Poddar and Rathor (1983), the thickness of the weathered zone varies slowly and therefore most of the anomalies are generated by rapid variations in its conductivity alone and can be mistaken for anomalies due to underlying conductive fractured zones. The VLF method can be used effectively for weathered zone characterization and mapping. The resistivity of the weathered zone can be estimated quickly. They further point out that in basaltic terrain the presence of a thin highly conductive layer at the base of the weathered zone of maximum thickness 25 m, totally obscures the information from deeper layers, because of the high frequencies used.

Both the responses attenuate with increasing depth of conductor and the rate of attenuation is more for a conductive host medium. With depth inphase response attenuates more than quadrature response (McNeill and Labson, 1991) and therefore the slope of inphase curve at the inflection point is also indicative of depth to the conductor (Wright, 1988). A steep inphase curve indicates the presence of a comparatively shallow conductor. Qualitatively the depth to the top of a steeply dipping conductor of large strike length and depth extent can be approximated as equal to half the distance between the main positive and negative peaks of the inphase anomaly. In the case where strike length and depth extent of conductor are limited, its depth equals the peak to peak distance (Wright, 1988). Since the inphase negative peak on the down-dip side of the steeply dipping conductor is poorly defined, Olsson (1980) proposes a graphical approach to compute the depth from a plot of the average of negative and positive peak amplitudes versus distance between the positive peak and cross-over point for different dips and overburden resistivities. Saydam (1981) shows that responses increase with an increase in conductance of the target

As a part of a qualitative interpretation, the cross-overs of inphase and quadrature curves in VLF profiles are located. The conductor is located below the cross-over or the inflection point where the tilt angle α (inphase component curve) changes sign (Fig. 10.3). The ellipticity ε (quadrature component curve) also changes sign as the profile passer over the conductor. When the target dips, as expected, the anomaly is not symmetrical. The effect of dip is better reflected in the in-phase curve with distinct larger anomaly amplitude on down dip side. The dip of the body can be estimated from the amplitude of negative and positive peaks of the prominent inphase anomaly for shallow, high conductivity targets. Vozoff and Madden (1971), Olsson (1980) and Wright (1988) present anomaly diagrams over a range of sheet model and readers may consult these models and anomaly shapes to interpret VLF data.

As such, from visual inspection of the inphase anomaly the following qualitative assessment can be made: location of conductive fractured zone directly below cross-over point of inphase curve, its dip from the asymmetric nature of anomaly and its depth from the gradient of the anomaly curve at inflection or cross-over point. A steep gradient at the inflection point with appreciable peak amplitude indicates a near surface conductor. Where multiple conductors are present locating cross-over points may be difficult and the centres of steepest gradient parts of the anomalies are the locations of conductors (Geonics, 1979).

The VLF anomalies are affected by topographic relief. A false anomaly can be produced by the sloping relief itself and may be mistaken for a conductor or it may superimpose on the conductor anomaly. In the case where a long profile crosses a hilly area or valley with appreciable topographic relief, corrections are to be applied to the data to remove the effect of topographic relief so that the anomaly due to

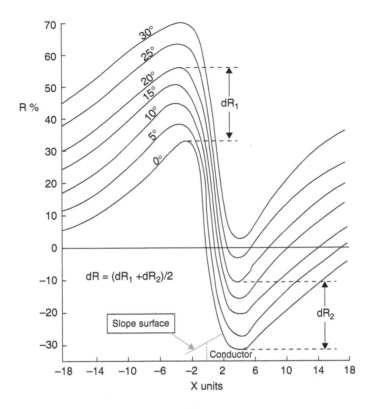

Figure 10.9 Effect of sloping (5° to 30°) traverse surface on VLF inphase (R) anomaly over a vertical conductor (Source: Abdul-Malik *et al.*, 1985).

the conductor alone is obtained. The effect of topographic relief is a function of skin depth, so to apply correction, terrain resistivity should be known (McNeill and Labson, 1991). In the case where the skin depth is much higher compared to the relief, i.e., the terrain is resistive, the effect of relief on the anomaly is, in general, less except for the conductive patches. Abdul-Malik *et al.* (1985) indicate that when the measurement profile and magnetic field are at right angles to the strike of a vertical conductor, i.e., in E-polarization mode, the shape of inphase anomaly does not change much with slope of topography (Fig. 10.9). However for a dipping conductor topographic slope affects the conductor anomaly considerably and dominates when the dip is small. For correcting the inphase data, Baker and Myers (1980) prepare a simple graph (Fig. 10.10) between surface slope angle and dR, where dR is the difference between the maximum on the measured R% curve for a given slope and the maximum on the measured R% curve when the surface is horizontal. The average value of dR for a particular surface slope is added to or subtracted from the observed anomaly. Abdul-Malik *et al.* (1985) further demonstrate by model experiments that Baker and Myers method is not effective for slopes more than 15° and could be generalised by considering the angles between the strike of the structure and the measurement profile line and the magnetic field

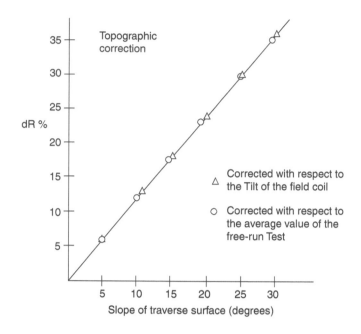

Figure 10.10 Relationship between the angle of traverse surface slope and dR as explained in figure 10.9 (Source: Baker and Myers, 1980).

as proposed by Eberle (1981). For a general qualitative but practical consideration the topographic heights should be plotted along the measurement line as suggested by Whittles (1969) referred in McNeill and Labson (1991) and one should check the normal polarity anomaly close to topographic peak and reverse polarity anomaly near the valley bottom.

The semi-quantitative interpretation of VLF data can be attempted using filters developed by Fraser (1969) and Karous-Hjelt (1983). Fraser gave a unique approach of interpretation through data presentation by a simple one-dimensional filtering of the raw in-phase data which shows cross-over response in profiles while traversing vertical or steeply dipping conductors. It uses the difference between successive values of in-phase data and computes the horizontal first derivative. The filter simplifies the interpretation by amplifying and improving the resolution of the local anomalies. It converts the inflections and cross-overs into 'peaks and troughs' by a 90° phase shift and reduces sharp noise. To apply the Fraser filter the data are to be at a regular interval of 10 to 15 m collected for a single transmitter frequency. The filtering is quite easy and can be done manually during field operation by tabulating the in-phase data in a specified format with proper sign (−/+) and adopting the simple procedure given by Frazer (1969). The positive values obtained from filtering, actually representing the cross-overs, are contoured to get the trend of the conductive body.

Karous and Hjelt (1983) use linear filter theory to calculate equivalent causative current source at depth for the measured magnetic field and produce subsurface current distribution to identify conductor geometry. This is quite useful. Before using this

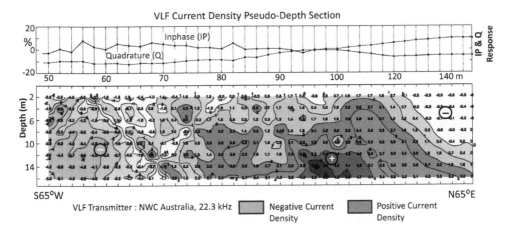

Figure 10.11 VLF current density pseudo-depth section obtained through Karous-Hjelt filtering of inphase anomaly (Blue: positive current density, red: negative current density).

filter technique, the inphase data is smoothened. The current density pseudo-depth section can be obtained by filtering the in-phase data either manually or using available software. The derived current-density pseudo-section manifests the apparent current concentrations at different depths (Fig. 10.11). The current density maxima lie within the conductor. So, the contoured pseudo-section approximates the spatial disposition of the conductor in the X-Z plane. It simply approximates the disposition and does not present the true current density distribution. The values of the current density do not indicate conductivity except that the higher positive current density values indicate better conductivity. The trend of positive current density contours indicates approximately the dip and geometric trend of the conductor. Since, the available filter expansion depends on the length of profile, i.e., the number of data points, which in turn indicates the depth of investigation, larger profiles are required for deeper information. Ogilvy and Lee (1991) use the Karous-Hjelt filter for computing theoretical current density pseudo-depth sections for a variety of conductive bodies and evaluated the performance of the filter. According to them, it can provide diagnostic information. For a vertical to sub-vertical plate conductor of limited cross-sectional area in a homogeneous resistive host medium the contouring of current concentrations can reliably approximate its geometry and dip. The presence of multiple conductors at different depths distorts the pattern by overlapping of anomalies and deteriorates the vertical resolution. However, the target locations can be identified. The presence of a conductive overburden reverses the inphase response and in such cases the pseudo-section is to be used cautiously. The current density maxima for a horizontal slab conductor may not occur within it and lead to erroneous depth estimation. McNeill and Labson (1991) indicate that filtering can slightly displace the anomaly peak along the profile line and also it alters the anomaly shape. Further, they point out that the amplitude of the filtered anomaly is relatively insensitive to the conductivity-thickness product of a vertical conductor. All these indicate that though Karous-Hjelt filtering provides a basic useful interpretational technique and the insight, it is necessary that the raw VLF data are also analysed.

Djeddi *et al.* (1998) develop a linear filtering technique to interpret anomalies due to 3-dimensional bodies. They also consider the real (in-phase) data and show that the dimension of the body can be determined even without considering its conductivity. Djeddi *et al.* indicate that the width of a target can be determined when it is buried at a depth less than or equal to its width. Djeddi *et al.* (1998) evaluate the applicability of the Fraser filter in a quantitative interpretation of 3-d bodies and indicate its effective use in estimating depth, width and extension of target.

To measure near surface resistivity variations through VLF-R measurements, the induced horizontal electric field component perpendicular to the primary magnetic field is measured which gives a surface impedance value to determine the apparent resistivities and phase. If the ground is homogeneous within the depth of investigation, apparent resistivity will be equal to true resistivity of the layer. For a layered subsurface, measurements yield apparent resistivity. In VLF-R resistivity measurements two parameters are obtained at each observation point. They are apparent resistivity, obtained through Cagniard's equation and the phase shift. This data set of resistivity and phase shift for a single frequency cannot yield information on a layered earth. Qualitatively a phase value more than 45 degrees indicates a conductive substratum and less than 45 degrees indicates a resistive substratum. Poddar and Rathor (1983) and Hjelt *et al.* (1985) present nomograms for quantitative interpretation by transforming these data into two-layered earth models at each point of observation, i.e., first layer or overburden resistivity (ρ_1), second layer or bed rock resistivity (ρ_2) and overburden thickness (h_1). It is assumed that horizontal variations in resistivity are small and smooth. For deriving a 2-layer model one of these three parameters is approximated which can be through a few resistivity soundings. It can be either first or second layer resistivity. In this way a two-layer resistivity cross-section presenting variations in resistivity and thickness of the overburden or weathered zone can be obtained. Instead of making VLF-R measurements by capacitive electrodes the apparent resistivity can be derived from VLF-EM measurements by using a filter developed by Chouteau *et al.* (1996). This is preferred in areas with fairly contrasting resistivity where moderate to strong variations in the vertical magnetic field are observed. Further, they suggest VLF-R measurement for resistive environment where VLF-EM measurement is not sensitive to small changes in resistivity and observed normalized vertical magnetic field is very small resulting in a noisy profile.

The thin vertical conductors – the fractured zones at depth in a resistive host medium can be located by carrying out surveys using two transmitters, one in the strike direction of the conductor and other perpendicular to it (Fischer *et al.*, 1983). This is possible when transmitters radiate a field at frequencies which are close to each other and can be considered as a single frequency. Like direct current resistivity survey, the apparent resistivities obtained in two orthogonal directions through these transmissions are different. The apparent resistivity obtained in E-polarization mode, when the along-strike transmitter is considered, is much higher than that obtained in H-polarization mode with transmitter perpendicular to the strike or survey profile. With these, the plot of apparent resistivity polar diagram at each station along the survey line reveals the apparent resistivity anisotropy variations. The maximum variation will be at stations over the conductor. Hence, the conductor can be located from the polar diagram.

10.7   CASE STUDIES

There are several references on VLF surveys for groundwater exploration in hard rocks of India, Africa and south America (Palacky *et al.*, 1981; Chandra *et al.*,1994; Sharma and Baranwal, 2005; Bernard and Valla,1991; Castelo Branco *et al.*, 2004, Lawrence and Ojo, 2012). Relevant application in delineating the gelogical formation contacts was attempted by Aina and Emofurieta (1991) in Nigeria. The main issues in hard rock are weathered zone mapping and fractured zone detection. The VLF technique can be used effectively in weathered zone mapping (Poddar and Rathor,1983). Secondly, the groundwater targets such as fractured zones, which can be considered as conductors but less conductive, can be detected, if they are reasonably long along the primary electric field, i.e., in-line with the VLF transmitter, because of the large magnetic field due to galvanic current flow (McNeill, 1985). It reduces as the fractured zone deviates from the electric field. The response also depends on the resistivity ratio of the host medium to that of the fractured zone and the resistivity and thickness of the overlying weathered material.

Reconnaissance VLF surveys were carried out in granites, volcanics and metasediments in the eastern part of India (Chandra *et al.*, 1994) using 22.3 kHz remote VLF transmitter at NWC, Australia. These were followed by detailed resistivity and controlled source frequency domain electromagnetic surveys. Some of the VLF results from the State of Jharkhand, India are discussed below. The results reveal that the VLF survey is useful for reconnaissance and delineation of shallow conductors and support from other geophysical methods may be required for confirmation.

10.7.1   Granitic terrain

A VLF profile at a site in granitic terrain was conducted over a stretch of about 200 m in an E-W direction (Fig. 10.12). It indicates thickening of the weathered column and the presence of dipping fractures underneath. A borehole was drilled on a positive peak obtained from Fraser filtering of the IP anomaly. It encountered 6 fractures at depths of 32.11, 35.11, 58.97, 63.59, 107.31 and 109.31 m. The cumulative discharge from a 2 m thick saprolite at 17 m depth and the fractures was 5.93 lps during drilling. Another borehole drilled 30 m away from the first one had a discharge of 1.99 lps only.

The VLF profile at another site in granitic terrain is shown in figure 10.13. It shows 2 cross-overs. The shape of the VLF anomaly and the central cross-over indicate the presence of a deeper resistive body. The filtered VLF pseudo-section reveals the presence of deeper conductors towards the SW end of the profile and at a point 200 NE of it separated by a resistive block. The reversal in inphase-quadrature polarity at point a is due to the presence of conductive overburden. With the support of gradient resistivity and FEM profiling two boreholes a and c and a test borehole b were drilled. The borehole a had a cumulative discharge of 6.4 lps from saprolite in the depth range of 12.61 to 25.61 m and fractured zone at 176.94 to 178.94 m respectively. The borehole c had a cumulative discharge of 5.6 lps from saprolite and fractured zone in the depth range of 14–23 m and 33.32 to 35.47 m respectively.

A VLF profile at a site in dyke infested granitic terrain of Jharkhand, India was conducted in N21°E orientation. It indicates phase rotation and enhancement of anomaly

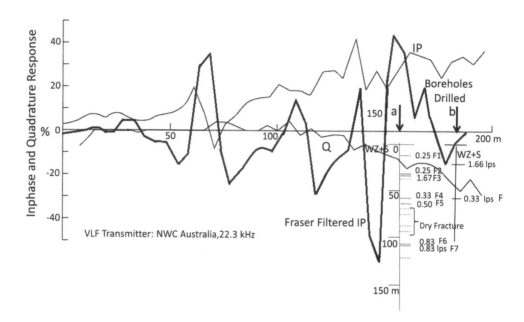

Figure 10.12   Fraser filtered VLF profile at site 1 in granitic terrain. Borehole 'a' was drilled on a Frazer filtered positive inphase anomaly and borehole 'b' was drilled 30 m away. Cumulative discharge from borehole 'a' was 5.93 lps and that from 'b' was 1.99 lps only (Source: Chandra et al., 1994).

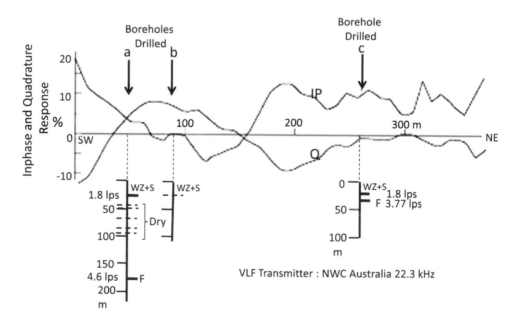

Figure 10.13   VLF profile at site 2 in granitic terrain (Source: Chandra et al., 1994).

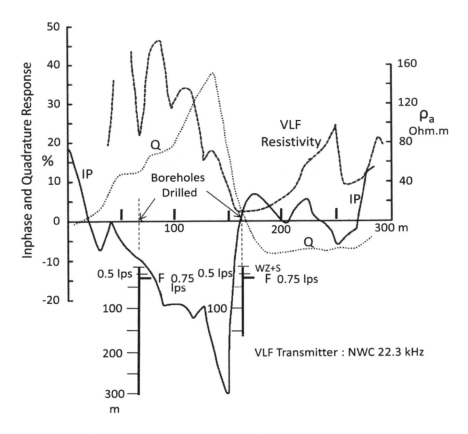

Figure 10.14 The VLF profile in dyke infested granitic terrain showing the response of shallow fractures within 24 m depth encountered in the borehole and anomaly enhancement due to conductive overburden (Source: Chandra *et al.*, 1994).

due to the presence of a conductive (10 ohm.m) overburden, revealed by VLF-R measurements. The skin depth would be about 10–20 m. A NE dipping shallow conductor was expected (Fig. 10.14). The contact of the quartz vein with granitic country rock forms the conductor. Two boreholes were drilled up to 300 and 160.82 m depth of which one was located at the IP/Q cross-over. The boreholes encountered almost identical subsurface conditions, a weathered zone up to 15 m depth followed by saprolite in the depth range of 15–20 m and a fracture in the depth range of 23 to 24 m. The discharges from the saprolite and fractured zones in the both the wells were 0.5 and 0.75 lps respectively.

10.7.2 Metasediments

A 3.5 km long reconnoitory VLF profile was conducted in metasediments along a road in NE-SW direction as shown in figure 10.5. A prominent anomaly was observed near its eastern end. Due to inaccessibility the VLF anomaly was not taken up for

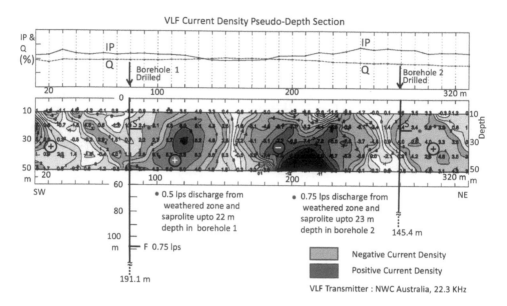

Figure 10.15 The Karous-Hjelt filter based VLF current density pseudo-section from metasediments, Jharkhand, India (Source: Chandra *et al.*, 1994).

further ground follow-up. Subsequently, the eastern end was traversed by another profile in NE-SW direction. It is almost in the strike direction to traverse across a lineament. The VLF current density pseudo-section (Fig. 10.15) reveals the presence of conductors with positive current density (blue coloured) towards the SW and NE parts of the profile separated by a resistive block of negative current density (red coloured). The NE conductor appears to be dipping in the NE while in the SW there are two conductors intersecting. The borehole drilled on SW anomaly encountered weathered zone down to 18 m, saprolite between 18 and 22 m (0.5 lps) and a productive deep fracture in the depth range of 107 to 109 m (0.75 lps) in the compact rock. A borehole could not be drilled on the NE conductor. However, a borehole was drilled at a site 30 m SW of the NE high current density contours (conductor). It encountered a weathered zone down to 15 m and saprolite between 15 and 23 m depth with 0.75 lps discharge followed by compact rock. The borehole missed the fractured zones.

10.7.3 Basic dyke and quartz reef

The results of VLF surveys across a basic dyke in granitic terrain exposed over 100 m in the NNE-SSW, i.e. almost parallel to the primary magnetic field generated by VLF transmitter NWC Australia are presented here. Three parallel VLF profiles, one towards the north and other two towards the south of the dyke were placed in N33° W-S 33°E direction crossing the ends of the dyke exposure (Fig. 10.16). The negative inphase anomaly towards the east of the dyke is well developed and the quadrature anomaly is positive and small indicating the presence of a good conductor in a resistive

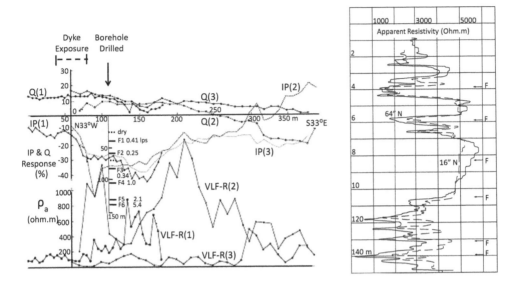

Figure 10.16 The VLF profiles across a basic dyke in granitic terrain, Jharkhand India and the borehole drilled on the IP anomaly in proximity with the dyke. Fractured zones encountered in the borehole and the resistivity log are also shown (Source: Chandra *et al.*, 1994).

host rock. For the survey the transmitter NWC, Australia was used, i.e., the transmitter is in the SE direction. The dyke being in the NNE-SSW direction, the response is mainly due to the "gathered current" and a strong resistivity anomaly because of the electric field variations is seen on VLF profile 2 which is closest to the structure (Gurer *et al.*, 2009). The highly resistive host rock is revealed by the resistivity log of the borehole. The VLF-R resistivity over the anomaly is more than 300 ohm.m and the skin depth is about 30 m. The filtered VLF pseudo-section along the profiles also reveals the presence of a conductive body immediate southeast of the dyke. However, the conductor dips towards the NW, i.e., towards the dyke. The weathered zone and saprolite being dry as revealed by the borehole drilled into the fractured zone underlying it was picked up. The borehole drilled over the IP anomaly towards the northern end of the dyke encountered productive fractures at depths of 39, 59, 83, 105, 131 and 140 m. Out of these the bottom 3 fractures were high yielding (between 1 and 5.4 lps). The VLF response obtained is only from the first occurrence of the conductor at 39 m depth.

Figure 10.17 presents the results of VLF surveys across a quartz reef near the above mentioned dyke. The quartz reef trends N43°W, i.e., almost perpendicular to the dyke and could be taken as in line with the NWC transmitter. Two VLF profiles were placed along the NE-SW across the quartz reef. The VLF profiles indicate cross-overs of the inphase and quadrature anomalies over the quartz reef. The depth to the conductor was estimated at around 28 m and resistivity of the overburden at around 500 ohm.m and therefore the skin-depth was taken to be about 40 m. The shape of the anomaly indicates the presence of a southwest gently dipping conductor overlain by a conductive

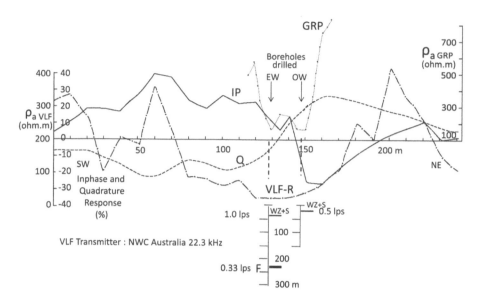

Figure 10.17 The VLF profile across quartz reef in granitic terrain along with borehole results (Chandra *et al.*, 1994).

overburden. Keeping in view of the cross-overs and the lows on the gradient resistivity profile two boreholes (EW and OW) were drilled on either side of the quartz reef. The EW drilled on southern side of the quartz reef encountered highly productive fractured quartz veins with a cumulative yield about 1.5 lps of which maximum contribution (1 lps) was from the saprolite zone up to 40 m depth.

REFERENCES

Abdul-Malik, M.M., Myers, J.O. & McFarlane, J. (1985) Model studies of topographic noise in VLF-EM data: Accounting for the direction of morphological strike relative to survey line and magnetic field directions. *Geoexploration*, 23 (2), 217–225.

Aina, A. & Emofurieta, W.O. (1991) VLF anomalies at contacts between Precambrian rocks in southwestern Nigeria. *Geoexploration*, 28 (1), 55–65.

Baker, H.A. & Myers, J.O. (1980) A topographic correction for VLF-EM profiles based on model studies. *Geoexploration*, 18 (2), 135–144.

Bayrak, M. (2002) Exploration of chrome ore in Southwestern Turkey by VLF-EM. *Journal of Balkan Geophysical Society*, 5 (2), 35–46.

Benson, A.K., Payne, K.L. & Stubben, M.A. (1997) Mapping groundwater contamination using DC resistivity and VLF geophysical methods-a case study. *Geophysics*, 62 (1), 80–86.

Bernard, J. & Valla, P. (1991) Groundwater exploration in fissured media with electrical and VLF Methods. *Geoexploration*, 27 (1–2), 81–91.

Castelo Branco, R.M.G., Cunha, L.S., Pineo, T.R.G. & de Castro, D.L. (2004) A combined FEM, VLF and Dipole-Dipole electrical imaging for ground water exploration in Irauçuba-Fumo region, Ceará State, NE Brazil. In: *Proceedings of the XXXIII Congress of IAH – Conference on Groundwater Flow Understanding: From Local to Regional Scale, Joint Conference IAH/ALHSUD, 11–15 October, 2004 Zacatecas, Mexico*.

Chandra, P.C., Reddy, P.H.P. & Singh, S.C. (1994) *Geophysical studies for groundwater exploration in Kasai and Subarnarekha River basins* (UNDP Project) Central Ground Water Board Tech. Report, Min. of Water Resources, Govt. of India

Chouteau, M., Zhang, P. & Chapellier, D. (1996) Computation of apparent resistivity profiles from VLF-EM data using linear filtering. *Geophysical Prospecting*, 44 (2), 215–232.

Djeddi, M., Baker, H.A. & Tabbagh, A. (1998) Interpretation of VLF-EM anomalies of 3D structures using linear filtering techniques. *Annali di Geofisica*, 41 (2), 151–163.

Eberle, D. (1981) A method of reducing terrain relief effects from VLF-EM data. *Geoexploration*, 19 (2), 103–114.

Fischer, G., Le Quang, B.V. & Muller, I. (1983) VLF ground surveys, a powerful tool for the study of shallow two-dimensional structures. *Geophysical Prospecting*, 31 (6), 977–991.

Fraser, D.C. (1969) Contouring of VLF-EM data. *Geophysics*, 34 (6), 958–967.

Geonics (1979) *Operating manual for EM 16 VLF-EM*. Geonics Ltd., Ontario, Canada.

Gomes, A. & Chambel, A. (2000) VLF method applied to hydrogeological prospecting in the gneissic-migmatitic hard rock aquifer of Evora, South Portugal. In: Sililo, O. *et al.* (eds.) Proceedings *of the XXX IAH Congress on Groundwater: Past Achievements and Future Challenges, 26 November-1 December, 2000, Cape Town, South Africa*. Rotterdam, The Netherlands, A.A. Balkema. pp. 153–157.

Gurer, A., Bayrak, M. & Gurer, O.F. (2009) A VLF survey using current gathering phenomena for tracing buried faults of Fethiye Burder Fault Zone, Turkey. *Journal of Applied Geophysics*, 68 (3), 437–447.

Gurer, A., Bayrak, M., Gurer, O.F. & Sahin, S.Y. (2008) Delineation of weathering in the Catalca granite quarry with the very low frequency (VLF) electromagnetic method. *Pure & Applied Geophysics* 165 (2), 429–441.

Hjelt, S.E., Kaikkonen, P. & Pietila, R. (1985) On the interpretation of VLF resistivity measurements. *Geoexploration*, 23 (2), 171–181.

Hutchinson, P.J. & Barta, L.S. (2002) VLF surveying to delineate longwall mine-induced fractures. *The Leading Edge*, 21 (5), 491–493.

Jeng, Y., Lin, M.-J. & Chen, C.-S. (2004) A very low frequency-electromagnetic study of the geo-environmental hazardous area in Taiwan. *Environmental Geology*, 46 (6–7), 784–795.

Kaikkonen, P. (1979) Numerical VLF modelling. *Geophysical Prospecting*, 27 (4), 815–834.

Kaikkonen, P. (1980) *Interpretation nomograms for VLF measurements*. Acta Universitatis Ouluensis, Series A, Scientiae Rerum Naturalium No.92, Physica No. 17 Univ. of Oulu.

Karous, M. & Hjelt, S.E. (1983) Linear filtering of VLF dip-angle measurements. *Geophysical Prospecting*, 31 (5), 782–794.

Lawrence, A.O. & Ojo, T.A. (2012) The use of combined geophysical survey methods for groundwater prospecting in a typical basement complex terrain: Case study of Ado-Ekiti, Southwest Nigeria. *Research Journal of Engineering & Applied Sciences*, 1 (6), 362–376.

McNeill, J.D. (1985) *The galvanic current component in electromagnetic surveys*. Technical Note TN 17, Geonics Ltd., Canada.

McNeill, J.D. & Labson, V.F. (1991) Geological mapping using VLF radio fields. In: Nabighian, M.N. (ed.) *Electromagnetic Methods in Applied Geophysics, Vol.2: Applications, Part B*, Chapter 7, Tulsa, Oklahoma, U.S.A., Society of Exploration Geophysicists. pp. 521–640.

McNew, E.R. & Arav, S. (1995) Surface geophysical surveys of the freshwater-saltwater interface in a coastal area of Long Island, New York. *Ground Water*, 33 (4), 615–626.

Ogilvy, R.D. & Lee, A.C. (1991) Interpretation of VLF-EM in-phase data using current density pseudosections. *Geophysical Prospecting*, 39 (4), 567–580.

Okonkwo, A.C. & Ezeh, C.C. (2012) A ground integrated geophysical exploration for sulphide ore deposits, case study: EPL A40 Mine Field, Lower Benue Trough, Nigeria. *International Research Journal of Geology & Mining*, 2 (8), 214–221.

Olsson, O. (1980) VLF anomalies from a perfectly conducting half plane below an overburden. *Geophysical Prospecting*, 28 (3), 415–434.

Paal, G. (1965) Ore prospecting based on VLF-radio signals. *Geoexploration*, 3 (3), 139–147.

Paal, G. (1968) Very low frequency measurements in northern Sewden. *Geoexploration*, 6 (3), 141–149.

Palacky, G.J., Ritsema, I.L. & De Jong, S.J. (1981) Electromagnetic prospecting for groundwater in Precambrian terrains in the Republic of Upper Volta *Geophysical Prospecting*, 29 (6), 932–955

Paterson, N.R. & Ronka, V. (1971) Five years of surveying with the Very Low Frequency Electromagnetic method. *Geoexploration*, 9 (1), 726.

Phillips, W.J. & Richards, W.E. (1975) A study of the effectiveness of the VLF method for the location of narrow-mineralized fault zones. *Geoexploration*, 13 (1–4), 215–226.

Poddar, M. & Rathor, B.S. (1983) VLF survey of the weathered layer in southern India. *Geophysical Prospecting*, 31 (3), 524–537.

Roberts, R.L., Hinze, W.J. & Leap, D.I. (1989) A multi-technique geophysical approach to landfill investigations. In: *Proceedings of the Third National Outdoor Action Conference on Aquifer Restoration, Ground Water Monitoring & Geophysical Methods, 22–25 May 1989, Orlando, Florida*. Dublin, Ohio, National Water Well Association, pp. 797–811.

Saydam, A.S. (1981) Very low-frequency electromagnetic interpretation using tilt angle and ellipticity measurements. *Geophysics*, 46 (11), 1594–1605.

Sharma, P.V. (1997) *Environmental and Engineering Geophysics*. Cambridge University Press.

Sharma, S.P. & Baranwal, V.C. (2005) Delineation of groundwater-bearing fracture zones in a hardrock area integrating Very Low Frequency electromagnetic and resistivity data. *Journal of Applied Geophysics*, 57 (2), 155–166

Shivaji, A. & Gnaneshwar, P. (1999) *VLF-EM and in situ conductivity measurements in Schirmacher range, East Antarctica*. Fifteenth Indian Expedition, Antarctica, Scientific Report, Dept. Ocean Dev. Tech. Publ. No. 13, pp. 227–240.

Smith B.D. & Ward, S.H. (1974) On the computation of polarization ellipse parameters. *Geophysics*, 39 (6), 867–869.

Stewart, M. & Bretnall, R. (1986) Interpretation of VLF resistivity data for groundwater contamination surveys. *Groundwater Monitoring and Remediation*, 6 (1), 71–75.

Tabbagh, A., Benderitter, Y., Andrieux, P., Decriaud, J.P. & Guerin, R. (1991) VLF resistivity mapping and verticalization of the electric field. *Geophysical Prospecting*, 39 (8), 1083–1097.

Vozoff, K. (1971) The effect of overburden on vertical component anomalies in AFMAG and VLF exploration: A computer model study. *Geophysics*, 36 (1), 53–57.

Vozoff, K. & Madden, T.R. (1971) *Selected plots from the VLF model suites*. 17, Winthrop Road, Lexington, Massachusetts, USA.

Whittles, A.B.L. (1969) *Prospecting with radio frequency EM-16 in mountainous regions*. Western Miner., pp. 51–56.

Wright, J.L. (1988) *VLF interpretation manual*. V4379/A/d19.01.88R0/M94, EDA Instruments, Canada.

Chapter 11

The time domain electromagnetic method

11.1 INTRODUCTION

The Time-Domain Electromagnetic (TDEM) or Transient Electromagnetic (TEM) method, like frequency domain (FEM), is a controlled source EM induction method. It has been widely used in metal ore prospecting and to some extent for deeper explorations over the past several decades. With advancement in instrumentation, portability and data acquisition technology, applications have widened and it is being used for near-surface mapping also. There has been considerable growth in applications of TEM for groundwater exploration where it has been mainly used for sounding depth-wise variations in electrical conductivity within 100 to 200 m depth. These variations can be translated into lithological layers with hydrogeological characteristics, such as unconsolidated and consolidated formations, sand and clay beds, presence of fresh and saline groundwater and conductive contamination plumes etc. It can also be used to assess weathered zone thickness, map resistive bed rock and approximate the presence of conductive shallow fractured zones. Being more responsive to conductive beds like saturated clays and saline water aquifers it has been mostly used in sedimentary terrain for hydrogeological investigations and environmental studies.

The TEM has advantages over FEM. In FEM measurement, small secondary field is measured in the presence of a primary field continuously emitted from a transmitter. Sometimes resolving the secondary from the primary field becomes difficult (Nabighian and Macnae, 1991) and more so when the target is deep. In TEM the primary magnetic field source is not continuous and measurement is made when it is off. TEM induction into the ground is caused by abruptly turning off the current in an ungrounded transmitter loop. The decay of the resulting secondary magnetic field is measured at sequentially increasing time gates when the primary field is off. In TEM deeper investigation with less surface spread compared to FEM is possible by increasing the length of recording time and magnetic moment if there is no late time noise creeping in. Unlike FEM, in areas with thick moderately conductive overburden, TEM sounding can be usefully conducted.

Compared to FEM and DC resistivity (VES) surveys, TEM survey is fast and cost-effective. The advantage of TEM over DC resistivity is, it requires much less space for field lay-out for an equivalent depth probing. An area of 150 m × 150 m taken by TEM transmitter square loop can yield equivalent depth information obtained from a VES spread (AB) over 2 km (Flores and Velasco, 1998). Of course it depends on subsurface layer-resistivities as shown by Pellerin *et al.* (1996). With relatively less space covered by the transmitter loop, no sequential change in transmitter-receiver position i.e., no static shift, TEM is less affected by topographic variation as well as lateral changes in resistivity, i.e., geological noise (Frischknecht *et al.*, 1991). The VES curves are associated with shifts and distortions due to changes in potential electrode positions, lateral inhomogeneities and orientation with respect to geological strike, whereas TEM decay curves are generally affected by electromagnetic noise. At late time gates signal to background noise (S/N) ratio becomes low. The presence of a highly resistive surface layer poses difficulty in current injection for VES but it is not so in TEM induction. Everett and Meju (2005) point out the merit of EM methods over DC resistivity. Flores and Velasco (1998) make a comparative analysis and detail the merits of TEM over VES. TEM can be conducted in open parks in urbanized areas if interference from electrical and other surface and subsurface structures is less. In an area, compared to VES a larger number of TEM soundings can be conducted, say up to 15 per sq km or more, in much less time, obviously rendering a better picture on aquifer dispositions economically. TEM sounding can replace VES at places. A constraint with TEM is that, while VES is responsive to conductive as well as resistive layers, TEM is more responsive to conductive layers and best suited for identifying them. The resolution in TEM depends on absolute resistivity while in VES it depends on resistivity contrast (Auken *et al.*, 2006). These could however be used as merit in reducing the ambiguities. Also, in TEM the induced voltage decay curve does not present a conceivable picture of resistivity distribution as readily obtained from DC resistivity sounding.

One of the early applications of TEM is by Anderson *et al.* (1983) for deep investigation at Medicine Lake volcano, California. Fitterman and Stewart (1986) and Hoekstra and Blohm (1990) evaluate the applications of TEM sounding in groundwater investigations. The TEM method has been widely used for saline water mapping, sea water intrusion and coastal fresh water aquifers at several places. Some of the references are Stewart and Gay (1986), Fitterman (1987), Mills *et al.* (1988), Goldman *et al.* (1991), Chen (1999), Fitterman and Deszcz-Pan (2004) on TEM applications in coastal tracts of West Central Florida, California, Israel, Taiwan and South Florida to identify sea water intrusion and fresh water aquifers. A novel approach is made by Hurwitz *et al.* (1999) to identify saline water in the sediments underneath a fresh water lake by using surface marine TEM measurements at the lake surface. A large scale TEM for groundwater has been carried out in parts of Denmark using modern techniques like High Moment TEM (HiTEM) and Pulled Array TEM (PATEM) by Auken *et al.* (2003), Christiansen (2003), Danielsen *et al.* (2003) and Sorensen *et al.* (2005). Harris (2001) conducts TEM measurements to delineate palaeo-channels and assess the groundwater resource. Jorgensen *et al.* (2003) and Oldenborger and Brewer (2014) use TEM to demarcate buried valleys. Albouy *et al.* (2001) reveal the merit of jointly interpreting TEM and VES in coastal areas. Buselli *et al.* (1990), Fitterman *et al.* (1990) and Buselli and Lu (2001) use TEM along with other methods in groundwater seepage and contamination studies and indicate its effective use in understanding subsurface hydrogeological conditions. Meju *et al.* (2001) and Meju (2005b) use single and multi-component TEM in hard rock to delineate deep weathering and the fractured zones.

11.2 BASICS

In TEM, a high steady current is passed through an ungrounded transmitter wire loop placed on the ground surface. After a certain time period the current is abruptly turned-off. During the short 'turn-off' time there is a rapid change in the primary magnetic field from its initial value to zero which induces a flow of eddy current in the conductive ground. Immediately after turn-off, i.e., at 'early time stage' the induced current is concentrated near the surface in the vicinity of the transmitter loop, imaging the loop shape and size. It flows in horizontal closed path below the transmitter loop in the same direction as that of the current in the transmitter loop. With lapse of time transmitter loop shaped eddy current ring descends in the subsurface and its area expands (Fig. 11.1). It diffuses downward and outward with decrease in its amplitude. The process continues. Since the induced current ring descends and expands it resembles a system of 'smoke-rings' (Nabighiyan, 1979). According to Hoversten and Morrison (1982), it is a single smoke ring system distorted by layering and not in the form of separate smoke rings in each layer. The diffusion or dissipation of the eddy current smoke ring is controlled by the electrical conductivity of the medium. The eddy current is prominent in conductive layers. The diffusion produces weak secondary magnetic field proportional to the strength of diminishing eddy currents (Everett and Meju, 2005). For horizontal layering the secondary magnetic field is vertical at the centre of the transmitter loop. The rate of change of fading secondary magnetic field with time induces a voltage (V(t) equivalent to |dBz/dt|; Bz: z or vertical component of secondary field) in the receiver coil measured in the absence of a primary magnetic field. The amplitude-characteristics of decaying secondary magnetic field (or induced voltage) at the receiver are measured at several successive time gates (Fig. 11.2). The induced voltage response varies with time and can be divided into three time stages, viz. early, intermediate and late stages. At late time stage voltage induced in the receiver becomes proportional to $t^{-5/2}$ and $\sigma^{3/2}$ or $\rho^{-3/2}$, where σ or ρ: conductivity or resistivity of the deepest layer (Fitterman and Stewart, 1986) and the relations (Barrocu and Ranieri, 2000) are

$$V(t) = \frac{k_1 M \sigma^{3/2}}{t^{5/2}} \quad \text{and} \quad \rho_a(t) = \frac{k_2 M^{2/3}}{V(t)^{2/3} t^{5/3}}$$

where V(t): output voltage from a single turn receiver coil of 1 m², k_1 and k_2: constants and M: transmitter moment, product of transmitter current and area and t is the time after transmitter current turn-off. It indicates that dB_z/dt or the voltage in the receiver coil varies with $\sigma^{3/2}$ indicating more sensitivity to variation in conductivity (Barrocu and Ranieri, 2000). The decay of induced current is faster in the resistive layer and slower in the conductive layer as shown in Figure 11.3 (Fitterman and Stewart, 1986).

This voltage response normalized to transmitter current or transmitter moment is measured as a function of time at selected time intervals (gates) during the primary field turn-off. The measurement of voltage induced in the receiver coil at successive time intervals reveals induced current flows at successive times and hence at different depths. The graph plot can be between V(t) and t, ρ_a and t or ρ_a and $\sqrt{t}$. Because the response depends on conductivity, early time measurements yield the conductivity or resistivity of near-surface layers while late time measurements, when the induced

(a) Eddy current immediately after transmitter current turn-off

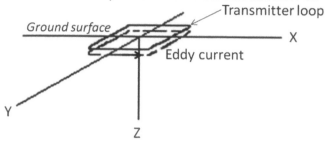

(b) Eddy current at later times $(t_1<t_2<t_3<t_4)$ after transmitter current turn-off

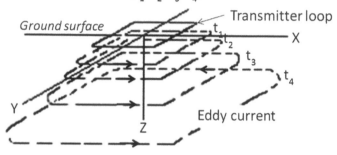

Figure 11.1 The flow of eddy current in the subsurface at early and late times after turning-off the current in the transmitter loop.

current has diffused to deeper horizons, yield the resistivity of deeper layers. Thus, the depth-wise layer resistivity characteristics of the subsurface – the TEM sounding is obtained.

The shape of transient curve (voltage decay vs time in second or inverted to apparent resistivity vs sq. root of time in second) does not exactly represent depth wise resistivity variations as obtained from conventional DC, apparent resistivity vs AB/2 VES curve, though depth of exploration is a function of time. It does not depend on transmitter-receiver separation or transmitter frequency. Since, resolution decreases (bulk physical property averaged over increasingly larger volume) and noise increases with depth because of diffusion, the signal to noise (S/N) ratio at late times defines the depth of investigation (DOI). Decay being fast during early times, the early time gates are kept very close-spaced compared to late time gates and for this TEM measurements are made at logarithmic time gates, i.e., the length of the time gates is proportional to the delay time (Christiansen, 2003).

In time-domain measurements the skin depth concept also prevails. It is known as diffusion depth- a qualitative measure of depth of penetrating electromagnetic energy in the subsurface (Fitterman and Labson, 2005) within a definite time. The skin depth δ is expressed as $(2t/\sigma\mu_0)^{1/2}$ where σ is conductivity and μ_0 is magnetic permeability and t is time. It is similar to frequency domain where skin depth is defined as $(2/\sigma\mu_0\omega)^{1/2}$ with

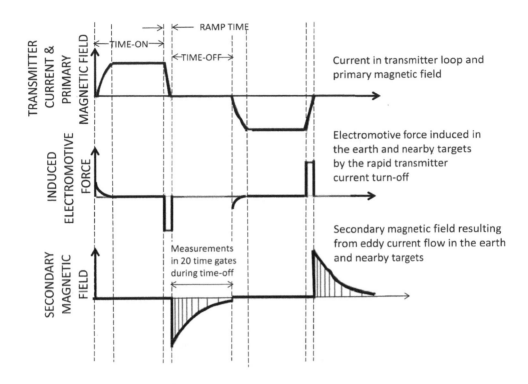

TRANSMITTER CURRENT & PRIMARY MAGNETIC FIELD

Current in transmitter loop and primary magnetic field

INDUCED ELECTROMOTIVE FORCE

Electromotive force induced in the earth and nearby targets by the rapid transmitter current turn-off

SECONDARY MAGNETIC FIELD

Secondary magnetic field resulting from eddy current flow in the earth and nearby targets

Figure 11.2 The time domain electromagnetic waveform and measurement time gates.

ω as angular frequency. The depth is proportional to inverse of sq. root of frequency in frequency domain and to sq. root of time in time domain. For further details reader may consult Fitterman and Stewart (1986), Nabighian and Macnae (1991), Spies and Frischknecht (1991), Auken *et al.* (2003) and other references given.

11.3 INSTRUMENT

A TEM system comprises a transmitter loop and a horizontal receiver loop. The transmitter is in the form of a horizontal square loop of insulated wire of different sizes placed on the ground surface. The dimensions of this ungrounded horizontal loop vary from a few tens of metres to hundreds of metres having a single insulated wire. The small loop acts as a magnetic dipole. The magnetic moment of the loop is equivalent to the product of current strength, number of turns of wire and its area. For deeper penetration the transmitter magnetic moment is increased by increasing the current, say from 1 to 4 A or more and also the loop size. Generally, TEM instruments with dual moments are required. The low moment transmitter is for near surface investigations with higher resolution whereas the higher moment is for late-delay time acquisition of data with least noise required for deeper investigations. The time constant (L/R) of the receiver coil should be much less than the earliest time of measurement (Spies and

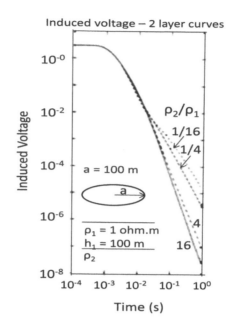

Figure 11.3 Induced voltage-time curves for a suite of two-layer models showing slow decay rate for conductive substratum and fast decay rate for resistive substratum. The first-layer thickness and transmitter loop radius are 100 m. The transmitter current is 1 A, and the receiver loop moment is 1 m² (Source: Fitterman and Stewart, 1986).

Frischknecht, 1991). A TEM instrument uses a constant current waveform consisting of equal periods of time-on and time-off. The TEM instruments are available with a stacking facility. The measurement can range between time as early as 6 μs and up to 100 ms after switching off and for groundwater exploration measurements may be required up to 10 ms.

11.4 FIELD PROCEDURES

The TEM can be employed for sounding as well as profiling. The field lay out is shown in figure 11.4. Three types of transmitter-receiver configurations are used, viz., central loop, loop-loop and coincident loop as shown in figure 11.5. In central loop configuration the receiver coil is placed at the centre of the transmitter loop and in the loop-loop or offset configuration the receiver coil is placed outside the transmitter loop. In coincident loop configuration the transmitter and receiver loops are of the same dimensions, spatially coincident and separated (Nabighian and Macnae, 1991). For this the transmitter loop can also be used as the receiver loop. In view of resolution and field efficiency required in groundwater exploration, Auken *et al.* (2003) advocate TEM soundings with central loop configuration. They conducted several thousand TEM soundings with central loop configuration using 40 m × 40 m size transmitter

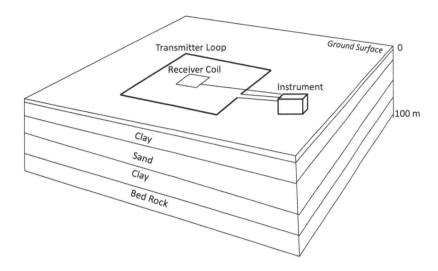

Figure 11.4 Field layout of transient electromagnetic sounding.

loop and 3 A current, measured transients over a time range of 9 µs to 9 ms and could investigate to a maximum depth around 130 m due to higher noise level beyond 2 ms. The variation in noise level ranges from 0.1 nV/m² in uninhabited areas to 10 nV/m² on outskirts of urban areas (Christensen and Sorensen, 1998). The central loop configuration is preferred as it is most stable for early-time measurements and loop layout errors are negligible (Auken *et al.*, 2006). Data scattering is not observed in central loop sounding and it is least affected by near-surface lateral variations in resistivity as induced current flows in rings around the receiver. The resolution is high in shallow central loop soundings. The transmitter loop size for central loop layout depends on the depth to be explored, decided for the objective and is based on field testing and noise level at late times. Auken *et al.* (2003) made a novel experiment of sounding with mixed configuration to get deeper information. They carried out low moment sounding with the central loop configuration for early times and with high moment loop-loop configuration for late times to avoid saturation of receiver amplifiers due to high moment transmitter in central loop configuration. Forward modeling based on geological, borehole lithological and geophysical information can also help in selecting the desired configuration.

The level of background noise at late time defines the limit of latest time of measurement and hence the depth. Approximating the noise and subsurface resistivity distribution, size of transmitter loop or amount of current to be sent can be estimated for roughly achieving the required depth sensitivity or DOI. For estimating this, the diffusion depth is calculated in terms of transmitter current (I), area (A) and noise level (V) for late-time voltage (Shu *et al.*, 2009). It is expressed as $\delta \approx 0.55(IA/\sigma V)^{\frac{1}{5}}$. A high amperage current, say 20 A and large loop size could be used for enhancing signal to noise ratio required in deeper exploration. The length of side of a square-loop transmitter can be varied from 25 to 200 m. The loop sizes of 40 m × 40 m or 50 m × 50 m

a. Central or In-Loop (receiver loop at the centre of the transmitter loop)

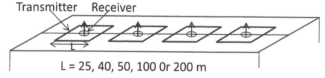

L = 25, 40, 50, 100 Or 200 m

b. Loop-Loop (offset - receiver loop out side the transmitter loop)

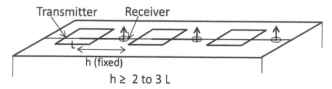

h ≥ 2 to 3 L

c. Coincident Loop (transmitter and receiver loops are spatially coincident)

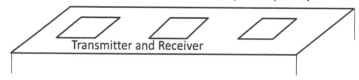

Figure 11.5 The different transmitter-receiver configurations (adapted from Hoekstra and Blohm, 1990).

are quite convenient. By these loop sizes depths up to 150–200 m can be explored. For laying the square loop, at first corners of the square are marked, e.g., for a 40 m side square, ends of the two orthogonal diagonals of length 56.6 m of the 40 m × 40 m square, intersecting at the mid-point are marked on the ground.

According to Spies and Frischknecht (1991) maximum S/N ratio and resolution are obtained when loop size and depth to be probed are of the same order. Hence, for shallow investigations the transmitter loop size has to be small. For smaller loops the turn-off time is smaller which makes measurements possible at early time gates required in shallow investigations. It gives better resolution and higher signal at early time. The effect of transmitter turn-off time on TEM sounding for resistive and conductive top layer with different thicknesses is discussed by Fitterman and Anderson (1987). A very small loop size of 10 m × 10 m is not preferred as in central loop system the distance between transmitter loop and receiver becomes too small and current leakages in transmitter coil affect response at the receiver. Large loop size at later times provides better signal. A peak current of 2 A can be sent through a loop of 40 m × 40 m for shallow exploration. Auken *et al.* (2002) show that noise can be reduced as well as the resolution of the model can be enhanced by increasing the transmitter current (Fig. 11.6).

It is always desired that the instrument is calibrated at a known test site and also checked for repeatability (Foged *et al.*, 2013) prior to its use in the field. Alternatively, if a calibrated instrument is available, Podgorski *et al.* (2013) recommends comparison

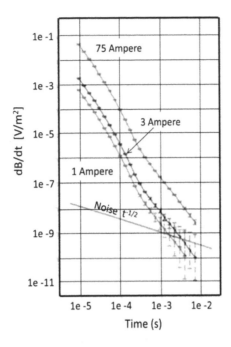

Figure 11.6 Enhancing the transmitter current for noise reduction (Source: Auken *et al.*, 2002).

of the performance of the deployed instrument with it. For better results, it is necessary that a few TEM soundings are conducted near existing boreholes and synthetic data through a forward model are compared with the measured data. Besides calibration, it helps data editing and a check on inversion of data from the area. Fitterman and Labson (2005) advocate that no measurement in the field should start without forward modeling.

The measurements can start at 6 to 10 μs after switching off and shallow zones can be investigated. The latest time could be up to 10–30 ms and can go up to 100 ms depending on the level of noise. The unit of measurement is volt per square metre per ampere of transmitter current. The minimum detectable signal (time derivative of secondary magnetic field: dB/dt) ranges from 10^{-6} to 10^{-12} V/m^2. Transient EM noise at later time stage restricts the length of time during which the transient can be sampled and thus deeper information cannot be obtained unless the transmitter moment is increased. The data can be collected at two base frequencies. Measurement gates are kept logarithmically spaced with about 10 gates per decade of time. After laying out the square loop at a site, measurements take only 10–15 minutes. A team of two to three persons as crew can carry out on average 10–15 TEM soundings in a day if the sites are located in close proximity. The site is selected in such a way that measurements are least affected by man-made structures, power-lines and buried pipes. A convention is to keep the transmitter loop at least one or two loop lengths away from the interfering structure (Anderson *et al.*, 2003).

11.5   PROCESSING OF DATA

The processing of TEM data is a vital step prior to meaningful inversion for generating reliable model parameters. It involves filtering of data for removal of spikes and careful elimination of very early time contaminated data and late time noisy data while preserving the shallow and deeper information. The first few early time gate responses get contaminated due to the flow of decaying residual current in the transmitter loop for a few microseconds after the current is shut-off. Kamenetsky and Oelsner (2000) compute the minimum delay time for coincident loop TEM system for different sizes of transmitter loop and half-space resistivities. For a 40 m × 40 m transmitter loop over a 100 ohm.m resistivity half-space, the minimum delay time is 11.2 μs. The effect of residual current at early time gates is removed by studying the change in slope of dB/dt (Podgorski et al., 2013). The late time gate data could be noisy and they are also eliminated. Such processing should be initiated for the data collected at a known point, viz., near a borehole so that effect of data elimination on model parameter estimation can be assessed.

11.6   INTERPRETATION

The prerequisite for interpretation is high-quality data and the availability of good computer code for interactive forward modeling and inversion. Forward modeling gives an idea of uncertainty in parameter estimation and those parameters which will be poorly estimated. Meju (1998) gives a simple preliminary method of analyzing the data for estimating effective resistivities at various sampling depths which can be used for model identification. Fitterman and Stewart (1986) analyze the applicability of TEM sounding under different hydrogeological conditions and present TEM sounding curves computed for various combinations of layer parameters. Out of these, the 2-layer curves expected from hard rock terrain are reproduced in figure 11.7.

The interpretation may start with forward modeling based on information obtained from existing boreholes in the area. Generally the inversion of TEM data is done using computer code for smoothest model using Occam's inversion scheme (Constable et al., 1987) which does not require any initial guess. The smooth model inversion uses constant-thickness layers and only resistivities vary. For areas without subsurface hydrogeological complexities, it is a good tool to visualize the resistivity distribution (Spies and Macnae, 1997) without any bias. The target of limited lateral extents may not give a good match in inversion. While inverting the data, it is to be also realized, as pointed out by Constable et al. (1987), that the diffusive nature of energy propagation in TEM sounding smears out the real earth structure and does not allow detection of sharp interfaces or 'thin' layers. So keeping this in mind, the smoothest model and extraneous information such as geology, resistivity sounding or imaging and shallow seismic data or borehole logs, if available can be used to guess the amount of layering and carry out inversion iteratively for simple or few-layer model with reduced range of equivalent models. An interpreted field TEM curve from granitic terrain of Hyderabad India taken from Baltassat et al. (2006) is reproduced in figure 11.8.

In a case where VES is available at the TEM sounding location the apparent resistivity data can be compared as graphs using the relation $t = \frac{\pi}{2}\mu\sigma L^2$ or $L = 711.8\sqrt{\rho}.\sqrt{t}$ given by Meju (2005a), where t is transient time in second, μ is magnetic permeability equal to μ_0, L is AB/2 in metre and ρ is resistivity of homogeneous subsurface in ohm.m which is obtained only after inversion of apparent resistivity data and therefore approximated by apparent resistivity. This type of joint viewing of TEM and VES data help resolve the layers or allow introducing a layer visible in the VES curve in TEM sounding interpretation. Also it helps in normalizing the segment shifts in VES curves through TEM data.

Auken *et al.* (2002) analyze TEM data for selected 3-layer K and H type geoelectrical layer models to illustrate the limitations in interpretations. In the K-type, the intermediate resistive layer may not be resolved except that resolution improves with increased transmitter moment. For example, the resistivity of the resistive fresh water aquifers underlying thick clay overburden will not be accurately determined. Auken *et al.* (2007b) define the upper limit of sensitivity to layer resistivity around 100 ohm.m and therefore values of resistivity for highly resistive bed obtained from TEM may not be reliable. Such a situation may be encountered in the basalt-flow terrain where the highly resistive massive basalt layers are overlain and underlain by moderate to low resistivity vesicular/fractured basalts and intertrappeans. The resistivity of the massive basalt layer may remain underestimated. When intermediate layer is conductive (less resistive) – H type model, generally obtained in granitic terrain, its resistivity gets resolved. Also, they indicate that the top of the intermediate conductive layer is resolved better than its bottom. That is, top of the weathered zone aquifer (saturated thickness) can be resolved but the depth to its contact with the underlying resistive bed rock, which may be transitional, may not get defined accurately. Also, if there are two conductive horizons the thicker one will be picked up. Besides, a good estimate of the resistivity of the top layer and substratum may not be possible. The TEM data interpretation has to be essentially validated by at least one or two known geological control points or borehole lithologs and geophysical logs available in the area.

Like DC resistivity, inversion of TEM also suffers from non-uniqueness and poor resolution at depth. To reduce non-uniqueness in TEM sounding inversion, the data and results obtained from other surface geophysical methods if available can be used to arrive at a realistic model closely representing the true subsurface condition. In this way ambiguities in interpretations can be reduced. It is done in two ways, either the inverted results of other geophysical method are further used to invert TEM data or both the data are combined in the inversion process. The former is known as sequential inversion while the latter is known as joint inversion. Jupp and Vozoff (1975) and Vozoff and Jupp (1975) conduct joint inversion of VES and MT soundings to get a model and show its applicability in resolving thin resistive layer at depth. Meju (1996) attempts joint inversion of TEM and magnetotelluric (MT) data to constrain the interpretation of MT apparent resistivity curves. Raiche *et al.* (1985) opine that since DC and AC fields behave in different fashions, a relatively thin conductive layer underlying a thick highly resistive layer may not be picked up in DC resistivity sounding, whereas in TEM it should not be a problem. They apply joint inversion to coincident loop TEM and Schlumberger soundings to improve the resolution of conductive thin layer and indicate that it is possible, but by using correct ramp turn-off time in TEM. It may be

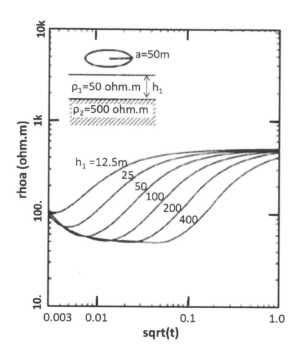

Figure 11.7 TEM sounding curves computed for 2-layer earth with varied overburden (first layer) thickness (h_1). In the time range plotted, the left hand descending branch is not present. With increasing thickness of the first layer the minimum of the curve becomes broad and flat and approaches the first-layer resistivity (Source: Fitterman and Stewart, 1986).

difficult to resolve thin layers in a sequence of alternate high and low resistivity layers (Carrasquilla & Ulugergerli, 2006) or a sequence of thin layers with increasing or decreasing resistivities and like VES, TEM also gives an average resistivity for the entire column of thin layers. Leite *et al.* (2013) however show the usefulness of joint inversion of TEM and VES in successfully locating conductive layers at depth within the resistive basalts. Yang and Tong (1999) attempt joint inversion of three types of data, viz., DC, TEM and MT and show its obvious superiority over joint inversion of two types of data mentioned above. An important and practical advantage of joint inversion is the minimum or no necessity of additional *a priori* information in deriving the realistic thickness-resistivity model (Schmutz *et al.*, 2000). While doing joint inversion separate inversions may also be conducted to visualize the improvement.

Similar to joint inversion but more robust, Auken *et al.* (2001) develop a mutually constrained inversion (MCI) scheme to concurrently invert two mutually constrained data sets to generate two closely related models which can be independently evaluated to get the best resolved parameters. Further improvement in joint interpretation of TEM and DC resistivity can be achieved by using first the laterally constrained inversion (LCI) among the TEM and DC resistivity data separately and then merging these through MCI (Christiansen *et al.*, 2007). It helps resolve the poorly resolved parameters.

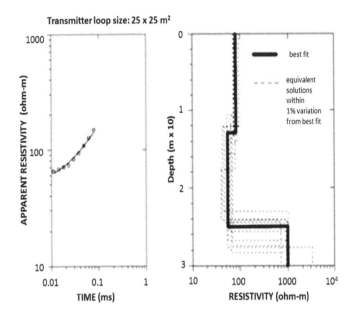

Figure 11.8 Interpreted TEM sounding curve from granitic terrain of Hyderabad, India. The transmitter loop is 25 × 25 sq.m. The interpretation shows equivalent models and best fit model (Source: Baltassat *et al.*, 2006).

11.6.1 Equivalence in electromagnetic sounding

Like DC resistivity sounding EM sounding interpretation also suffers from equivalence which cannot be removed but can be minimized. The equivalence in intermediate layers becomes a serious problem when the layers are thin and occur at deeper horizons, providing a wide range of equivalent layer-models. Equivalent models can generate the same field data or the field data can yield a number of equivalent models. It becomes difficult to select a representative model or narrow down the range of equivalence without support from other information constraining the range either through conductivity or thickness of the intermediate layer. Equivalence in layered-earth models for FEM is studied by Verma (1980) and Sharma and Kaikkonen (1999) and for TEM by Mallick and Verma (1979). Fitterman *et al.* (1988) compare the equivalence in VES, FEM and TEM soundings.

The basic objective here is to resolve parameters with least ambiguity from the response which is either a product or ratio of thickness and resistivity. Compared to VES and FEM, Fitterman *et al.* (1988) indicate a better resolving power of TEM for S equivalence where the intermediate layer is conductive (h_2/ρ_2 in H-type curve, h_2: intermediate layer thickness and ρ_2: layer resistivity). For T equivalence ($h_2\rho_2$ in K-type curve) FEM and TEM can resolve h_2 but not ρ_2. This can be achieved to a great extent by joint inversion of compatible VES and FEM or TEM sounding using selective resolving potency of each method (Raiche *et al.*, 1985). McNeill (1990) indicates the superiority of VES in resolving the resistive intermediate layer, but advocates joint inversion of VES and TEM. According to Sharma and Kaikkonen (1999), though the EM method is insensitive to resistive layer, by joint inversion uncertainty in parameters of intermediate resistive layer (K-type) is reduced more as compared to that of a conductive layer; the reason being the sensitivity of EM to thickness of the resistive layer and that of DC to the transverse resistance of the layer and hence EM constrains the thickness variations in joint inversion. They point out that both the methods being sensitive to conductance of a conductive intermediate layer (H-type), as compared to K-type, the H-type is poorly resolved by joint inversion. Further, in a multi-layer sequence the resistive layer is resolved better, if it overlies a conductive bed rather than underlies it. To resolve equivalence or suppression and ascertain the limitation, the best approach is to constrain the depth, information on which can be derived either from a high resolution method such as seismic or a few borehole geophysical logs. It is also necessary to generate a forward model from available subsurface geological information and by varying the layer parameters to see which layer can be best resolved.

11.6.2 Detectability and depth of investigation

In the case of electromagnetic sounding the definitions of depth of investigation (DOI) and detectability differ from those defined conventionally in DC resistivity. Detectability is basically the ability to resolve a layer-parameter under certain hydrogeological conditions (in terms of conductivity contrast) for a given survey design and instrument capability. A target is detectable if the combination of its parameters creates a minimum 10% variation in response with respect to background. In TEM, generally the detectability of a conductive layer is more than the resistive layer. As mentioned by Fitterman (1989), it can be determined in simplest form by studying the difference between the forward model response generated for a set of expected layer-parameters and that for another model by varying one of the parameters. Spies (1989) conducts sensitivity analysis for central loop response by perturbing the conductivity of a thin layer (thickness one-tenth of depth; conductivity twice the background) in uniform half-space and concludes that for near-zone sounding (a/d <1, where, a is radius of the loop and d is depth to the thin conductive layer) the maximum contribution originates from a depth equivalent to 0.35 times the diffusion depth. Further, he shows that in a 2-layer model the bottom layer can be detected with upper layer thickness equal to diffusion depth and detectability increases with an increase in resistivity contrast. Obviously, the layer parameters are affected by equivalence and hence are non-unique. In TEM measurements detectability is also limited at late delay times, where expected variation in response due to parametric changes is not dependable due to masking by high noise (Fitterman, 1989).

In TEM sounding theoretically speaking there is no specific depth below which there is no information (Christiansen and Auken, 2012). Practically DOI is controlled by transmitter moment, sensitivity of instrument, the way resistive and conductive layers occur in depth sequence and the noise level defined in terms of standard deviation of dB/dt at late times. DOI, therefore can be more or can be less than the 'skin depth' or 'depth of diffusion'. In TEM the earliest time gate indicates the minimum depth or shallowest DOI where response can be resolved and the maximum delay time up to which noise free (or with adequate signal to noise ratio) transients can be measured indicates maximum DOI. Spies (1989) defines maximum DOI as the thickness of over-burden for which TEM signal equals the noise level irrespective of transmitter moment and subsurface resistivity. It can also be defined as the depth below which TEM data are insensitive to the variations in layer parameters. According to Spies (1989) noise level may vary from 0.1 to 10 $\eta V/m^2$. Since noise limits the DOI, one of the ways of increasing DOI is to enhance the S/N ratio. This can be done by increasing the transmitter moment through increased current input rendering a higher S/N ratio at late delay times. Spies (1989) shows that to double the DOI, the transmitter moment has to be increased by several orders which may not be feasible. So stacking and filtering is necessary to reduce and filter out the noise. Fitterman (1989) computes DOI on the basis of maximum delay-time where a signal of sufficient strength can be measured. For this purpose the signal strength for a geoelectrical layer model can be computed and compared with the anticipated noise. Using the definition of late-stage voltage apparent resistivity he derives a formula for latest time (t) of measurable signal given as

$$t = \mu \left[\frac{\left(\frac{M_t}{N_m} \right)^2}{400(\pi \rho_a^L)^3} \right]^{1/5}$$

where μ is magnetic permeability, M_t is transmitter moment (area of transmitter loop multiplied by current), N_m is noise level in the transient system and ρ_a^L is late-stage voltage apparent resistivity and computes the detectability envelop for a value of $N_m = 1 \times 10^{-9} V/m^2$. Since the latest measurement time is inversely related to late-time voltage apparent resistivity, DOI depends on resistivity of layers at depth. Figure 11.9 computed by Fitterman for two-layer models reveals the detection limits in terms of delay time or DOI for different transmitter moments and bottom layer resistivities. The DOI is higher with shallow resistive layers compared to shallow conductive layers. Also, the deeper conductive layers are better resolved than the deeper resistive layers. The constant M_t lines across the curve indicate the delay time limits. Towards the left of the line the signal is measurable for a particular magnetic moment and bottom layer resistivity, while towards the right it is not. Also it indicates that for a given magnetic moment and noise level late time measurements are much better for the conductive bottom layer than the resistive bottom. As suggested by Fitterman (1989) for TEM survey design and transmitter moment to be used in an area, estimation of latest delay-time measurement and hence DOI can be made through generation of TEM apparent resistivity curve for anticipated geoelectrical layer model based on hydrogeological conditions or electrical resistivity log.

DOI has been investigated through various other approaches also (Olderburg and Li, 1999, Minsley, 2011, Christiansen and Auken, 2012 and Schamper et al., 2012). Similar to the approach by Oldenburg and Li, Christiansen and Auken define DOI for any 1D DC or EM data as *"a measure of the capability of the measured data and their uncertainty to resolve the given model"* and assign *"an absolute threshold value with DOI for the minimum amount of information required for something to be resolved"*. This definition appears quite logical and practical. The DOI is computed on the layer-model obtained from inversion of observed data. It is done by calculating the sensitivity of an observed data point for a layer parameter. An empirically defined threshold value for sensitivity is considered to get the value of DOI. Schamper et al.

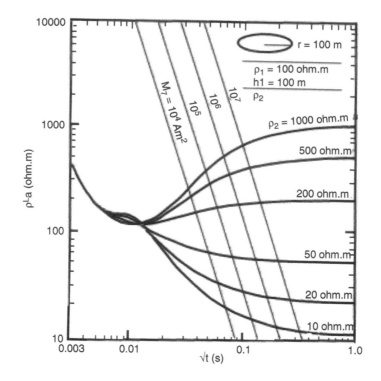

Figure 11.9 Plot of late-time stage apparent resistivity vs square root of time for geoelectrical two-layer models. The detectability limits for different transmitter moment are shown (Source: Fitterman, 1989).

(2012) compute the DOI index as a function of depth for FEM data and a value more than 0.1 indicates that it has been achieved. It is observed that DOI depends on subsurface resistivity distribution. In this computation the sensitivity of substratum is not defined. Also, the DOI computation is model specific and equivalence is not considered.

For a logical estimation of DOI, Fitterman and Labson (2005) attempt computation of sensitivity of TEM measurements to changes in 2- to 4-layer resistivities for a defined noise level and transmitter moment. According to them DOI depends on noise level and ability to discriminate the layered-models based on differences in apparent resistivities in a certain time range. Finally they concluded that estimating DOI is an "inexact" process.

11.7 DELINEATION OF FRACTURED ZONES IN HARD ROCK

In FEM, the detection of deeper fractured zones is constrained by the presence of a conductive weathered zone/overburden which does not pose much problem in TEM. The measurement of the Z-component of the secondary field is generally done in TEM for horizontal layering in sedimentary area. It is not effective in delineating deeper fractured zones underlying the weathered zone, as these zones behave as elongated sheet or plate-like vertical to sub-vertical discrete conductors. In this regard acquisition of multi-component TEM data can prove quite useful. Macnae (1984) suggests acquisition of horizontal component data which may also help the interpretation process. Meju (2005b) proposes measurements of components other than the Z-component to get the details of heterogeneous conductive targets. Theoretically speaking, measurement of the horizontal component of the secondary magnetic field, along a profile perpendicular to the strike of vertical to sub-vertical conductor/ fractured zone, strongly coupling with it can be used to locate it. Generally, the component along the strike of the target is taken as Y-component and that perpendicular to it as the X-component. The Y-component is supposed to be zero for a layered earth and any variation in Y-component can be used in identifying the off-line lateral inhomogeneities (Smith and Keating, 1996). The X-component response is also weak and contains much more noise as compared to the Z-component. Kirkegaard *et al.* (2012) analyze the sensitivity functions of X and Z components and found that while Z component sensitivity function is positive, the X component sensitivity function has multiple sign reversals which can differentiate the lateral conductivity variations. They also indicate that X-component sensitivity to lateral conductivity variations is highest at early times, i.e., shallow vertical to sub-vertical targets can be detected by X-component measurements. As reported by Kirkegaard *et al.* (2012) for SkyTEM heliborne system the Z/X component ratio is of the order of 5:1 at early times and 10: 1 at late times therefore it is most essential to reliably measure the X-component particularly at early times. The X-component profile across the strike of the fractured zone presents a maximum value near it with a possible sign reversal over the middle of it (Meju, 2005b) and thus can help define better the fractured zone location. Smith and Keating (1996) indicate the use of all the three components in characterizing depth, dip and strike of the discrete conductor.

The BRGM, France in collaboration with CSIR-NGRI, India conducted a large number of close grid TEM soundings in granitic terrain of Hyderabad India (Baltassat *et al.*, 2006). It was observed that 25 m × 25 m size transmitter loop is sufficient for shallow investigations and estimation of depth to the bed rock with reasonable accuracy as shown through TEM sounding interpretation and range of equivalent models in figure 11.8. However, it was not possible to distinguish the fractured zone from the underlying bedrock because of high values of resistivity and low resistivity contrast. Meju *et al.* (2001) carry out TEM along with FEM (HLEM) in granitic terrain of northeast Brazil to overcome the limitation of FEM in delineating fractured zones underneath thick conductive saprolite overburden. According to them, ideally the thick saprolite zone should be associated with a single-peak, broad and high amplitude early-time single-loop TEM anomaly, while the deep fractured zone is expected to be with twin-peak late-time TEM anomaly (Fig. 11.10).

11.8 AIRBORNE ELECTROMAGNETIC SURVEYS

Airborne electromagnetic (AEM) survey is of two types, viz., the frequency domain and time-domain. Airborne FEM as well as TEM have been in use for mineral exploration for several decades and remarkably contributed to the discovery of a large number of

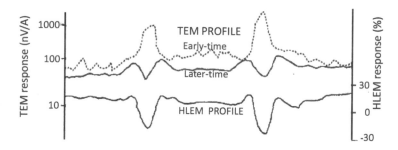

Figure 11.10 The schematic pattern of TEM early-time single-peak anomalies over zones of thickened saprolite overburden (upper dashed line) and later-time twin-peak anomalies over the steep fractured zones (middle solid line), the expected HLEM anomaly of steep fractured zones (bottom solid line) is also shown for comparison (source: Meju *et al.*, 2001).

deposits (Palacky and West, 1991; Holladay and Lo, 1997; Allard, 2007). The varied applications of airborne electromagnetic surveys are given in the USGS Bulletin-1925 (1990) edited by David V Fitterman. Conventionally, airborne frequency domain EM survey is carried out for shallow investigations while airborne time domain EM is conducted for relatively deeper investigations (Sorensen and Auken, 2004). A great advantage of AEM survey is that it can present a comprehensive view of the subsurface 3d conductivity distribution at a regional scale with adequate resolution (Palacky and West, 1991). Besides, it gives fast and dense data coverage and accessibility to inaccessible areas and rugged terrains. AEM was used for geological mapping in hard rocks (Palacky, 1981). One of the recent high resolution applications is mapping palaeochannels for uranium exploration (Reid and Viezzoli, 2007). For groundwater and environmental studies application of AEM is relatively new and sporadic. Sattel and Kgotlhang (2004) report the application of AEM and aeromagnetic for groundwater exploration in parts of Botswana. Aquifer disposition and salinity over a large area can be obtained through resistivity mapping and delineation of subsurface structures using high resolution surveys by helicopter at low altitude and narrow flight-line spacing. There are quite a good number of published papers on AEM case studies and some of them are referred here. References of early investigations mentioned by Palacky and West (1991) are of AEM surveys by Collett (1966) over Winkler aquifer, Manitoba and Baudoin *et al.* (1970) for fresh-saline interface in Rhone Delta. Sengpiel and Meiser (1981) from BGR, Germany carried out helicopter borne single frequency EM surveys (HFEM) in the late seventies to delineate fresh water- sea water interface in a small North Sea Island (referred in Sengpiel, 1986). This was followed by HFEM surveys around 1987–88 to delineate fresh aquifers in the deserts of Pakistan (Sengpiel and Flusche, 1992). Paterson and Bosschart (1987) and Bromley *et al.* (1994) report the applications of HFEM in delineating fractured zones in hard rocks. Lipinski *et al.* (2008) report HFEM surveys for large scale hydrogeological mapping and monitoring within 50 m depth and selecting the optimum location for impoundment of water. Shirzaditabar *et al.* (2011) report the use of HFEM and magnetic to image spatial resistivity distribution and 3d geological structures and validate the geological map. Cook (1992) attempts defining groundwater recharge rates through electrical conductivity variations mapped by HFEM. Applications of HFEM surveys are reviewed by Siemon *et al.* (2009). Steuer and Siemon (2005) and Steuer *et al.* (2009) state the limitation of HFEM in penetrating through conductive clays which was overcome through helicopter borne time-domain electromagnetics (HTEM).

The TEM airborne surveys started in the early 1960s (Holladay and Lo, 1997). Wynn (2006) carries out airborne TEM survey for aquifer delineation up to a depth of 400 m in Upper San Pedro River basin, Cochise County, South Eastern Arizona using aircraft at a nominal terrain-clearance of 120 m. Now, HTEM with terrain-clearance of 30 m have been developed. The near surface high resolution HTEM survey with early-time measurements is being widely used in several countries for aquifer mapping, sea water intrusion and aquifer vulnerability to contamination etc. The information obtained from HTEM can be used for further detailing on the ground wherever required. Pfaffhuber *et al.* (2010, 2011) use it for mapping clays associated with tunneling hazards and active rock slides in western Norway. Viezzoli *et al.* (2009) study the HTEM data of Lower Murray region of South Australia for inter-aquifer connectivity, surface water-groundwater interaction and floodplain salinization. HTEM surveys have been carried out for salinity distribution in coastal aquifers in many areas (Kirkrgaard *et al.*, 2011; Wiederhold *et al.*, 2010; Kok *et al.*, 2010, Jorgensen *et al.*, 2012). While Street (1992), Anderson *et al.* (1993) and Street and Anderson (1993) present the studies through precision airborne TEM surveys for soil salinity mapping, Odins *et al.* (1995) and Paine and Collins (2003) report the use of HTEM for groundwater quality study up to a depth around 120 m. d'Ozouville *et al.* (2008) conduct HTEM over a volcanic island for delineating perched aquifers and salt water wedges. Besides mapping buried valley structures, Sorensen *et al.* (2004) use it for aquifer vulnerability evaluating clay contents up to 30 m depth. Ahmed (2014) uses a heliborne SkyTEM system to map aquifers in different representative hydrogeological terrains of India including granites, basalts, alluvium, desert sands and coastal sediments. Smith and Annan (1997) present the advances in airborne TEM with a good number of references. Smith *et al.* (2007) summarize airborne electromagnetic surveys for hydrologic studies carried out by USGS. The readers may like to refer to the papers mentioned above and those by Jacobsen and Edsen (2005), Auken *et al.* (2007a), Auken *et al.* (2009) and Pugin *et al.* (2014). Overall, the applicability of HTEM survey in groundwater investigations is established. It is observed that the dual moment HTEM system with early time gate ($\approx$6 to 10 μs) measurement facility could prove quite effective and economic in delineating shallow and deep aquifers with varying water quality. The resolution of near surface layers has been a serious issue because of erroneous early time data. Very recently Schamper *et al.* (2014) reported on developing a HTEM system that can start measurement within a few μs after the transmitter turn-off and resolve near-surface layers precisely, thereby opening scope for its wider applications in near surface hydrogeology particularly in hard rock and vadose zone characterization.

There are several systems developed for airborne TEM surveys by fixed wing aircraft and helicopter. For groundwater exploration and environmental studies a helicopter-borne system-SkyTEM, which is a central loop system, has been introduced by the Hydro-Geophysics Group of University of Aarhus, Denmark in 2004. In this the transmitter coil frame is kept suspended 30 m below the helicopter flown at a height of 60 m, i.e., with 30 m ground clearance. The flight lines can be spaced at 100 to 500 m depending on the objective and the resolution required. The height of the transmitter above ground depends on ground installations and buildings etc. The flight lines are in general kept perpendicular to the strike of geological formations or across structures of interest. Generally, flight speed is kept around 45 to 50 km per hour. The response is recorded 10–12 μs after the transmitter is turned off.

In AEM the received responses are of relatively low magnitude as compared to ground TEM and therefore the S/N ratio related to several factors including man-made is generally low. For getting a high resolution subsurface resistivity picture, much care is required in acquisition, processing and filtering of data. The data are processed and corrected at stages for altitude variations, transmitter tilt, geographic locations and for removing unwanted effects of man-made structures and power-lines so as to avoid erroneous interpretations (Auken, et al., 2009). The presence of a highly conductive overburden can prevent detection of underlying resistive layers or reduce the DOI considerably. The deeper resistive target such as bedrock topography with a highly conductive near-surface layer may have low resolution as it contributes very less to the response having much noise at late times. The inversion of data can be done involving the concept of spatially constrained inversion (SCI) developed by Viezzoli et al. (2008). The results are presented either as images or conductivity-depth sections. Preparation of a DOI map is quite useful as it manifests areas with higher DOIs indicating better late time data with higher signal to noise ratio, thereby rendering information on deeper aquifers/hydrogeological conditions.

11.8.1 Airborne TEM survey for groundwater in hard rock

In hard rock groundwater exploration the basic requirement is to accurately map the weathered and saprolite zones of resistivity within an approximate range of 30 to 200 ohm.m and thickness upto 60 m. The resistivity can be much less with increased altered clay content or salinity of groundwater. The resistivity can be even higher than 200 ohm.m, if the weathered zone is dry. Secondly, it is to delineate underlying shallow fractured zones in compact rock which may be either inter-connected and connected to the weathered zone or isolated. The saturated fractured zone though represents a relatively conductive body in resistive host rock, its resistivity is generally higher than that of the overlying saturated weathered zone and saprolite. Auken et al. (2007b) indicate a better performance of TEM with layer resistivities less than 100 ohm.m. Thus, the fractured zone response will not be as high as obtained from highly conductive targets in mineral exploration and its detection warrants an airborne TEM system with higher sensitivity and signal to noise ratio and comparable resolution at shallow as well as deeper levels (Sorensen and Auken, 2004). The deeper compact resistive formations will be associated with lower signal level and hence with low signal to noise ratio at late times. This further require a high resolution system which can record the secondary field decay from very early times around 10 μs yielding shallow information on the weathered zone as well as noise free reliable data at late times rendering information upto about 200 or 300 m depth. For high resolution shallow information a transmitter of low magnetic moment is required whereas for deeper information it is to be with high moment. Meju (2005b) shows the applicability of measuring three spatial components of the secondary field instead of a single component in TEM profiling to delineate the fractured zones underlying conductive weathered material. It comprises recording in three coordinate directions (x,y,z) and is termed multi-component transient EM profiling. As discussed above the X component measurements by helicopter surveys in hard rock areas can help define the fractured zone location if its amplitude is adequate and the noise level is comparatively low.

Prior to the interpretation of HTEM data from hard rocks and preparation of maps it is essential that the geology of the area and presence of structures like faults, folds, dikes, quartz reefs etc. and associated photo-lineaments are studied on a large scale map preferably at 1:50,000. Also, the geomorphology of the area is studied to infer the possible impact of geologic structures on surface features such as drainage pattern and possible concealed subsurface extensions of the structures. These apparently help define areas of interest. Besides, it helps recognize HTEM anomalies, associate them with hydrogeological attributes – probable types of groundwater target viz., horizontal conductive layer or vertical to sub-vertical conductor etc and rule out unwanted anomalies. Overall, the process helps conceptualize subsurface hydrogeological model to constrain the hydrogeological transformation of heliborne data which is very crucial and the ultimate. It has to be invariably followed by correlation of HTEM inversion results with lithological and geophysical log information of existing boreholes and other ground geophysical survey results so that the inversion is hydrogeologically constrained as well as its limitations are known.

Generally, the quantum of HTEM sounding data generated is enormous and practically it is not possible to analyze and interpret individual data sets. Therefore, after conducting LCI, various trend maps are prepared. To start with, a DOI map is prepared from HTEM data that helps identifying stripes and areas with higher DOI for further analysis within the zone of interest, e.g., near a fault, dyke or confluence and criss-crossing of such structures or on the axis of a fold where bedrock conductors-the deeper fractured zones can be expected. Once an area or stripe with higher DOI is selected and data are confirmed for adequate late time higher signal to noise ratio, the generation of inverted model using Occam's inversion can be attempted for that area or stripe. It does not need *a priori* assumption on number of layers. Also, the few-layer inversion of TEM soundings along multi-directional transects across that area or spot can be attempted. It is for inferring depth-wise lateral resistivity variations and also for possible occurrence of a less resistive fractured zone impregnated in highly resistive host medium. It is better to select areas with a linear stripe of higher DOI than spots with higher DOI values. Besides, resistivity maps for different depth ranges can also be analyzed for the purpose.

The probability of saturated fractured zones in hard rock increases near geological structures and the confluence and intersection of structures. Therefore, it is better to combine low altitude precision heliborne magnetic (HMAG) with HTEM. It is quite useful and is almost essential if objective is to delineate regional and local fractured zone aquifers. The HMAG data can be collected, processed and interpreted at a small additional cost. The correlation and coincidence of aeromagnetic linears and anomalies of varied wavelengths with favourable HTEM results can support the inferences drawn on fractured zones. In hard rock, analysis should be initiated through HMAG data so that structures are delineated first and confirmed and prioritized with the help of a geological map and hydrogeological information. It is to be followed by analysis of X-component HTEM data preferably along and across the structures.

In case the X-component data are not available, the DOI of conventional Z-component HTEM data acquired can be analyzed for broadly identifying the fractured zones. Such an attempt was made by CSIR-NGRI, Hyderabad in the hard rocks of Karnataka, India (Chandra, personal communications). The DOI will be small in areas with near surface compact formation without fractured zones. The saturated fractured zones in compact granite gneiss being relatively conductive compared to surrounding, the DOI is expected to increase with the presence of shallow saturated fracture zones. Maps for different DOIs say from 50 to 100 m may be prepared and correlated with the HTEM derived resistivity distribution maps at these depths and the local geomorphology, drainage pattern, photo and magnetic lineaments, saturated weathered zone thickness, depth to compact rock, structures and the geology to approximate the regional fractured zone network. Prior to utilizing such fractured zone related DOI maps for exploration purpose they are to be checked through existing borehole information and essentially validated through a few boreholes drilled at selected points. These maps may not indicate the presence of individual fractures but help demarcate the areas or narrow down the zone of interest having maximum possibility of fracture occurrences.

Also, the HTEM DOI cross-sections along and across the structures inferred from HMAG can be studied and stripes or areas with higher DOI near or on the structures can be selected for fractured zone delineation. A spatially linear long stripe of increased DOI is preferred. The next step should be to study the generated TEM soundings in high DOI areas or stripes through Occam's inversion and pinpoint the best site for drilling to possibly tap the fractured zone. The fractured zones though relatively conductive, have less contrast in conductivity with the surrounding compact formation, and being thin, may not get reflected as a conductive layer, however, inferences can be drawn through the range of resistivities and their local lowering. Finally, as mentioned above a few boreholes are to be drilled at selected sites pinpointed through ground EM or resistivity profiling across high DOI zones to corroborate the HTEM findings, modify interpretations, standardize hydrogeophysical transformation and refine existing geological and hydrogeological maps.

As per requirement, maps can be prepared on top soil resistivity, weathered zone thickness, saturated thickness of unconfined aquifer, bed rock topography, local and regional fractured zones, preferential orientation and types of geological structures for groundwater, extents of contaminated aquifers, aquifer protection and vulnerability and areas suitable for various types of artificial recharge structures.

REFERENCES

Ahmed, S. (2014) A new chapter in groundwater geophysics in India: 3D aquifer mapping through heliborne transient electromagnetic investigations. *Journal of Geological Society of India*, 84 (4), 501–503.

Allard, M. (2007) On the origin of the HTEM species. In: Milkereit, B. (ed.) *Exploration 07: Proceedings of the Fifth Decennial Int. Conf. on Mineral Exploration, Sept 9 to 12 2007, Toronto, Canada*. pp. 355–374.

Albouy, Y., Andrieux, P., Rakotondrasoa, G., Ritz, M., Descloitres, M., Join, J.-L. & Rasolomanana, E. (2001)Maping coastal aquifers by joint inversion of DC and TEM soundings-three case histories. *Ground Water*, 39 (1), 87–97.

Anderson, A., Dodds, A.R., McMahon, S. & Street, G.J. (1993) A comparison of airborne and ground electromagnetic techniques for mapping shallow zone resistivity variations. *Exploration Geophysics*, 24 (3 & 4), 323–332.

Anderson, W.L., Frischknecht, F.C., Bradley, J.A., Grette, R., & Grose, C. (1983) *Inversion results of time-domain electromagnetic soundings near Medicine Lake, California (Part 2).* U.S.Geological Survey Open-File Report 83-910, U.S.G.S.

Anderson, M.L., Hart, D.J. & Alumbaugh, D.L. (2003) Use of the Time-domain electromagnetic method for determining the presence and depth of aquitards. Presented at: *The 27th Annual Meeting, American Water Resources Association (AWRA), Wisconsin Section, 27–28 February 2003, Lac du Flambeau, Wisconsin.*

Auken, E., Jorgensen, F., & Sorensen, K.I. (2003) Large-scale TEM investigation for groundwater. *Exploration Geophysics,* 34 (3), 188–194.

Auken, E., Nebel, L., Sorensen, K., Breiner, M., Pellerin, L. & Christensen, N.B. (2002) EMMA-A geophysical training and education tool for electromagnetic modeling and analysis. *Journal of Engineering & Environmental Geophysics,* 7 (2), 57–68.

Auken, E., Pellerin, L., Christensen, N.B. & Sorensen, K. (2006) A survey of current trends in near-surface electrical and electromagnetic methods. *Geophysics,* 71 (5), G249–G260.

Auken, E., Pellerin, L. & Sorensen, K.I. (2001) Mutually constrained inversion (MCI) of electrical and electromagnetic data. Presented at: *The SEG International Exposition and 71st Annual Meeting, 9–14 September, 2001, San Antonio, Texas, USA.*

Auken, E., Westergaard, J., Christiansen, A.V. & Sorensen, K. (2007a) Processing and inversion of SkyTEM data for high resolution hydrogeophysical surveys. Presented at: *ASEG 2007: The 19th Geophysical Conference& Exhibition, 18–22 November 2007, Perth, Australia.*

Auken, E., Foged, N., Christiansen, A.V. & Sorensen, K. (2007b) Enhancing the resolution of the subsurface by joint inversion of X- and Z-component SkyTEM data. Presented at: *ASEG 2007: The 19th Geophysical Conference & Exhibition, 18–22 November 2007, Perth, Australia.*

Auken, E., Foged, N., Christiansen, A.V., Westergaard, J.H., Kirkegaard, C., N. Foged & A. Viezzoli, (2009) An integrated processing scheme for high-resolution airborne electromagnetic surveys, the SkyTEM system. *Exploration Geophysics,* 40 (2), 184–192.

Baltassat, J.M., N.S. Krishnamurthy, J.F. Girard, S. Dutta, B. Dewandel, S. Chandra, M. Descloitres, A. Legchenko, H. Robain, V. Ananda Rao & S. Ahmed, (2006) *Proton magnetic resonance technique in weathered – fractured aquifers,* IFCPAR project 2700-W1, BRGM/RP-54538-FR, pp 33–42.

Barrocu, G. & Ranieri, G. (2000) TDEM: a useful tool for identifying and monitoring the fresh-saltwater interface, *Proc., 16th Salt Water Intrusion Meet (SWIM), Delft, Wolin Island, Poland, pp 12–15*

Baudoin, P., Durozoy, G. & Utard, M. (1970) *Study of a fresh water-salt water contact in the Rhone Delta using airborne electromagnetic methods.* In: Morley, L.W. (ed.) *Mining and Groundwater Geophysics, 1967;* Geol. Surv. Canada, Econ. Geol. Rep. 26, 627–637 (in French)

Bromley, J., Mannstrom, B., Nisca, D. & Jamtlid, A. (1994) Airborne Geophysics: Application to a ground-water study in Botswana. *Ground Water,* 32 (1), 79–90.

Buselli, G., Barber, C., Devis, G.B. & Salama, R.B. (1990) Detection of groundwater contamination near waste disposal sites with transient electromagnetic and electrical methods. In: Ward, S.H. (ed.) *Geotechnical and Environmental Geophysics, Vol. II: Environmental and Groundwater.* Tulsa, Oklahoma, USA, Society of Exploration Geophysicists. pp. 27–40.

Buselli, G. & Lu, K. (2001) Groundwater contamination monitoring with multichannel electrical and electromagnetic methods. *Journal of Applied Geophysics,* 48 (1), 11–23.

Carrasquilla, A.A.G. & Ulugergerli, E. (2006) Evaluation of the transient electromagnetic geophysical method for stratigraphic mapping and hydrogeological delineation in Campos Basin, Brazil. *Brazilian Journal of Geophysics,* 24 (3), 333–341.

Chandra, S. Senior Scientist, CSIR-National Geophysical Research Institute, Hyderabad, India (Personal Communications, May–June, 2015).

Characterization using Advanced Geophysical Techniques in Representative Geological Terrains of India – AQKAR, Tumkur, Karnataka, sponsored by Central Ground Water Board, Ministry of Water Resources, River Development & Ganga Rejuvenation, Govt. of India and funded by The World Bank under HP II.

Chen, Chow-Son (1999) TEM investigations of aquifers in the, southwest coast Taiwan. *Ground Water*, 37 (6), 890–896.

Christiansen, A.V. (2003) Application of airborne TEM methods in Denmark and layered 2D inversion of resistivity data. *Ph. D Thesis, Dept. Earth Sci., University of Aarhus, Denmark*.

Christiansen, A.V. & Auken, E. (2012) A global measure for depth of investigation. *Geophysics*, 77 (4), WB 171-WB 177.

Christiansen, A.V., Auken, E., Foged, N. & Sorensen, K.I. (2007) Mutually and laterally constrained inversion of CVES and TEM data: A case study. *Near Surface Geophysics*, 5 (2), 115–123.

Christensen, N.B. & Sorensen, K.I. (1998) Surface and borehole electrical and electromagnetic methods for hydrogeological investigations. *European Journal of Environmental & Engineering Geophysics*, Vol. 3, pp. 75–90.

Collett, L.S. (1966) *Airborne electromagnetic survey over the Winkler aquifer, Manitoba*. In: Blackadar, R.G. (ed.) *Report of Activities, November 1965 to April 1966*. Geol. Survey of Canada, Paper 66-2. pp. 6–9.

Constable S.C., Parker, R.L. & Constable, C.G. (1987) Occam's inversion: A practical algorithm for generating smooth models from electromagnetic sounding data. *Geophysics*, 52 (3), 289–300.

Cook, P.G. (1992) A helicopter-borne electromagnetic survey to delineate groundwater recharge rates. *Water Resources Research*, 28 (11), 2953–2961.

CSIR-NGRI (2015) *Final report on Pilot Aquifer Mapping Project (AQUIM): AQKAR, Tumkur District, Karnataka*. CSIR-National Geophysical Research Institute, Hyderabad, India report on Aquifer

Danielsen, J.E., Auken, E., Jorgensen, F., Sondergaard, V. & Sorensen, K.I. (2003) The application of the transient electromagnetic method in hydrogeohysical surveys. *Journal of Applied Geophysics*, 53 (4), 181–198.

d'Ozouville, N., Auken, E., Sorensen, K., Violette, S., de Marsily, G., Deffontaines, B. & Merlen, G. (2008) Extensive perched aquifer and structural implications revealed by 3 D resistivity mapping in a Galapagos volcano. *Earth and Planetary Science Letters*, 269 (3–4), 518–522.

Everett, M.E. & Meju, M.A. (2005) Near-surface controlled-source electromagnetic induction: background and recent advances. In: Rubin, Y. & Hubbard, S.S. (eds.) *Hydrogeophysics: Water Science and Technology Library, Vol. 50, Chapter 6*. The Netherlands, Springer. pp. 157–183.

Fitterman, D.V. (1987) Examples of transient sounding for ground-water exploration in sedimentary aquifers. *Ground Water*, 25 (6), 685–692.

Fitterman, D.V. (1989) Detectability levels for central induction transient soundings. *Geophysics*, 54 (1), 127–129.

Fitterman, D.V. & Anderson, W.L. (1987) Effect of Transmitter Turn-Off Time on Transient Soundings. *Geoexploration*, 24 (2), 131–146.

Fitterman, D.V. & Labson, V.F. (2005) Electromagnetic Induction Methods for Environmental Problems. In: Butler, D.K. (ed.) Near-Surface Geophysics, Part 1, Chapter 10, SEG Investigations in Geophysics Series: No. 13, Tulsa, Oklahoma, USA, Society of Exploration Geophysicists. pp. 301–355.

Fitterman, D.V. & Deszcz-Pan, M. (2004) Characterization of saltwater intrusion in South Florida using electromagnetic geophysical methods, *18 SWIM, Catagena, Spain*, pp. 405–416.

Fitterman, D.V., Frischknecht, F.C., Mazzella, A.T. & Anderson, W.L. (1990) Example of transient electromagnetic soundings in presence of oil field pipes. In: Ward, S.H. (ed.) *Geotechnical and Environmental Geophysics, Vol. II: Environmental and Groundwater.* Tulsa, Oklahoma, USA, Society of Exploration Geophysicists. pp. 79–88.

Fitterman, D.V., Meekes, J.A.C. & Ritsema, I.L. (1988) Equivalence behavior of three electrical sounding methods as applied to hydrogeological problems. Presented at: *The 50th Annual Meeting and Technical Exhibition of the European Association of Exploration Geophysicists, 6–8 June 1988, The Hague, The Netherlands.*

Fitterman, D.V. & Stewart, M.T. (1986) Transient electromagnetic sounding for groundwater. *Geophysics,* 51 (4), 995–1005.

Flores, C. & Velasco, N. (1998) A comparative analysis between transient electromagnetic soundings and resistivity soundings in the Tres Virgenes geothermal zone, Mexico. *Geofisica International,* 37 (3), 183–199.

Foged, N., Auken, E., Christiansen, A.V. & Sørensen, K.I. (2013) Test-site calibration and validation of airborne and ground-based TEM systems. *Geophysics,* 78 (2), E95–E106.

Frischknecht, F.C., Labson, V.F., Spies, B.R. & Anderson, W.L. (1991) Profiling methods using small sources. In: Nabighian, M.N. (ed.) *Electromagnetic Methods in Applied Geophysics, Vol. 2: Applications, Part A,* Chapter 3, Tulsa, Oklahoma, U.S.A., Society of Exploration Geophysicists. pp. 105–270.

Goldman, M., Gilad, D., Ronen A. & Melloul, A. (1991) Maping of seawater intrusion into the coastal aquifer of Israel by the time domain electromagnetic method. *Geoexploration,* 28 (2), 153–174.

Harris, B.D. (2001) Transient electromagnetic methods and their application to the delineation and assessment of groundwater resources in the Eastern Goldfields, Western Australia. *Ph.D Thesis, Curtin University of Technology, Australia.*

Hoekstra, P. & Blohm, M.W. (1990) Case histories of time domain electromagnetic soundings in environmental geophysics. In: Ward, S.H. (ed.) *Geotechnical and Environmental Geophysics, Vol. II: Environmental and Groundwater.* Tulsa, Oklahoma, USA, Society of Exploration Geophysicists, pp. 1–15.

Holladay, S. & Lo, B. (1997) Airborne Frequency-Domain EM-Review and Preview. In: Gubins, A.G. (ed.) *Proceedings of Exploration 97: Fourth Decennial International Conference on Mineral Exploration, 14–18 September, 1997, Toronto, Canada,* GEO F/X Div. of AG Inf. Sys. Ltd. Canada, pp. 505–514.

Hoversten, G.M. & Morrison, H.F. (1982) Transient fields of a current loop source above a layered earth. *Geophysics,* 47 (7), 1068–1077.

Hurwitz, S., Goldman, M., Ezersky, M. & Gvirtzman, H. (1999) Geophysical (time domain electromagnetic model) delineation of a shallow brine beneath a freshwater lake, the Sea of Galilee, Israel. Water Resources Research, 35 (12), 3631–3638.

Jacobsen, L.H. & Edsen, N.A. (2005) *1D LCI-interpretation applied to large scale SkyTEM data sets.* Presented at: *EAGE Near Surface 2005: The 11th meeting of environmental and engineering geophysics, Near Surface division, 4–7 September 2005, Palermo, Italy.*

Jorgensen, F., Sandersen, P.B.E. & Auken, E. (2003) Imaging buried Quaternary valleys using the transientelectromagnetic method. *Journal of Applied Geophysics,* 53 (4), 199–213.

Jorgensen, F., Scheer, W., Thomsen, S., Sonnenborg, T.O., Hinsby, K., Wiederhold, H., Schamper, C., Burschil, T., Roth, B., Kirsch R. & Auken, E. (2012) Transboundary geophysical mapping of geological elements andsalinity distribution critical for the assessment of future sea water intrusion in response to sea level rise. Hydrol. *Earth Syst. Sci.,* 16 (7), 1845–1862.

Jupp, D.L.B. & Vozoff, K. (1975) Stable Iterative Methods for the Inversion of Geophysical Data. *Geophys. Journal of Royal Astr. Soc.,* 42 (3), 957–976.

Kamenetsky, F. & Oelsner, C. (2000) Distortions of EM transients in coincident loops at short time-delays. *Geophysical Prospecting*, 48 (6), 983–993.

Kirkegaard, C., Foged, N., Auken, E., Christiansen, A.V. & Sorensen, K. (2012) On the value of including X-component data in 1 D modeling of electromagnetic data from helicopterborne time-domain systems in horizontally layered environments, *Journal of Applied Geophysics*, 84, 61–69.

Kirkegaard, C., Sonnenborg, T.O., Auken, E. & Jorgensen, F. (2011) Salinity distribution in heterogeneous coastal aquifers mapped by airborne electromagnetic. *Vadose Zone Journal*, 10 (1), 125–135.

Kok, A., Auken, E., Goren, M., Ribeiro, J. & Schaars, F. (2010) Using ground based geophysics and airborne transient electromagnetic measurements (SkyTEM) to map salinity distribution and calibrate a groundwater model for the island of Terschelling-The Netherlands. In: *Proceedings of Twentyfirst Salt Water Intrusion Meeting (SWIM)*, 21–25 June 2010, Azores, Portugal.

Leite, D.N., Porsani, J.L., Bortolozo, C.A. & Couto Jr, M.A. (2013) Geoelectrical characterization by VES/TEM joint inversion at Urupês region, Paraná basin, Brazil. Presented at: *Near Surface Geoscience 2013,The 19th European Meeting of Environmental and Engineering Geophysics*, 9–11 September, 2013, Bochum, Germany.

Lipinski, B., Sams, J., Smith, B. & Harbert, W. (2008) Using Helicopter Electromagnetic Surveys to Evaluate Coal bed Natural Gas Produced Water Disposal in the Powder River Basin, Wyoming. *Geophysics*, 73 (3), B77–B84.

Macnae, J.C. (1984) Survey design for multicomponent electromagnetic systems. *Geophysics*, 49 (3), 263–273.

Mallick, K. & Verma, R.K. (1979) Time-domain electromagnetic sounding computation of multi-layer response and the problem of equivalence in interpretation. *Geophysical Prospecting*, 27 (1), 137–155.

McNeill, J.D. (1990) Use of electromagnetic methods for groundwater studies. In: Ward, S.H. (ed.) *Geotechnical and Environmental Geophysics, Vol. I: Review and Tutorial*. Tulsa, Oklahoma, USA, Society of Exploration Geophysicists, pp. 191–218.

Meju, M.A. (1996) Joint inversion of TEM and distorted MT soundings: some effective practical considerations. *Geophysics*, 61 (1), 56–65.

Meju, M.A. (1998) A simple method of transient electromagnetic data analysis. *Geophysics*, 63 (2), 405–410.

Meju M.A. (2005a) Simple relative space–time scaling of electrical and electromagnetic depth sounding arrays: implications for electrical static shift removal and joint DC-TEM data inversion with the most-squares criterion. *Geophysical Prospecting*, 53 (4), 463–469.

Meju, M.A. (2005b) *Non-invasive characterization of fractured crystalline rocks using a combined multicomponent transient electromagnetic, resistivity and seismic approach*. In: Harvey P.K., Brewer, T.S., Pezard, P.A. & Petrov V.A. (eds.) Petrophysical Properties of Crystalline Rocks, Special Publications 240, London, The Geological Society, pp. 195–206.

Meju, M.A., Fontes, S.L., Ulugergerli, E.U., La Terra, E.F., Germano, C.R. & Carvalho, R.M. (2001) A joint TEM-HLEM geophysical approach to borehole siting in deeply weathered granitic terrains. *Ground Water*, 39 (4), 554–567.

Mills, T., Hoekstra, P., Blohm, M. & Evans, L. (1988) Time domain electromagnetic soundings for mapping sea-water intrusion in Monterey County, California. *Ground Water*, 26 (6), 771–782.

Minsley, B.J. (2011) A trans-dimensional Bayesian Markov chain Monte Carlo algorithm for model assessment using frequency-domain electromagnetic data. *Geophysical Journal International*, 187 (1), 252–272.

Nabighian, M.N. (1979) Quasi-static transient response of a conducting half-space – An approximate representation. *Geophysics*, 44 (10), 1700–1705.

Nabighian, M.N. & Macnae, J.C. (1991) Time domain electromagnetic prosecting methods. In: Nabighian, M.N. (ed.) *Electromagnetic Methods in Applied Geophysics, Vol. 2: Applications, Part A*, Chapter 6, Tulsa, Oklahoma, U.S.A., Society of Exploration Geophysicists. pp. 427–520.

Odins, J.A., Beckham, J. & O'Neill, D.J. (1995) Supplementary ground geophysics for airborne electromagnetic salinity survey over Jemalong-Wyldes Plains area. *Exploration Geophysics*, 26 (2–3), 195–201.

Oldenborger, G.A. and Brewer, K. (2014) *Time-domain electromagnetic data for the Spiritwood valley aquifer, Manitoba.* Geological Survey of Canada, Open File 7593. doi:10.4095/293700.

Oldenburg, D.W. & Li, Y. (1999) Estimating depth of investigation in DC resistivity and IP surveys. *Geophysics*, 64 (2), 403–416.

Paine, J.G. & Collins, E.W. (2003) Applying airborne electromagnetic induction in ground-water salinization and resource studies, West Texas. In: *Proceedings of the Symposium on the Application of Geophysics to Environmental and Engineering Problems, 6–10 April 2003, Texas, USA.* Environmental and Engineering Geophysical Society, pp. 722–738.

Palacky, G.J. (1981) The airborne electromagnetic method as a tool of geological mapping, *Geophysical Prospecting*, 29 (1), 69–88.

Palacky, G.J. & West, G.F. (1991) Airborne electromagnetic methods. In: Nabighian, M.N. (ed.) *Electromagnetic Methods in Applied Geophysics, Vol. 2: Applications, Part B*, Chapter 10, Tulsa, Oklahoma, U.S.A., Society of Exploration Geophysicists. pp. 811–879.

Paterson, N.R. & Bosschart, R.A. (1987) Airborne geophysical exploration for groundwater. *Ground Water*, 25 (1), 41–50.

Pellerin, L., Johnston, J.M. & Hohmann, G.W. (1996) A numerical evaluation of electromagnetic methods in geothermal exploration. *Geophysics*, 61 (1), 121–130.

Pfaffhuber, A.A., Grimstad, E., Domaas, U., Auken, E., Foged, N. & Halkjaer, M. (2010) Airborme EM mapping of rockslides and tunnelling hazards. *The Leading Edge*, 29 (8), 956–959.

Pfaffhuber, A.A., Bazin, S., Domaas, U. & Grimstad, E. (2011) Electrical Resistivity Tomography to follow up an airborne EM rock slide mapping survey – Linking rock quality with resistivity. Presented at: *The Twelfth International Congress of the Brazilian Geophysical Society, 15–18 August 2011, Rio de Janeiro, Brazil.*

Podgorski, J.E., Auken, E., Schamper, C., Christiansen, A.V., Kalscheuer, T. & Green, A.G. (2013) Processing and inversion of commercial helicopter time-domain electromagnetic data for environmental assessments and geologic and hydrologic mapping. *Geophysics.* 78 (4), E149–E159.

Pugin, A.J.-M., Oldenborger, G.A., Cummings, D.I., Russell, H.A.J. & Sharpe, D.R. (2014) Architecture of buried valleys in glaciated Canadian Prairie regions based on high resolution geophysical data. Quaternary Science Review, 86, 13–23.

Raiche, A.P., Jupp, D.L.B., Rutter, H. & Vozoff, K. (1985) The joint use of coincident loop transient electromagnetic, and Schlumberger sounding to resolve layered structures. *Geophysics*, 50 (10), 1618–1627.

Reid, J. & Viezzoli, A. (2007) High-resolution near surface airborne electromagnetic – SkyTEM survey for uranium exploration at Pelle Range, WA, *ASEG 2007 – Nineteenth Geoph. Conf., extended abstract, Conference Handbook.*

Sattel, D. & Kgotlhang, L. (2004) Groundwater Exploration with AEM in the Boteti area, Botswana. *Exploration Geophysics*, 35 (2), 147–156.

Schamper, C., Jorgensen, F., Auken, E. & Efferso, F. (2014) Assessment of near-surface mapping capabilities by airborne transient electromagnetic data – an extensive comparison to conventional borehole data. *Geophysics*, 79 (4), B-187-B199.

Schamper, C., Rejiba, F. & Guérin, R. (2012) 1D single-site and laterally constrained inversion of multifrequency and multicomponent ground-based electromagnetic induction data-application to the investigation of a near-surface clayey overburden. *Geophysics*, 77 (4), WB19–WB35.

Schmutz, M., Albouy, Y., Guerin, R., Maquaire, O., Vassal, J., Schott, J.-J. & Descloitres, M. (2000) Joint electrical and time domain electromagnetism (TDEM) data inversion applied to the super sauze earthflow (France). *Surveys in Geophysics*, 21 (4), 371–390.

Sengpiel, K.P. (1986) New possibilities for groundwater exploration using airborne electromagnetics, *Proc., 9th Salt Water Intrusion Meet (SWIM), Delft, The Netherlands* pp. 671–683.

Sengpiel, K.P. & Meiser, P. (1981) Locating the fresh water/salt water interface on Spiekeroog island by airborne EM resistivity/depth mapping. *Geologisches Jahrbuch C 29*. Hannover, Germany.

Sengpiel, K.P. & Fluche, B. (1992) Application of airborne electromagnetic to groundwater exploration in Pakistan. *Z. dt. geol. Ges. (Hannover)*, 143, 254–261.

Sharma, S.P. & Kaikkonen, P. (1999) Appraisal of equivalence and suppression problems in 1D EM and DC measurements using global optimization and joint inversion. *Geophysical Prospecting*, 47 (2), 219–249.

Shirzaditabar, F., Bastani, M. & Oskooi, B. (2011) Imaging a 3D geological structure from HEM airborne magnetic and ground ERT data in Kalat-e-Reshm area, Iran. *Journal of Applied Geophysics*, 75 (3), 513–522.

Shu, Y., Xian-Xin, S. & Ming-Sheng, C. (2009) The probing depth of transient electromagnetic field method. *Chinese Journal of Geophysics*, 52 (3), 693–703.

Siemon, B., Christiansen A.V. & Auken, E. (2009) A review of helicopter-borne electromagnetic method for groundwater exploration. *Near Surface Geophysics*, 7 (5 & 6), 629–646.

Smith, R.S. & Annan, A.P. (1997) Advances in airborne time-domain EM technology. In: Gubins, A.G. (ed.) *Proceedings of Exploration 97: Fourth Decennial International Conference on Mineral Exploration, 14–18 September, 1997, Toronto, Canada*, GEO F/X Div. of AG Inf. Sys. Ltd. Canada, pp. 497–504.

Smith, B.D., Grauch, V.J.S., McCafferty, A.E., Smith, D.V., Rodriguez, B.R., Pool, D.R., D-Pan, M. & Labson, V.F. (2007) Airborne electromagnetic and magnetic surveys for groundwater resources: A decade of study by the U.S. Geological Survey. In: Milkereit, B. (ed.) *Exploration 07: Proceedings of the Fifth Decennial Int. Conf. on Mineral Exploration, Sept 9 to 12, 2007, Toronto, Canada.* pp. 895–899.

Smith R.S. & Keating, P.B. (1996) The usefulness of multicomponent, time-domain airborne electromagnetic measurements. *Geophysics*, 61 (1), 74–81.

Sorensen, K.I. & Auken, E. (2004) SkyTEM – a new high-resolution helicopter transient electromagnetic system. *Exploration Geophysics*, 35 (3), 194–202.

Sorensen, K.I., Sorensen, B., Christiansen, A.V. & Auken, E. (2004) Interpretation of a hydrogeological survey- data from high resolution SkyTEM system. *Paper presented at EAGE 10th European Meeting of Environmental Engineering & Geophysics*.

Sorensen, K.I., Auken, E., Christensen, N.B. & Pellerin, L. (2005) An integrated approach for hydro-geophysical investigations: new technologies and a case history. In: Butler, D.K. (ed.) *Near-Surface Geophysics, Part 2, Chapter 21, Series: Investigations in Geophysics: No. 33*, Tulsa, Oklahoma, U.S.A., Society of Exploration Geophysicists.

Spies, B.R. (1989) Depth of investigation in electromagnetic sounding methods. *Geophysics*, 54 (7), 872–888.

Spies, B.R. & Frischknecht, F.C. (1991) Electromagnetic sounding. In: Nabighian, M.N. (ed.) *Electromagnetic Methods in Applied Geophysics, Vol. 2: Applications, Part A, Chapter 5*, Tulsa, Oklahoma, U.S.A., Society of Exploration Geophysicists. pp. 285–422.

Spies, B.R. & Macnae, J.C. (1997) Electromagnetic Trends—Spatial, Temporal and Economic, *electrical and electromagnetic method*. In: Gubins, A.G. (ed.) *Proceedings of Exploration 97: Fourth Decennial International Conference on Mineral Exploration, 14–18 September, 1997, Toronto, Canada*, GEO F/X Div. of AG Inf. Sys. Ltd. Canada, pp. 489–496.

Steuer, A. & Siemon, B. (2005) A comparison of TEM, HEM and SKYTEM measurements at a buried valley in Northern Germany. In: Near Surface 2005: *Proredings of the Near Surface 2005, Italy, Eleventh European Meeting of Env. & Eng. Geop, 4–7 September, Palermo, Italy*. Near Surface Geoscience Div. EAGE.

Steuer, A. Siemon, B. & Auken, E. (2009) A comparison of helicopter-borne electromagnetic in frequency- and time-domain at the Cuxaven valley in Northern Germany. *Journal of Applied Geophysics*, 67 (3), 194–305.

Stewart, M. & Gay, M.C. (1986) Evaluation of transient electromagnetic soundings for deep detection of conductive fluids. *Ground Water*, 24 (3), 351–356.

Street, G.J. (1992) Airborne geophysical surveys: Applications in land management. *Exploration Geophysics*, 23 (1&2), 333–338.

Street, G.J. & Anderson, A. (1993) Airborne electromagnetic surveys of the regolith. *Exploration Geophysics*, 23 (3 & 4), 795–800.

USGS (1990) *Developments and applications of modern airborne electromagnetic surveys*. Fitterman, D. (ed.) U.S. Geological Survey Bulletin 1925. U.S.G.S.

Verma, R.K. (1980) Equivalence in electromagnetic (frequency) sounding. *Geophysical Prospecting*, 28 (5), 776–691.

Viezzoli, A., Christiansen, A.V., Auken, E. & Sorensen, K. (2008) Quasi-3D modeling of airborne TEM data by spatially constrained inversion. *Geophysics*, 73 (3), F105–F113.

Viezzoli, A., Auken E. & Munday, T. (2009) Spatially constrained inversion for quasi 3D modelling of airborne electromagnetic data-an application for environmental assessment in the Lower Murray Reion of South Australia. *Exploration Geophysics*, 40 (2), 173–183.

Vozoff, K. & Jupp, D.L.B. (1975) Joint inversion of geophysical data. *Geophysical Journal of the Royal Astronomical Society*, 42 (3), 977–991.

Wiederhold, H., Siemon, B., Steuer, A., Schaumann, G., Meyer, U., Binot, F. & Kühne, K. (2010) Coastal aquifers and saltwater intrusions in focus of airborne electromagnetic surveys in Northern Germany. *SWIM21-21st Salt Water Intrusion Meeting, 21–25 June 2010, Azores, Portugal*.

Wynn, J. (2006) *Mapping ground water in three dimensions – An analysis of airborne geophysical surveys of the Upper San Pedro River Basin, Cochise County, Southeastern Arizona*. U.S. Geological Survey Professional Paper-1674. U.S.G.S.

Yang, C.-H. & Tong, L.-T. (1999) A study of joint inversion of direct current resistivity, transient electromagnetic and magnetotelluric sounding data. *Terrestrial, Atmospheric and Oceanic Sciences,Diqiu Kexue Jikan (TAO)*, 10 (1), 293–301.

Chapter 12

The borehole geophysical logging methods

12.1 INTRODUCTION

The in-hole continuous depth-wise measurement of physical properties of geological formations immediately surrounding the borehole is known as borehole geophysical logging and is within the domain of subsurface geophysics. In hard rock groundwater exploration geophysical logging is carried out to recognize different geological formations, the presence of fractured zones and assess their hydrological characteristics. According to Keys (1979), the identification or confirmation, orientation and characterization of fractured zones in terms of their water transmitting capabilities is one of the most important results desired from logging.

The surface geophysical methods such as electrical and electromagnetic presented so far sense the subsurface conditions remotely and help identify the fractured zones which are prominent. The resolving power of these methods decreases with depth. The combined use of these methods and other methods, though they increase the resolution and minimize the ambiguities, may not resolve an individual fracture – a thin planar feature at depth, which could be highly productive and may still remain undetected. While in a borehole, providing direct access to the subsurface for *in situ* high density measurements, the thin horizontal or dipping fractured zones intersected at depths can be defined with a far better resolution. The advantage of geophysical logging is that it gives a continuous depth-wise record at a scale constant with depth. If required sampling of measured parameters at specific depths is also possible which cannot be obtained from surface geophysical methods where deeper information is averaged out and the dimension and physical property contrast of the target at depth has to be large enough to get a response at the surface. Obviously, to reduce the number of test boreholes and the level of uncertainty or well failure and to select borehole sites, where productive fractures can be encountered, a surface geophysical survey is essential and the only economic alternative.

Geophysical logging methods were primarily evolved for oil wells and continue with well developed instrumentation and techniques. The water wells compared to oil wells involve a different objective, much less depth, lesser diameter generally varying from 6 to 12 inches, a relatively fresh water hydrogeological environment and are constructed in a different way. Accordingly, the logging methods adopted are used with portable machines, purpose-driven modifications in probe size, electronics, cable type and length, recording and data processing systems and interpretational techniques. The early developments and various applications are given in Keys (1967).

After a borehole is drilled for groundwater, its geophysical logging is conducted for well construction. It is essential for water wells drilled in sedimentary terrain. This is to determine properties of the water bearing formation viz., lithology, vertical extent, quality of water (in terms of electrical conductivity), confining clays, porosity, flow etc. and design well-assembly for construction of a production well. An idea about the yield and hydraulic properties of aquifers can also be obtained. In hard rock which differs from sedimentary terrain by its composition, compaction and fracture opening, the aim of logging is mainly to obtain depth-wise precise information on hydrogeologically significant, open (yielding or receiving) thin fractured zones, groundwater flow from these zones, groundwater quality, contaminated and non-contaminated ones and if required identify depth zones for hydraulically induced fracturing. Also, it can be used to understand the interconnectivity of fractures and their role in contaminant transport.

Besides, the geophysical log is also used to validate and standardize surface geophysical measurements and correct the borehole lithological log. In fact it adds value to surface and borehole lithological data. The correlation of geophysical logs of boreholes in an area and also assessing the response of a well through logging due to the pumping of another, help define lateral continuity of aquifers/fractured zones and evaluate the sub-surface hydrogeological condition. The most significant use of fracture characterization through logging is in assessing their groundwater transport capability and planning aquifer protection as fractures are vulnerable paths for contaminant movement. Time-lapsed logging helps monitor water quality and contamination levels in fractured zones. Also logging of existing wells helps assess the casing wall and screen conditions required for aquifer remediation and well rehabilitation.

The logging methods can be grouped as active and passive. The passive methods include measurements of physical properties without exciting the formations i.e., without putting a source in the borehole. Such measurements are conventional spontaneous or self-potential, temperature, natural gamma radioactivity and also magnetic susceptibility. Besides, the caliper and optical televiewer and flowmeter record the borehole wall condition and groundwater ambient flow without using any active source for excitation. In active methods formations are excited by electrical, electromagnetic, radioactive or seismic source which moves through the borehole depth and the response of the surrounding formations is recorded. The conventional logging, either active or passive, involves measurements in a single borehole and generates 1d information. Some of the measurements can also be made involving more than one borehole to generate a subsurface image and the cross-borehole fracture flow pattern which is generally referred as 'cross-borehole'. Also, there could be measurements such as surface to borehole and borehole to surface depending on the location of transmitter (source) and receiver, i.e., source on surface or in borehole. These are used for solving specific groundwater problems related to fracture dimension and contaminant movement and not in regular groundwater exploration. Generally, the borehole is logged cost-effectively by a selected logging method or combination (multi-parametric) logging methods involving simultaneous or sequential measurement of more than one physical property. Since all the methods have some limitations in characterizing the subsurface, the multi-parametric logging approach is adopted that helps remove or minimize the ambiguities in interpretations.

While logging a borehole, measurements are made by different probes or sonde lowered into or pulled out of the borehole through a cable on winch located on the

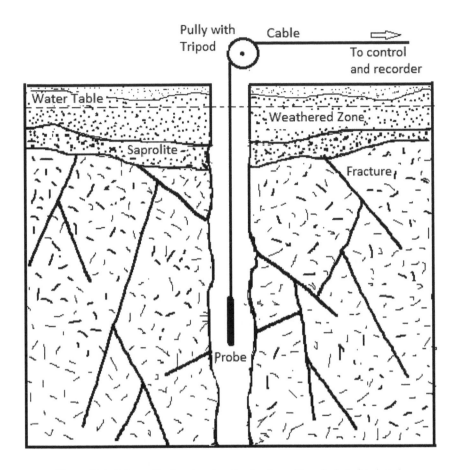

Figure 12.1 Schematic diagram showing logging of borehole in hard rock.

ground surface. The data transmitted are recorded and stored at the surface through an analog or digital logging console (Fig. 12.1). The noise or spurious effects are examined and filtered. Digital logging is preferred as it facilitates easy storage and access and computerized filtering, processing and interpretations using software. The system can be portable or installed in a vehicle. While interpreting the fractured zone from geophysical logs, they are integrated with the lithological log and drill-time log. These are borehole logs showing depth-wise variations in lithology and the time taken in drilling through the formations, i.e., the variations in rate of drilling (Fig. 12.2). The drilling through fractured zones takes less time compared to that through a compact formation. It could be an indicator of change in the compactness of the formation or lithology as well as the occurrence of fractured zones. Also, it is necessary to record the depths of water strikes and discharges.

The geophysical logging for groundwater was initiated through electrical logging. It comprised only single point resistance (SPR) and self potential (SP). The interpretation was mainly qualitative to decipher aquifers, non-aquifer zones, interfaces and

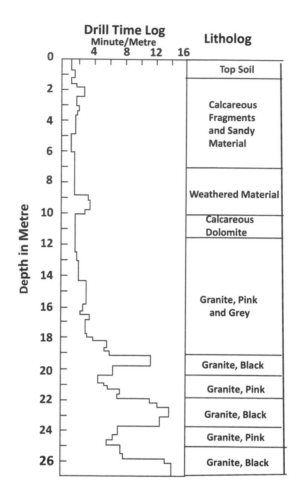

Figure 12.2 Drill-time and lithological log of a borehole in granitic terrain (Radhakrishna *et al.*, 1979).

water quality. Subsequently, a variety of methods like electrical resistivity with varied electrode spacing, nuclear, fluid conductivity, temperature, caliper, flowmeter and acoustic were added. In addition to these there are logging methods using electromagnetic induction and magnetic susceptibility. The other methods used are acoustic and optical borehole televiewers, radar and nuclear magnetic resonance. In hard rock wells fractures encountered are generally quite thin about a few centimeters only and may hold water of different quality and also altered material. So, the variation in resistivity across a fracture measured through the most commonly used conventional resistivity log alone may not yield the authentic information with desired resolution unambiguously and constrain the hydrogeological interpretation. Therefore, multiparametric logging combining some of the methods mentioned above is used to study other fracture related parameters, minimize the ambiguity and get adequate information on fractured zones, lithology and fracture flow. For example, a combination of resistivity low, increased borehole diameter, change in temperature gradient and flow at depth in a borehole indicates the presence of an open fracture. Some of the important references are Keys (1979, 1988), Crosby and Anderson (1971), Nelson *et al.* (1982), Soonawala (1983), Davison *et al.* (1984), Paillet *et al.* (1987b), Paillet (1994), National Academy of Sciences (1996), Stumm *et al.* (2001), Williams *et al.* (2002), Schmitt *et al.* (2003) and Paillet & Ellefsen (2005) on identification of fractured zones combining the methods of natural gamma radioactivity, single point resistance, caliper, electromagnetic induction, SP, resistivity, fluid temperature, flow, acoustic and optical televiewer and radar. Most of the logs respond to the presence of fractures but all of them may not be conclusive. However, it is not a common practice in water well construction in hard rock to use all logging methods as cost of equipment and expertise increases. Therefore, a combination of selected logs is preferred. The standard combination is of conventional log like electrical resistivity, caliper, temperature-fluid conductivity and flowmeter. The maximum combination of logs has been used in ensuring 'no-fracture', required for underground tunnel construction or nuclear waste disposal and 'fractures' for hydrocarbon exploration.

12.2   SPONTANEOUS POTENTIAL

Spontaneous Potential or SP logging is the oldest, simplest, most-used, cost-effective logging method run in mud or water filled open borehole to ascertain lithological boundaries, clay layers and quality of formation water (in terms of EC). SP is the natural potential developed in a water and mud filled borehole which pierced through geological formations holding interstitial fluid of electrical conductivity different from borehole fluid. SP of the order of tens to hundreds of milliVolts develops. The potential is a sum of potentials due to chemical activity originating from the difference in salinity of mud filtrate and formation water, known as diffusion or 'liquid-junction' potential and potential at the contacts between non-granular and granular zones known as 'membrane potential'. Because of this, a natural flow of current takes place through the borehole, invaded zone, non-invaded part of the granular layer and the non-granular layers. The lithology of individual beds and difference in water quality play an important role in the development of SP. If formation water is more saline compared to borehole fluid, SP development is negative otherwise positive. Positive potentials are indicated by shifting of the curve towards the right while negative potentials are towards the left. The point of inflection on the SP curve indicates lithologic contact. Besides this, movement of ions with liquid through mud cake and invaded zone into the pores of permeable formations due to difference in hydrostatic pressures also gives rise to a potential called 'electrokinetic' or 'streaming' potential. Generally, its contribution is negligible but becomes considerably high when the pressure difference between borehole fluid and formation water is high. Streaming potential also develops when formation water is at a high pressure and gushes into the borehole.

In hard rock, sedimentary conditions for development of SP are not present and therefore SP log is not the usual one as obtained in sedimentary terrain with consistency and reliability in its expected character. It is generally difficult to interpret the developed potential, if any, in terms of fractured zone occurrences or assign it a specific reason. It may be even totally featureless. Because of the electrokinetic potential developed due to groundwater in-flow or out-flow from fractured zones- gaining or losing zones (Keys and MacCary, 1971) there could be a minor development of negative potential or sudden oscillation imposed as 'noise' or spikes on the SP curve against these zones with considerable flow. Thus, the depth zones of inflowing or losing water can possibly be picked up. However, the polarity of developed potential will depend on the electrical conductivities of water in the borehole and that in the fractured zone. Also, potential can develop due to mineralization and secondary filling of fractures. These potentials are at times erratic, departing from typical response (Keys and MacCary, 1971) and mostly non-conclusive. According to Stumm *et al.* (2001) SP results are not significant as far as fracture delineation is concerned. Davison *et al.* (1984) say that in hard rock the SP log does not provide any additional information which cannot be obtained from other logs. The lithological changes could possibly be picked up by the development of electrochemical potentials as reported by West *et al.* (1975) and Houtkamp and Jacks (1972). Crosby and Anderson (1971) indicate the inconsistency in character of electrical logging in the Columbia Group of basaltic lava terrain and stress the need of sonde contact with the borehole wall. Overall, the SP log can only support other log responses and cannot be used independently for fracture detection as its interpretation is subjective. The SP log is usually obtained with resistivity logs and no extra log run is involved. Figures 12.3a and 12.3b show two examples of SP logs from metasediments of Jharkhand, India (Chandra *et al.*, 1994). The borehole (a) drilled up to 145.4 m did not encounter any fractured zone. It had a discharge of 0.75 lps only from saprolite in the depth range of 15–23 m. The prominent negative SP at 102 m depth though associated with a resistivity low, is not associated with any fractured zone. A fractured zone was encountered at 107–109 m depth in a nearby borehole shown with dotted horizontal lines in figure 12.3a. Another borehole (b) drilled up to 160.1 m depth intersected fractured zones at 23.4–24.4 m, 31.02–32.0 m and 46.26–47.0 m depths with discharges 0.5, 0.75 and 1.25 lps respectively. The weathered and saprolite zone up to 23 m depth was dry. The fractured zone at 46.26–47 m depth is associated with a sharp negative SP as well as a prominent resistivity low (a discrepancy in depth measurement is observed).

12.3 SINGLE POINT RESISTANCE

A Single Point Resistance (SPR) log is run along with SP and the same electrodes (borehole and surface) are used, each acting as potential as well as current electrode. SPR and SP logs taken together can bring out useful information on fractured zones. Like SP, SPR also cannot be run in cased and dry boreholes. It is a qualitative log used for identifying lithologic boundaries. SPR log measures the resistance offered by formations encountered in the borehole. The resistance is measured between two electrodes, one of which moves in the borehole and the other is placed on the ground surface near the borehole. It investigates only 5 to 10 times the electrode diameter, i.e., the volume of investigation is quite low. Other than conventional SPR, differential SPR measurement can be made by measuring the resistance between two very closely spaced electrodes separated by an insulator, placed on the probe. In this the surface electrode is placed on the probe at a very small separation from the other electrode. It provides better resolution than conventional SPR probe. Interpretation of SPR log is

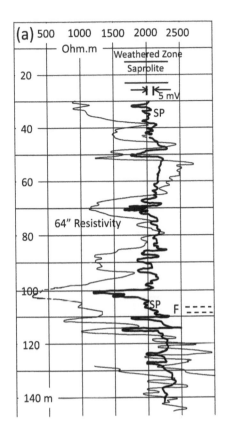
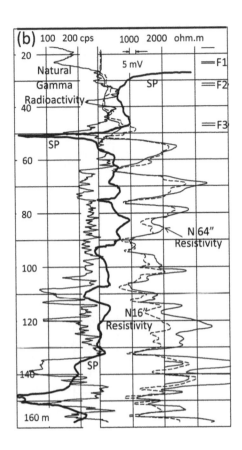

Figure 12.3 SP log (dark black continuous curve) from metasediments, Jharkhand India; (a) borehole drilled up to 145.4 m did not encounter any fractured zone but a prominent negative SP at 102 m depth and in a nearby borehole a fracture was encountered at 107–109 m depth shown by horizontal dotted lines, (b) in a borehole drilled up to 160.1 m the fractured zone at 46.26–47 m depth is associated with a sharp negative SP (Source: Chandra *et al.*, 1994).

quite simple. The changes in lithology are defined effectively by the inflection points on log curve. Against formation with higher resistivities SPR curve always shows higher resistances. The reduction in resistance corresponds to clay filling, water saturation or mineralization of fractured zone. The resistance is further reduced by the enlargement in borehole diameter generally observed at the fractured zone. Having small volume of investigation, SPR is highly sensitive to variations in surrounding volume of borehole fluid due to changes in borehole diameter. For large diameter boreholes or boreholes with saline mud fluid the measured values are dampened.

In hard rock boreholes fractured zones are resolved by SPR log. Olsson et al. (1984) identify the fractures by SPR log in deep boreholes in granites and gneisses of Sweden (Fig. 12.4a) Stumm *et al.* (2001) demonstrate identification of fractured zone from SPR logs in granites, gneisses and schists in Northern Queens County, New York.

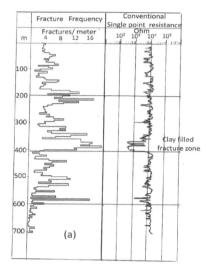

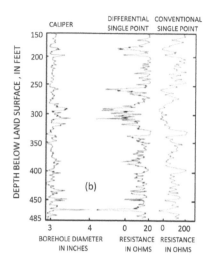

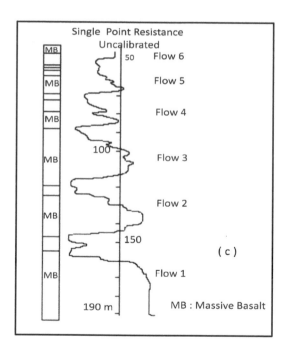

Figure 12.4 Single Point Resistance logs from hard rock terrains, (a) SPR log of a borehole in granite gneiss in Sweden, clay filled fractured zone shows a low (Source: Olsson *et al.*, 1984), (b) caliper, differential and conventional SPR logs of a well in fractured crystalline rock; differential SPR indicates the fractured zone with better resolution supported by diameter enlargement picked up in caliper log (Source: Keys 1988) and (c) uncalibrated SPR from basaltic terrain defining different basalt flows (Source: CGWB, 2004).

According to Davison *et al.* (1984) SPR and differential SPR both can detect major fractured zones, but not the minor ones, while Keys (1988) shows the superiority of differential SPR over conventional SPR log and caliper log in defining narrow fractures (Fig. 12.4b). The differential SPR with very small electrode separation is very sensitive to borehole diameter and hence picks up the fractured zones with better resolution (Keys, 1988; Howard, 1990). The SPR may not be useful in detecting steeply dipping fractures as the associated low resistivity zone will be near-parallel covering a larger depth interval (Keys, 1988). The differential SPR with an excellent thin bed resolution capability is the simplest method of delineating or confirming a fractured zone. The flow-stratigraphy in basaltic terrain comprising amygdaloidal, jointed, vesicular and massive basalts is defined by SPR log. The massive basalt units are associated with higher resistances (Fig. 12.4c, CGWB, 2004).

12.4 RESISTIVITY

Electrical resistivity is an important physical property to identify lithology, hydrologic characteristics and quality of water (in terms of EC) in geological formations intersected by the borehole. The probe consisting of current and potential electrodes injects current into the formation through borehole mud fluid or water. Resistivity logging can be done only in a conductive fluid filled borehole. It is not conducted in metal-cased boreholes. In PVC cased boreholes it can be run to confirm the slot positions as there will be formation related response against the slots only. The resistivity recorded against blank PVC cases will be high.

The probe is moved down or up in the borehole and the potential difference between the potential electrodes is measured to yield variations in resistivity with depth. A variety of electrode configurations are available for resistivity measurements. The most used in groundwater exploration is Normal configuration of electrodes. In Normal configurations one current and one potential electrode are on the probe and return the current electrode is on the cable and the reference potential is grounded. The spacing between active current and potential electrodes could be 4″, 8″, 16″, 32″ and 64″. The commonly used spacings are 16″ or 64″ known as N16″ and N64″ or Short Normal and Long Normal. The electrode at the bottom of a probe located 16″ below the current electrode and the one located at the top of probe, 64″ above current electrode are used for 16″ and 64″ potential measurements. While smaller spacing is useful in thin bed detection, larger spacing provides deeper information. Under normal borehole conditions and bed thickness the measured apparent resistivity by N64″ tends towards true resistivity of the bed.

The measured apparent resistivity of a bed depends on resistivities of the bed, shoulder (underlying and overlying) beds, invaded zone and borehole mud fluid. Also it depends on the diameters of the borehole and the invaded zone, the electrode configuration and bed thickness. Because of the greater effects of these factors the apparent resistivity values recorded by a smaller electrode separation of N16″ is generally less than that by N64″. The N16″ resistivity is greatly affected by mud fluid resistivity and only the variations in flushed or invaded zone resistivity is obtained. The N64″ resistivity, with greater depth of investigation, in qualitative terms, is dominated by the response from formations beyond the invaded zone surrounding the borehole wall.

The N16″ and N64″ resistivity response curves for beds intersecting the borehole are symmetrical with respect to the beds. The logs can be interpreted qualitatively by visual inspection. The approximate bed boundaries and zones of interest are directly interpreted from logs by identifying inflection points on the curve. For quantitative information departure curves are used which involve the effects mentioned above.

In sedimentaries, since the invasion of mud filtrate can occur only in porous and permeable formations, the zone of invasion in clay beds is negligible or thin. Therefore, the low resistivity of clay beds recorded by both the configurations almost overlaps. Against granular zones N16″ and N64″ curves depart. For moderate differences in formation water and borehole fluid resistivities, the N16″ and N64″ curves separate more for permeable than for less permeable ones. The separation can also be due to high contrast in formation water and borehole fluid resistivities where in N16″ it is affected most by borehole fluid resistivity and the curve mostly records the invasion zone resistivity. Also, the response gets dampened with increase in borehole diameter and decrease in borehole fluid resistivity as mentioned by Telford et al. (1990). For a resistive bed as thick as 3 to 4 times the current-potential electrode spacing, apparent resistivity keeps on increasing as the probe approaches from the bottom till the current electrode reaches the middle of the bed. Thereafter, it drops off symmetrically as the probe crosses the bed. The recorded resistivity is different from theoretical curves as all the details are rounded off. With decreasing thickness of beds vis-à-vis current-potential electrodes spacing, the resistivity peak keeps on shrinking in size and disappears altogether when the bed thickness equals or becomes less than the spacing. Rather resistive beds are shown as conductive. That is, thin resistive beds picked up as resistive in N16″ are seen as relatively conductive beds in N64″. This is known as 'resistivity reversal'. The apparent resistivity recorded may be equal, greater or less than the true resistivity depends on the influencing factors (CGWB, 2004). For resistive beds to appear as resistive in N64″ the bed thickness should be at least 1.5 times the electrode spacing, i.e., about 2.5 m.

In hard rocks, the resistivity log can be studied qualitatively. It is useful for studying the resistivity variations in the weathered zone where values are generally moderate and can be interpreted similarly to sedimentary areas. But it is seldom used, as the thickness of the aquifer zone between the water table and the bottom of the weathered zone is generally a few tens of metres only. For compact hard rocks, underlying the weathered zone, interpretation of the resistivity log is quite difficult mainly because of very high resistivity values, in thousands of ohm.m, which are mostly beyond the measuring scales of conventional logging machines and not dependable for any hydrologic parameter estimation (Keys, 1979, Davison et al., 1984). Also, when formation resistivity is very high, current flows mostly along the borehole through mud fluid and the apparent resistivity recorded does not represent the formation resistivity. Against a conductive fractured zone surrounded by compact highly resistive formations though the apparent resistivities recorded are always less than the surrounding, the values are higher than its true resistivity and much higher if the fracture is thin. The values cannot be corrected and used in quantitative interpretations. The fractured zones are qualitatively picked up from logs as resistivity lows. The identification of a fractured zone and its water quality by resistivity log are reported by Davison et al. (1984). The fractured zones can be identified by studying the differences in resistivities obtained through shallow and deep focused resistivity logs (Dual Laterolog) having varied radius

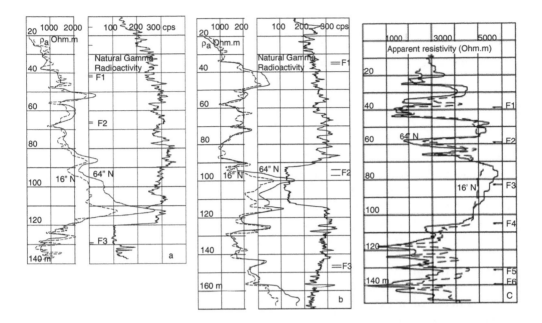

Figure 12.5 The N16″ and N64″ resistivity logs of boreholes in hard rock terrain of eastern India: boreholes (a) and (b) are located at a site in metasediments, the fractures are associated with relatively 'low' resistivities and borehole (c) is located very close to a basic dyke in granitic terrain, out of the 6 fractured zones (F1 to F6) detected, F3 with 0.34 lps discharge is not picked up; F4, F5 and F6 picked up in the log had discharges in the range of 1 to 5.9 lps (Source: Chandra *et al.*, 1994).

of investigation and high resolution (high density vertical and horizontal borehole wall coverage) microresistivity log or scanner which yield an image of micro-level resistivity variations in the borehole wall (Vasvari, 2011; Cull, 1988) but they are seldom used in shallow groundwater exploration in hard rock. Figures 12.5a, b and c present the N16″ and N64″ resistivity logs of boreholes in metasediments and granites. The fractured zones are associated with relatively 'low' resistivity values.

12.5 ELECTROMAGNETIC INDUCTION

Induction logging works on the principles of electromagnetic induction. It can be used in dry as well as water- and water based or oil based mud-filled borehole. It measures the electrical conductivity of rock formations intersected by the borehole. Like surface EM measurements induction logging is more sensitive to conductive beds than resistive beds. The salinity of drilling fluid, however, restricts its use and therefore induction logging works better in dry boreholes. The advantage of induction logging over conventional resistivity logging is that it can be deployed for monitoring in wells with PVC casing as well. EM induction logging is effective for contaminant detection (variation in electrolyte conductivity) and its monitoring through time-lapsed measurement of formation conductivity variations (Mack, 1993; Williams *et al.*, 1993; Metzger and Izbicki, 2013). It can be used for monitoring conductivity variations in the near surface unsaturated zone also. Besides, it can be used in post-hydraulic fracturing studies. A detailed description of induction logging is given by Ellis and Singer (2007).

The probe comprises a single transmitter and single or multiple receiver coils. The transmitter (T) emits alternating current at a constant frequency and amplitude. It induces eddy currents in formations surrounding the borehole which flows in a horizontal plane loop in the formations. The secondary magnetic field due to an eddy current induces a voltage in the receiver (R) coil crossing it. If the system operates at a low induction number (dimensionless ratio of T-R spacing to skin depth), i.e. T-R spacing is much less than skin depth, the amplitude of the secondary field varies linearly with conductivity of the surrounding formation (McNeill, 1980; McNeill *et al.*, 1990). The measured response includes contributions from the borehole fluid as well as the invaded zone. So, a true response is obtained when the bed is conductive and the invaded zone is relatively resistive. With increasing conductivity the skin effect becomes significant. To reduce the effect of the borehole and the invasion, additional focusing coils are placed on the probe so that eddy current flows in region further away surrounding the borehole. The T-R coil spacing dictates vertical resolution as well as depth of investigation. Smaller spacing gives better vertical resolution but a smaller depth of investigation. The probe is essentially calibrated before conducting measurements and correction for invasion and borehole diameter is applied.

The general order of resistivity for compact hard rock is around 1000 to 5000 ohm.m. The resistivity being quite high, the induction logging which senses conductivity better, has limited applicability. It may not give the resistivities accurately (Mwenifumbo *et al.*, 2009). Also the fractured zones which are quite thin and with resistivity mostly higher than 100 ohm.m may not get detected by the induction probe. However, Stumm *et al.* (2001) show an increase in electrical conductivity through focused induction measurements against the fractured zones. Mwenifumbo *et al.* (2009) suggest capacitive conductivity measurements.

12.6 FLUID CONDUCTIVITY

The borehole fluid conductivity logging is for continuous depth-wise measurement of electrical conductivity (EC) of fluid contained in the borehole. The unit in general used is micro Seimens per cm or micro mho per cm). It does not include any measurement related to formations and interstitial fluids surrounding the borehole. It is used to ascertain the quality of water in the borehole in terms of EC, locate transmissive fractured zones and monitor temporal changes in the depth-wise electrical conductivity of water in a borehole (Michalski, 1989). It helps identify zones with poor quality water.

The conductivity is measured by closely spaced electrodes positioned a few centimeters apart in the probe. The electrodes are placed in normal resistivity configuration comprising two current and two potential electrodes. The fluid can enter and come out of the probe tube through the openings as it moves through fluid in the borehole. The measurements inside the tube remain unaffected by the resistivity of formations surrounding the borehole. Measurements are done by alternating current to avoid the effect of polarization at the electrodes. The logs are recorded while going down the borehole at a low speed to get undisturbed fluid and avoid considerable air circulation above the fluid level. Also, low speed is required since fluid conductivity and temperature measurements are carried out simultaneously with a single probe.

The log chart directly gives borehole fluid conductivity or specific conductance in micro Siemens per centimeter. Borehole correction is not required. To compute total salt contents a conductivity probe is calibrated in the laboratory with standard solutions of known NaCl concentration for which charts are available. To correct for the presence of other ions conversion factors are used. Variations in temperature of the fluid affect its electrical conductivity considerably because of the changes in ionic mobility and therefore all measured conductivities are adjusted to a standard temperature generally taken as 25°C. That is, once the fluid conductivity and temperature are known, the equivalent NaCl concentration can be worked out using standard charts. To incorporate corrections for other ion concentrations separate chemical analysis of borehole fluid sample is required (Keys, 1988). The logging can also be conducted after the well is pumped to assess fracture inflow (Stumm et al., 2001).

The borehole fluid conductivity log can show a change against the screened zone and blank casing providing information on depths where the quality of water changed. The water yielding fractured zones in uncased boreholes are picked up by varied deflections in the conductivity log because fractured zones in general hold groundwater with different chemical compositions, ion concentrations and electrical conductivities. The magnitude of the deflections could be associated with transmissive capacity of the fractures (Michalski, 1989). Tsang et al. (1990) demonstrate a cost-effective use of time sequence conductivity logging in identifying fracture flow. For this the borehole is first washed out by deionized water up to its bottom to create a contrast between the fracture fluid and borehole fluid conductivities. A fluid conductivity log is first conducted without pumping the well to get a depth wise conductivity profile to be used for background values. The borehole is pumped at a low rate to stabilize the borehole fluid conductivity at a considerably low value which is monitored by measuring the conductivity of out-flowing water from the borehole. The logging is repeated several times at a gap of a few hours while continuously pumping the well at a constant low flow rate. The time sequence conductivity data is analyzed to identify inflow points and individual inflow rates. The inflows from fractures are associated with sharp peaks in conductivity logs while those from porous layers are flat topped. Tsang et al. (1990) confirm the fractures through borehole televiewer and temperature logs of which the latter was used to normalize the conductivity values for a standard temperature. Tsang and Doughty (2003), Doughty and Tsang (2005) and Doughty et al. (2005) further demonstrate the effective application of multi-rate fluid conductivity logging when well is being pumped to infer the inflow strength, salinity and transmissivity of fractures intersected by the borehole.

12.7 TEMPERATURE

The temperature probe measures depth-wise variation in borehole fluid temperature. The temperature of the fluid surrounding the probe is recorded. It may or may not represent the temperature in the formation surrounding the borehole (Keys and

MacCary, 1971). The temperature increases in the subsurface with depth at a constant rate. It is known as the geothermal gradient. On an average the near-surface thermal gradient (beyond 10–15 m depth with no seasonal temperature change) is a rise by 2.5 to 3°C per 100 m depth increment. The gradient depends on the thermal conductivity of geological formations and heat flow. If there is no vertical movement of fluid in the borehole and it is in thermal equilibrium with the surrounding formation the borehole fluid temperature log can yield a geothermal gradient (Keys and MacCary, 1971). The flow of water into or from an open fracture intersected in a borehole causes transfer of heat and perturbation in the geothermal gradient which can be used to detect the zone of fluid flow and mixing in the borehole. Whether the fracture is contributing or receiving water from the borehole, upflow or downflow the rate and amount of flow can be defined (Sorey, 1971). Besides locating permeable fractured zones, it is useful in locating zones of movement of recharge water (Keys and Brown, 1978). The temperature logs can be used to locate setting of fresh cement behind the casing (Keys, 1988). Temperature log is also run to incorporate temperature corrections in other logs which are affected by temperature.

Temperature logs can be run in two ways, either measuring absolute temperatures or differential temperature. In differential temperature the rate of change in temperature per unit depth is recorded. It is sensitive to changes in the slope of the absolute temperature record and therefore can identify even small changes in temperature vs depth slope caused by minor inflows. Figure 12.6 taken from Keys (1988) shows temperature and differential temperature logs and the caliper log. A differential temperature log is obtained by a probe having two temperature sensing elements kept a fixed small distance apart and recording the difference between the temperatures measured at these sensors simultaneously. This can also be done through a single sensor on the probe moving down the borehole and using electronic memory to record the changes in temperature at two points of times i.e. at two fixed depths. In this the constant speed of the probe is maintained carefully. The differential temperature log can be obtained even mathematically by getting the first derivative of the digitized absolute temperature vs depth record for selected depth spacings equivalent to different sensor spacings or time-lapses. As pointed out by Keys (1988) it maximizes the sensitivity. Keys and MacCary (1971) recommend that temperature and differential-temperature logs should be run simultaneously.

In a borehole, fluid movement disturbs the linear geothermal gradient mentioned above and a distorted temperature profile with abrupt variation or anomaly is obtained. With considerable vertical flow of water in the borehole it may not show an increasing temperature over a depth interval or can even show a decrease. With time the disturbance diminishes and the borehole fluid comes to thermal equilibrium with the surrounding rock under the prevailing geothermal gradient. The temperature probe is therefore run in undisturbed borehole condition for the first down-hole log at constant low speed in combination with fluid conductivity.

The temperature log of a borehole having open fractured zones shows perturbations in the temperature gradient. The anomaly depends on relative temperature of water coming in or going out of the fractured zone and flowing up or down the borehole. Identification of such anomalies to characterize the fractured zone has been in practice for a long time. To have an idea of the thermal signatures associated with fracture flow dynamics, Drury and Jessop (1982), Drury et al. (1984) and Drury

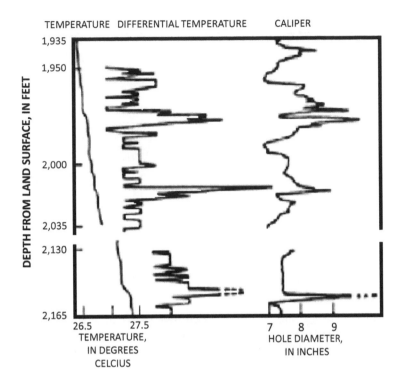

Figure 12.6 Temperature, differential temperature and caliper log of a well (modified Fig. 87 of Keys, 1988). In the differential temperature curve the spikes are well correlated with the increase in borehole diameter shown in caliper log against the fractures.

(1989) analyze the thermal signatures associated with different categories of fracture flows viz., fractures (a) contributing or accepting water from the borehole but not connected before drilling, i.e., the borehole connects these fractures, (b) not contributing after drilling but accepting circulation water during drilling and (c) transmitting water before and after drilling. Type (a) fractures are hydrogeologically quite significant in hard rocks. In an undisturbed normal temperature profile the possible perturbations caused by type (a) discrete fractures are schematically shown by Borner and Berthold (2009) and this is reproduced here (Fig. 12.7). Temperature logs may not show discrete steps and the upward and downward flow mixing are reflected as smooth convex upward and concave downward anomalies respectively. As observed the anomaly gets reversed with reversal of flow direction. Fracture type (b) produces a characteristic spike-like thermal anomaly that decays with time (Drury, 1989). Type (c) shows a smooth change in thermal gradient and there is a scope for ambiguity as a change in thermal conductivity in the vertical direction can also produce the same type of change in the thermal gradient (Drury, 1989).

To assess the extension of the fractured zone the temperature logs of boreholes in an area can be correlated. For improved understanding of fracture inter-connectivity Silliman and Robinson (1989) suggest an approach of studying the natural response of fluid temperature in an observation well due to pumping or injection in another well.

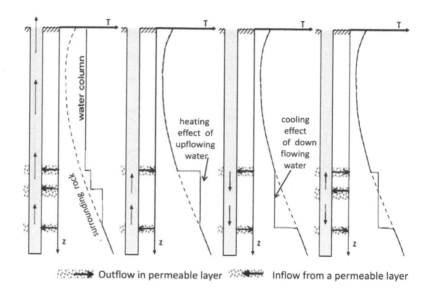

Figure 12.7 Temperature log anomalies or deviations from normal thermal gradient associated with upflow and different types of interflows due to hydrostatic pressure differences between fractures connected by the borehole (after Borner and Berthold, 2009).

The technique is constrained by certain combinations of flow geometry, flow rate and observation well diameter which is minimized by using a small diameter observation well and maximized flow rate. Klepikova *et al.* (2011) based on temperature profiles under ambient, single and cross-borehole pumping conditions attempt estimating the fracture hydraulic head, local transmissivity and inter-connectivity of fractures between nearby boreholes.

12.8 NATURAL GAMMA RADIOACTIVITY

Natural gamma radioactivity logging is the commonly used nuclear logging in groundwater exploration. It measures total gamma radiations emanating from formations in the immediate vicinity of the borehole wall that interact with the detector probe moving in the borehole. It is used to identify litho-units, primarily clays and shales in formations crossed by the borehole and correlate aquifers and other zones amongst boreholes. Ninety percent of gamma radiation recorded at the detector originates from formations within 15 to 30 cm of the borehole wall (Keys, 1988). Thus, the depth of investigation can be approximately taken as 30 cm. The presence of borehole metallic casing and cement grouting, increased borehole diameter, position of the probe in borehole, drilling fluid (mud), type of detector etc. affect the measurements. In large diameter boreholes the measured radiation is less than the actual and it is further reduced if the borehole is filled with heavy drilling mud. In hard rock generally wells are not cased except the top part and mud is not used for drilling. So these effects are not observed.

The quantitative value of radioactivity is not required in groundwater exploration and therefore gamma radiation measured in counts per second is not calibrated. The increase or decrease in gamma radiation incorporating borehole effects is sufficient to differentiate lithologies. The emission of gamma rays is random, even without any movement of the probe the count rate fluctuates. For accuracy in gamma radiation measurements the probe has to stay at a point in the borehole for a reasonable time and the count rate has to be fast enough so that sufficient counts of fluctuations are measured within the short stay-time. The measurement accuracy increases with higher count rate for a longer period or for a longer stay. It is achieved by a fixed low speed probe movement and a fixed time span for averaging the counts. The fixed time span in seconds is known as the time constant (t_c) and is required in analog recording. In digital logging it is replaced by sampling. The normally selected time constant values are 1, 2, 3 and 5 sec. Obviously, a probe moving with a large time constant reduces the vertical resolution. According to Repsold (1989), it is to be kept greater than 200/N, where N is the average gamma count per second to be measured in a borehole. The probe speed is kept less than $20/t_c$, observing that it does not exceed 5–6 metres per minute. Too small a time constant renders the log noisy and too fast a logging speed flattens the log record. A long time constant with slow logging speed is preferred for better vertical resolution. For example, if t_c is 1 second, the averaged data is recorded at every 1 sec during which the probe with a speed of 6 m/minute or 10 cm/second will move by 10 cm. That is, the data is recorded either at the top or at the bottom of the 10 cm thick zone. This is known as lag and obviously a layer thinner than 10 cm cannot be differentiated on the basis of gamma counts.

Usually, the basic igneous rocks are less radioactive than the acidic rocks (Keys, 1979). The natural gamma radioactivity log response is generally higher for acidic rocks containing feldspar and mica than the basic rocks. But the response is not uniquely related to lithology and therefore to be used in combination with other logs. As in sedimentary terrain the lithological correlation among the boreholes can be attempted. The natural gamma radioactivity logs of two boreholes in metasediments located 35 m apart shown in figures 12.5a and 12.5b reveal the presence of relatively less radioactive ($\approx$100 cps) bed compared to the surrounding rocks ($\approx$300 cps). The correlation indicates the dip (apparent) of the bed about 30°. The fractured zone F3 in borehole (a) can be correlated with fractured zone F2 in borehole (b). The gamma log can also be used effectively to identify the basalt flows (CGWB, 2004) as shown in figure 12.8 and define the flow dispositions in an area through log correlations which in turn help identify the possible extents of aquifers.

Independently, the natural gamma log response may not provide any conclusive information regarding groundwater occurrence in igneous and metamorphic rocks, nor can the gamma response resolve thin fractured zones which act as water producing zones. However, the concentration of natural radioactive isotopes in trace amounts along the fractures due to leaching or precipitation may produce differentiating gamma response and get them identified (Davison et al., 1984). Stumm et al. (2001) relate the variations in gamma response to changes in lithology and fractures in boreholes drilled through fractured crystalline bedrock in southwestern Manhattan Island, New York. Conversely any increase in gamma response need not be associated with fractures. An increase in gamma count may be associated with secondary clay minerals in fractured zones and pegmatite etc. Interpretation of gamma logs depends to a larger extent on

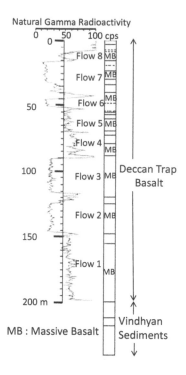

experience in a restricted geologic environment. The interpretation usually takes place on a qualitative basis in combination with lithological samples.

12.9 GAMMA-GAMMA (DENSITY)

Gamma-gamma logging, also known as density logging is used for *in situ* assessment of bulk density of formations in the vicinity of the borehole wall. With other logs it can be used to estimate porosity and differentiate the lithology. A low energy gamma source on the probe bombards gamma rays on formations surrounding the borehole. A detector on the probe shielded from the source records gamma rays, after they are attenuated and backscattered by their collision with the electrons in surrounding formations and mud fluid in the borehole. There can be two detectors on the probe to compensate the extraneous effects. The number of such collisions is directly related to the number of electrons in surrounding formations. Hence the attenuation of gamma radiation depends on electron density which is actually measured and is related to bulk density of formation. The intensity of gamma radiation back scattered to the detector is inversely related to density. Gamma-gamma logging can be conducted both in cased as well as uncased boreholes and it can be conducted in dry or fluid filled boreholes.

Gamma-gamma response is strongly affected by hole diameter, casing and borehole fluid. As the probe moves up or down the borehole at a constant low speed, the number of back-scattered gamma rays detected in unit time is recorded as a function of depth, producing the gamma-gamma log. The variations in gamma-gamma log response can be correlated with the presence of thick fractured zones but not conclusively as lithological changes also produce such variations. Thin fractures may not get detected. Besides, the extraneous effects like borehole diameter enlargement also produce changes in log response (Davison *et al.*, 1984). The probe holds an active radioactive source and therefore, radiation safe handling procedures are to be followed to avoid radiation exposure. Details of gamma-gamma logging can be obtained from Keys (1988).

12.10 NEUTRON

Neutron logging is used to determine porosity and lithology in combination with a gamma-gamma density log. The neutron probe holds a neutron source. It emits continuously high speed neutrons. A detector on the probe at a fixed separation from the source detects the resulting captured neutron product. Neutrons entering the borehole fluid and formations surrounding the borehole lose energy on collision with nuclei and get slowed down. The loss of energy on collision varies with elements. The maximum neutron capture occurs, when the mass of the neutron and that of the colliding nucleus is comparable or approximately the same, like nucleus of hydrogen atom holding one proton. The slow-down of neutrons indicates the hydrogen content of the formation. It indicates water in formation pores and quantifies the pore spaces and hence the porosity. The neutron logging can be conducted in uncased as well as cased boreholes with or without fluid. With calibration and corrections porosity values are obtained from the log. While carrying out neutron logging the standard safety norms for radioactive hazards are to be strictly observed.

Galle (1994) indicates that neutron logging for porosity determination in sedimentary formation is quite effective but in hard rock such as granite specifically with low porosity, the porosity determination is complex as the neutron matrix effect due to clays and micas and macroscopic capture cross-section of lithium can be significant. The neutron porosity related to rock matrix is much higher than the total free water porosity. Galle (1994) emphasizes that while evaluating the water content from neutron porosity in hard rock it is necessary to consider the matrix effect and also neutron tool calibration for the formation under study is made instead of the usual standard calibration in limestone. A comparison of neutron porosity log with acoustic porosity log may help in discriminating the fracture porosity. The probe with larger source and spacing is used to estimate the porosity of the formation adjacent to the borehole while a smaller source and source-detector spacing can be used in a small diameter borehole to record and monitor the moisture content profile of the vadose zone.

12.11 CALIPER

A caliper log is run for continuous measurement of depth-wise variations in borehole diameter and can also be called borehole geometry logging. It can be used in open or cased, water filled or dry boreholes. The caliper log is generally included in a multi-parametric logging programme to assimilate an important correction - the borehole diameter correction in other logs and also validate inferences obtained from them. Borehole diameter can change due to several reasons, viz., cavities formed, swelling of clays, mud cake, change in lithology and compactness (soft and hard formation), fragmentation of borehole wall due to fracturing etc. The variations in diameter obtained from the caliper log can also be used to calculate the volume of borehole, gravel and cement seal, get the precise depth of caving and the depth at which drill-bit was changed and accordingly information on casing lengths of different diameter is obtained. It can be used to get exact depth of parting, damage or collapse and corrosion in casing required for well rehabilitation etc. In hard rock the caliper log is run in an open hole to determine the increase in borehole diameter caused by rock fragmentation that can be related to fractured zones.

A caliper probe is a motor controlled mechanical device. The probe can have 1 to 6 arms. In water well drilling generally a 3-arm caliper is used which has equally spaced three spring loaded radiating arms from the probe. The arms get pressed against the borehole wall by spring action and can mechanically spread-out or spread-in depending on borehole diameter variations. The spreading of arms is converted to an electrical signal and recorded. Since in a conventional caliper probe there is no way to differentiate between the individual arm movement and azimuth of the arms, it gives only a mean value of the borehole diameter. The 4-arm caliper with 90° apart dual axis and each arm having independent measurement can provide information on the directional increase in borehole diameter.

The spring-loaded arms of the caliper probe are controlled by a motor within the probe. The arms can be opened or closed in the borehole at any depth from surface command. The probe is lowered to the bottom of the borehole with arms closed and are opened only during up-log. The caliper probe is calibrated prior to logging by a calibration jig. The log records variations in borehole diameter in inches compatible with drill bit sizes given in inches. The resolution depends on arm-wall contact. The vertical resolution depends on the contact-length of the arms with the borehole wall and the pressure exerted by the arms while horizontal resolution depends on the number of arms (Keys, 1988). The caliper probe gives an erroneous reading if the borehole is not vertical. A variety of caliper probes is available with multiple arms and a facility for knowing the arm orientations. Also caliper probes for simultaneous measurement of diameter at different depths of borehole are available. Besides, acoustic calipers using transducers give better estimates of borehole diameter.

12.12 FLOWMETER

The flow meter or flow logging constitutes depth-wise measurement of groundwater flow rates in a borehole. It can measure vertical flow under ambient conditions or flow in a well during pumping (Paillet *et al.*, 1987b) as shown in figure 12.9. It is a method to find the yield contribution of individual fractured zones or tapped aquifer zones intersected by a borehole, locate zones which are taking water out of the well, direction of flow, vertical variations in hydraulic conductivity (Molz *et al.*, 1989; Boman *et al.*, 1997) and also monitor the performance of slotted screen zones (Crowder

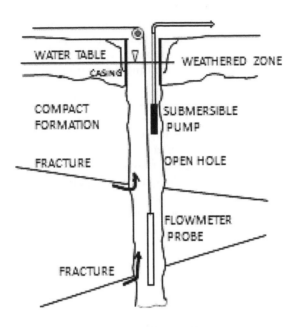

Figure 12.9 Schematic diagram showing flowmeter application while pumping a borehole.

and Mitchell, 2002). The oldest flow logging probe is a mechanical impeller or spinner. The 3 or 4 bladed impeller in metallic housing lowered in the borehole can measure the flow rate continuously while moving up or down at a constant speed of about 5–6 m/minute or kept stationary at a selected depth (Keys, 1988).

At low flow velocities, less than 2 to 3 m per minute the blades of the impeller flow meter do not rotate due to friction. Such low velocities may be observed from fractures encountered in hard rock wells. To measure very low flow velocities, a heat-pulse flowmeter is used (Hess, 1982). It can measure flow velocities ranging from 0.06 to 6.0 m/minute and can resolve velocity differences of 0.01 m/minute (Hess, 1986). Ito *et al.* (2006) improve the flow meter velocity detection limit up to 0.018 m/minute. By positioning the flowmeter probe at a specified depth or at a regular depth interval the ambient flow rates can be measured. It can also be used while pumping the well. The heat pulse flowmeter probe contains a horizontal wire grid heating element located between two thermistors, one above and one below it. The wire grid is heated by a short pulse of electric current to heat the water in its vicinity. The heated front of the water flows either towards upper or lower thermistor depending on the direction of flow, and get detected. From the time difference between electric current triggering and arrival of the heated water front at a thermistor the velocity and direction of flow is determined. The deflection of response trace to right or left indicates upward or downward flow respectively (Keys, 1988). Another type is electromagnetic flowmeter which measures the flow continuously. The electromagnet in the probe creates a magnetic field and water which is a conductor flowing through the magnetic field creates a voltage gradient proportional to average flow velocity. The voltage is detected by electrodes (Young and Pearson, 1995; Young *et al.*, 1998). The positive and negative voltages are taken as upward and downward flows respectively. Compared to the impeller type the EM flowmeter is better as it does not have any mechanical moving part which is affected by frictional forces and errors due to lateral flow generally observed in impeller flowmeter is not there. The EM and heat pulse flowmeter responses are of comparable quality and can be successfully used in estimating flows from fractured zones (Young and Pearson, 1995).

To identify fractured zones either heat-pulse or electromagnetic flow logging is conducted under ambient conditions as well as optimized (no dewatering) constant discharge pumping with stabilized water level; dewatering is often observed due to large drawdown. According to Cohen et al. (1996) heat-pulse flowmeter may detect more flowing intervals than the impeller flowmeter because of a much lower detection limit. They further mention that flow logging may respond differently if another nearby well is pumped and also the response may vary with time. Paillet (2004) describes the various corrections required in flow meter measurements for accuracy in interpretation. A computer code 'FLASH' developed by Day-Lewis et al. (2011) for analyzing vertical flow under ambient and stressed conditions can be used to estimate transmissivities.

12.13 ACOUSTIC

The acoustic logging is also known as sonic logging. It gives continuous depth-wise measurement of the travel time of the acoustic wave through formations intersected by the borehole. Acoustic log is used to determine porosity of formation, identify fractured zones and lithology. The acoustic wave emitted as energy pulses from a source transmitter on the probe reaches the borehole wall rock through mud fluid. The compressional wave also known as a longitudinal or P wave emitted by the transmitter travels in all directions through the borehole mud and hits the borehole wall. A part of the energy is reflected back from the borehole wall. The part of energy reaching the borehole wall at critical angle gets refracted along it and travels through the wall rock formation. The P wave travelling through the rock has higher velocity. It travels fast and again is critically refracted back to the borehole as a head wave arriving first at the receiver. The probe may hold more than one receiver. The receiver is isolated from the transmitter by a spacing generally more than 60 cm (Keys, 1988). The transmitter and receivers are acoustic transducers transforming the electrical energy into acoustic and vice versa. The refracted P waves traveling along the borehole wall rock reach these two receivers on the probe separated by a distance about 30 cm, taking more time to reach the distant receiver. The first arrivals at these receivers are picked up. The distance between the two receivers being known, the P wave velocity is determined from the difference in first arrival times i.e., time taken in travelling the distance between them. The log data records the interval transit time (Δt). It is the travel time taken per unit distance travelled. The unit is microsecond per foot or metre. The velocity of the P wave depends on the elastic properties of the rock and the fluid present in its pore spaces, in turn depends on porosity which can be estimated from the transit time. The acoustic energy source operates in a frequency range 10 to 30 kHz and could be monopole or dipole. A monopole source transmits energy equally in all directions and a dipole source emits energy in specified directions. For acoustic logging the borehole is to be uncased and water filled or mud filled. The probe speed is kept between 3 to 6 m/minute. For standard transmitter-receiver spacing, the depth of investigation depends on frequency of the source and acoustic velocity of the formation. According to Keys (1988), the theoretical depth of investigation for 20 kHz acoustic velocity probe is about 22 cm (0.75 ft) for unconsolidated material (acoustic velocity: 5000 ft/sec) and about 112 cm (3.75 ft) for consolidated hard rocks (acoustic velocity: 25,000 ft/sec). The vertical resolution is equivalent to the spacing between the receivers.

As the wave propagates its amplitude decreases depending on several factors. If attenuation is too strong, the first arrival cannot be detected at the far receiver and detection occurs at the second or third cycle arriving at this receiver. That is only one of the receivers is triggered. It will result in an abrupt increase in interval transit time. This is known as 'cycle skipping' (Serra, 1984). One of the probable factors causing cycle skipping is the presence of a fractured zone. The cycle skipping on the transit time log is seen as an abrupt, sharp high transit time peak that can be corroborated with other logs such as SPR, caliper, temperature, flowmeter, borehole televiewer or borehole radar to confirm the presence of fractured zone.

Besides this, the P wave striking at the borehole wall at less than a critical angle generates refracted shear waves also called transverse or S waves (Serra, 1984) but the reflection of S wave is not possible through borehole fluid. The S wave has higher magnitude than the P wave. Like the P wave the receiver can receive the generated shear wave traveling along the borehole wall rock if it is fast formations (hard formation). Fast formation means the S wave velocity in it is faster than the compressional velocity of borehole fluid ($V_p > V_s > V_{mud}$). So, in 'slow formation' where $V_p > V_s < V_{mud}$ an S wave cannot be recorded. In hard rocks both P and S wave velocities can be determined. Also, a pseudo-Rayleigh wave is generated at the borehole wall of fast formation. The pseudo-Rayleigh wave and the S wave have almost same velocity and cannot be distinguished. These waves are followed by a high amplitude Stoneley wave and a mud wave. A Stoneley wave travels along the fluid-borehole wall interface as a guided wave. Its velocity is less than the compressional velocity of the borehole fluid and the shear wave velocity. Its amplitude is large. At low frequency it is known as a tube wave. It is a guided wave propagating along the wall of a fluid filled borehole. It is sensitive to fractured zones and is strongly attenuated or eliminated at major fractured zones and sensitivity to fracture permeability is very high at low frequency (Tang et al., 1991). According to Hardin and Toksoz (1985) the effect of a guided wave may not exist if the formation shear wave velocity is much less than the borehole fluid velocity. By comparing the attenuation of tube wave energy against the fault or fractured zones their relative permeability can be assessed (Tang and Cheng, 1993; Pflug et al. 1997). The greater the attenuation the more permeable is the zone. Davison et al. (1984) correlate the tube wave amplitude with fracture permeability determined through packer test and conclude that tube wave attenuation correlates the best. In a tube wave study it is important to select the frequency band of the source which will generate a large amplitude tube wave. Also the borehole diameter should be small. At the fractured zone intersected by a borehole the tube wave can also be generated by incident compressional energy from a near surface source in proximity of the borehole. Since the in-hole and near surface tube wave method works at different scales of investigation, the hydraulic characteristic of the fractured zone obtained by these methods need not match (Paillet et al., 1987a). Huang and Hunter (1984) explain that compressional energy striking the fractured zone squeezes out water from it into the borehole, generating the tube

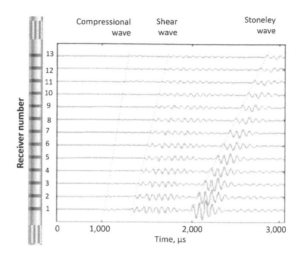

Figure 12.10 Typical waveforms from a monopole transmitter in a fast formation: compressional (P) wave, shear wave (S) and Stoneley wave. The pink dashed lines are arrival times. The acoustic logging tool receiver array is shown at left (Source: Haldorsen *et al.*, 2006).

wave radiating up and down the borehole. That is, permeable fractured zones can be identified by locating the tube wave generation sources in the borehole. Li *et al.* (1994) however, indicate that though tube wave generating sources in the borehole are fractured zones, but not all of them may be yielding.

The conventional acoustic log measures P wave travel time only. To get complete information on the travel time and amplitude of P, S and Stoneley waves mentioned above, which can be analyzed for fractured zones and their permeability, full waveform acoustic logging is conducted. It records the whole waveform from all the receivers. Compared to conventional acoustic, in full waveform logging, the frequency of transmitter used is relatively low, the transmitter-receiver distance is large and transmitters and receivers are more in number. Figure 12.10 shows the P, S and Stoneley waves arrivals at the receivers of a full waveform acoustic probe (Haldorsen *et al.*, 2006).

12.14 BOREHOLE TELEVIEWER

A borehole televiewer is used to obtain an image of the wall of a thin mud or water filled open borehole. It is used to locate the voids, joints, fractures, alteration, bedding planes, change in lithology and also ruptures, perforations and corrosion in the casing of cased wells. The dip of such intersecting structures can also be measured. Fractures are directly viewed by borehole televiewer. It provides better information than any other method. It is of two types, viz., acoustic televiewer and optical televiewer. The acoustic televiewer images the borehole wall or the casing by means of an acoustic transducer (low frequency rotating or high frequency fixed transducer with a rotating acoustic mirror) on the probe which sends a focused beam of ultrasound waves in a

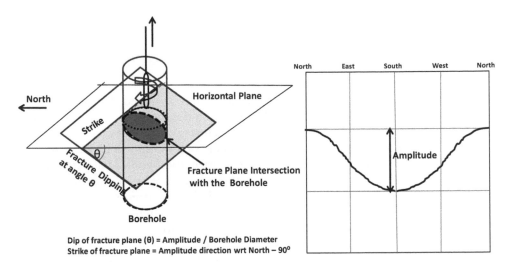

Figure 12.11 Schematic diagram showing response of acoustic televiewer on fracture intersected by borehole.

horizontal plane while the probe moves up or down in the borehole. The probe is run centralized in the borehole. The travel time of reflection of these waves from the borehole wall and the amplitude are converted to get a 360° digital image of the wall which is a record of acoustic energies reflected from the borehole wall. The amplitude of the reflected signal is high for smooth and low for mechanically damaged rough borehole wall which may represent the fractured zone. The resolution is quite high. The vertical resolution is in millimeters. It depends on the logging speed which varies from 1 to 3 m per minute. The circular resolution is of the order of millimeters and depends on the number of samplings per revolution. The travel-time log gives a high resolution 360° variation in borehole diameter and is used as a high resolution caliper log measuring enlargement in diameter at the fractured zone. The direction of the image is referred to magnetic north defined by a built-in magnetometer. The obtained log display chart gives a planar view of the borehole wall image presenting the inclined fracture as a dark sinusoidal trace, as indicated schematically in figure 12.11. The direction of maximum amplitude difference gives the direction of dip whose value is obtained by a simple calculation shown in figure 12.11. The hydraulic conductivity of a fracture is related to its width or aperture which can be obtained from the acoustic televiewer (Keys 1979).The advantage of an acoustic televiewer is that it can be used in thin mud fluid filled borehole.

In an optical televiewer a probe holding a high resolution digital CCD camera, a conical mirror and a light source is lowered in the borehole to capture 360° reflections of the borehole wall. In this also the magnetic north is defined by the magnetometer. The merit of the optical televiewer is that it gives a complete image of the borehole wall in natural colours and the demerit is that it requires that the fluid if present in the borehole is totally clean or that the borehole is dry. There are a good number of publications on borehole televiewer. Some of them are Williams *et al.* (2002), Williams and Johnson (2004), and Morin *et al.* (1997) presenting the applications of the acoustic and optical televiewers.

12.15   BOREHOLE RADAR

The ground penetrating radar (GPR), in general, has a limited depth investigating capability up to about 40–50 m and varied resolution depending on antenna frequency and conductivity. For information on deeper fractures, borehole radar (BHR) can be used. It has been in practice since 1985 (Olsson *et al.*, 1992a). BHR helps delineate fractured zones intersecting the borehole, their orientation and radial extent beyond the borehole (Williams *et al.*, 2002) as well as those which are located near the borehole but not intersected. The BHR characterizes the lithological heterogeneities also.

BHR uses electromagnetic wave propagation in the frequency range of 10 to 1000 MHz. The transmitter and receiver antennas are lowered in the open or PVC cased borehole. The transmitted radar pulse propagating into the surrounding rock formation, impinges on the inhomogeneity interface. A part of the energy gets reflected from the interface and a part propagates further till attenuated. The antennas are moved up or down in small steps (0.1 to 1.0 m) and a radar scan is taken at each location in the borehole to generate a radar reflection profile along the borehole (Singha *et al.*, 2000). Omni-directional and directional antennas are used. The omni-directional antenna radiates a pulsed signal and receives reflected energies from 360° directions and therefore the azimuth of the reflector cannot be determined (Liu and Sato, 2006). By using both type of antennas or cross-hole survey, the location, extent, orientation and dip and strike of the fracture planes can be obtained. In the Stripa granite, Sweden Olsson *et al.* (1992b) using 20–60 MHz radar transmitter frequency investigate approximately 100 m in single-borehole mode and 200–300 m surrounding the borehole in cross-borehole mode. As with GPR, in BHR also higher frequency provides better resolution but with a lesser range of investigation. Olsson *et al.* (1992b) indicate a broader application of BHR in understanding the complexities of structures, variations in physical properties and channeling of flow within fractures and fractured zones. Lane *et al.* (1998b) showed that it is difficult to get near-borehole structures due to large minimum transmitter-receiver offset and stressed the need for multi-offset data acquisition to enhance the near-borehole imaging and accurate estimation of radar propagation velocities required to determine the dip of the reflector and assess the near-hole hydrology.

The EM wave propagation velocity is sensitive to the presence of fluid while wave attenuation depends on conductivity. For non-magnetic good dielectric material the radar wave velocity in the subsurface (v), attenuation (α) and skin depth (δ) are given by the relations (Liu and Li, 2001).

$$v = \frac{c}{\sqrt{\varepsilon_r}}, \quad \alpha = \frac{188.5\sigma}{\sqrt{\varepsilon_r}}, \quad \delta = \frac{1}{\alpha}$$

where ε_r: dielectric constant, σ: electrical conductivity of the material and c is the radar wave velocity in vacuum In a highly conductive (saline water) environment the attenuation is strong. The amplitude attenuation by saline water is exploited by combining

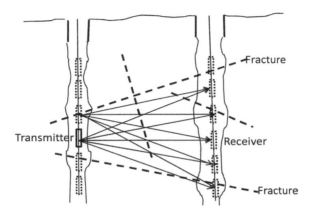

Figure 12.12 Cross-borehole radar tomography for fracture delineation.

the cross-borehole radar tomography with injection of saline tracer. It was initiated by Deadrick *et al.* (1981). Ramirez and Lytle (1986) name it 'Alterant geophysical tomography' and indicate that the measurement of changes in attenuation is more accurate compared to that of the absolute. The 'BHR difference-attenuation tomography' as it is known today, measures the changes and traces the preferential flow path in a fractured zone through time-lapse imaging of injected highly conductive saline tracer migration causing attenuation (Lane *et al.*, 1998a, Day-Lewis *et al.*, 2003). Day-Lewis *et al.* (2006) further demonstrate its usefulness in providing an insight into site specific heterogeneities in hydraulic conductivity and channeling through fractures. Zhou *et al.* (2001) develop a nonlinear tomographic inversion method to simultaneously reconstruct EM wave velocity and attenuation distribution. It uses information on first arrival travel time and amplitude spectra from cross-borehole radar measurement.

BHR can be used across multiple boreholes for cross-borehole measurement. It is now in regular use to generate 2d sections and to know the azimuth of the fracture plane. It is known as cross-hole tomography. The transmitter is located in one borehole and the receiver in the other and numerous measurements are made (Fig. 12.12). Either transmitter or receiver can be kept at a fixed location in the borehole. For each transmitter-receiver the ray path travel time and amplitude are computed to get cross-sectional images of 'radar slowness' and 'radar attenuation' respectively (Day-Lewis *et al.*, 2006). Singha *et al.* (2000) indicate the use of cross-borehole measurements in assessing lithological variations between the boreholes, identifying fractured zones and their connectivity, monitoring tracer tests and contamination remediation. Further, they reveal that BHR combined with other surface and borehole methods can be effective in distinguishing the transmissive fractures from lithological variations and closed fractures. Conventional logging can detect variations in formation properties in the direction of measurement i.e., along the borehole depth due to the features such as fractured zones intersecting the borehole. Since connectivity controls the sustained yield, assessing the fracture connectivity by cross-hole methods is more important than identifying open fractures through combination logs.

12.16 NUCLEAR MAGNETIC RESONANCE

The Nuclear Magnetic Resonance (NMR) logging is high-end logging used in oil exploration. In groundwater exploration it has been used only rarely. Nakashima and Kikuchi (2007) show its superiority over other conventional logs. They indicate the possible ambiguities in measuring fracture apertures by borehole wall resistivity imaging and neutron logging, while the former is not suitable for fractures filled with resistive water, the latter cannot distinguish water in open fractures from that in clays. The amplitude of the NMR signal increases linearly with the increase in aperture of a water-filled fracture (Nakashima and Kikuchi, 2007) and therefore the signal can be analyzed to estimate the fracture aperture. Since NMR responds only to water content and can distinguish between free and clay-bound water, the hydrologic properties of fractured zones can be better assessed.

The basics of NMR are discussed in the section on Magnetic Resonance Sounding (MRS) of Chapter 13 on Integrated Geophysical Survey. In NMR logging, instead of Earth's magnetic field as static field used in MRS, a strong static magnetic field is created in formations surrounding the borehole by permanent magnets on the probe. It is to magnetize the formation fluid. Because of which the randomly oriented proton magnetic moment of the hydrogen nuclei in the pore water of the surrounding formations gets aligned in the direction of the applied static field. Now, electromagnetic, i.e. oscillating magnetic field pulses at the required Larmor frequency are sent. The Larmor frequency being proportional to the static field, it is in radio frequency range as the applied static field is quite strong ≈ 500 to 1000 times stronger than the Earth's field. By this the protons get aligned to a plane perpendicular to the static magnetic field, start precessing in the direction of the oscillating field and produce a decaying magnetic field. The time taken by the protons to align with the static field is known as the build-up time T1 or longitudinal relaxation time and the time taken for the decay of the oscillating field generated by the protons is known as T_2 or the transverse relaxation time. It is a measure of amplitude vs time. The initial amplitude is proportional to the number of protons and hence related to the porosity and water content. A large number of T_2 measurements are made between the pulses, i.e., over a very short time, known as spin-echoes. The time interval between successive pulses is generally of the order of hundreds of microseconds. The details of NMR logging can be obtained from published papers and books e.g., Coates *et al.* (1999) and Dunn *et al.* (2002).

REFERENCES

Boman, G.K., Molz, F.J. & Boone, K.D. (1997) Borehole flowmeter application in fluvial sediments: methodology, results and assessment. *Ground Water,* 35 (3), 443–450.

Borner, F. & Berthold, S. (2009) Vertical flows in groundwater monitoring wells. In: Kirsch, R. (ed.) *Groundwater Geophysics, Chapter 13.* Springer, pp. 367–389.

Chandra, P.C., Reddy P.H.P. & Singh, S.C. (1994) *Geophysical studies for groundwater exploration in Kasai and Subarnarekha River basins* (UNDP Project) Central Ground Water Board, Min. of Water Resources, Govt. of India Tech. Report.

CGWB (2004) *Manual on geophysical logging of water wells* Central Ground Water Board, Min. Water Resources Govt of India Publication.

Coates, G.R., Xiao L. & Prammer, M.G. (1999) NMR logging: principles & applications *Halliburton Energy Services Publication H02308.*

Cohen, A.J.B., Karasaki, K., Benson, S., Bodvarsson, G., Freifeld, B., Benito, P., Cook, P., Clyde, J., Grossenbacher, K., Peterson, J., Solbau, R., Thapa, B., Vasco, D. & Zawislanski, P. (1996) *Hydrogeologic characterization of fractured rock formations: a guide for groundwater remediators.* Project Summary, U S EPA, National Risk Management Laboratory, EPA/600/S-96/001.

Crosby III, J.W. & Anderson, J.V. (1971) Some applications of geophysical well logging to basalt hydrogeology. *Ground Water,* 9 (5), 12–20.

Crowder, R.E. & Mitchell, K. (2002) Spinner flowmeter logging: A combination of borehole geophysics and hydraulics. *Well Design & Installation Workshop, Arizona Hydrological Society, 18 July, 2002, Phoenix Arizona.*

Cull, R. (1988) Formation microscanner tool imaging and laterolog resistivity measurements in interpretation of crystalline terrains. In: Boden, A. & Eriksson, K. G. (eds.) *Proceedings of the Int.. Symp.on Deep Drilling in Crystalline Bedrock, Volume 2: Review of Deep Drilling Projects, Technology, Sciences and Prospects for the Future.* Springer-Verlag, pp. 315–327.

Davison, C.C., Keys, W.S. & Paillet, F.L. (1984) *The use of borehole geophysical logs and hydrologic tests to characterize plutonic rock for nuclear fuel waste disposal.* Atomic Energy of Canada Ltd. Report No. AECL-7810, WNRE.

Day-Lewis, F.D., Johnson, C.D., Paillet, F.L. & Halford, K.J. (2011) A computer program for flow-log analysis of single hole (FLASH). *Ground Water,* 49 (6), 926–931.

Day-Lewis, F.D., Lane Jr, J.W., Harris, J.M. & Gorelick, S.M. (2003) Time-lapse imaging of saline-tracer transport in fractured rock using difference-attenuation radar tomography. *Water Resources Research,* 39 (10), 1290, SBH 10-SBH 14.

Day-Lewis, F.D., Lane, J.W., & Gorelick, S.M. (2006) Combined interpretation of radar, hydraulic, and tracer data from a fractured-rock aquifer near Mirror Lake, New Hampshire, USA. *Hydrogeology Journal,* 14 (1&2), 1–14.

Deadrick, F.J., Ramirez, A.L. & Lytle, R.J. (1981) Insitu fracture mapping using geotomography and brine tracers. *preprint DE82 002101, UCRL 85712, IEEE, Nuclear Science Symposium, 21-23 October 21 1981, San Francisco, California, USA.*

Doughty, C., Takeuchi, S., Amano, K., Shimo, M. & Tsang, C.-F. (2005) Application of multirate flowing fluid electric conductivity logging method to well DH-2, Tono site, Japan. *Water Resources Research,* 41 (10), W10401.

Doughty, C. & Tsang, C.-F. (2005) Signatures in flowing fluid electric conductivity logs. *Journal of Hydrology,* 310 (1-4), 157–180.

Drury, M.J. (1989) Fluid flow in crystalline crust: detecting fractures by temperature logs. In: *Beck, A.E., Garven, G. & Stegena, L. (ed.) Hydrogeological regimes and their subsurface effects: Geophysical Monograph 47, IUGG Vol. 2,* IUGG, American Geophysical Union.

Drury, M.J. & Jessop, A.M. (1982) The effect of a fluid-filled fracture on the temperature profile in a borehole. *Geothermics,* 11 (3), 145–152.

Drury, M.J., Jessop, A.M. & Lewis, T.J. (1984) The detection of groundwater flow by precise temperature measurements in boreholes. *Geothermics,* 13 (3), 163–174.

Dunn, K.-J., Bergman, D.J. & Latorraca, G.A. (2002) Nuclear magnetic resonance: Petrophysical and logging applications. In: Helbig, K. & Treitel, S. (ed.) *Handbook of Geophysical Exploration: Seismic Exploration, Vol. 32,* U.K., Elsevier Science Ltd.

Ellis, D.V. & Singer, J.M. (2007) *Well logging for earth scientists.* The Netherlands, Springer.

Galle, C. (1994) Neutron porosity logging and core porosity measurements in the Beauvoir granite, Massif Central Range, France. *Journal of Applied Geophysics,* 32 (2 & 3), 125–137.

Hardin, E. & Toksoz, M.N. (1985) Detection and characterization of fractures from generation of tube waves. In: *Tran. SPWLA 26th Annual Logging Symp.,* pp. 205–225.

Haldorsen, J.B.U., Johnson, D.L., Plona, T., Sinha, B., Valero, H.-P. & Winkler, K. (2006) Borehole acoustic waves. *Oilfield Review,* 18 (1), 34–43.

Hess, A.E. (1982) *A heat-pulse flowmeter for measuring low velocities in boreholes.* U.S. Geological Survey Open-File Report 82–699. U.S.G.S.

Hess, A.E. (1986) Identifying hydraulically conductive fractures with a slow-velocity borehole flowmeter. *Canadian Geotechnical Journal*, 23 (1), 69–78, (Abstract).

Houtkamp, H. & Jacks, G. (1972) Geohydrologic well-logging. *Nordic Hydrology*, 3 (3), 165–182.

Howard, K.W.F. (1990) Geophysical well logging methods for the detection and characterization of fractures in hard rocks. In: Ward, S.H. (ed.) *Geotechnical and Environmental Geophysics, Vol. I: Review and Tutorial*. Tulsa, Oklahoma, USA, Society of Exploration Geophysicists. pp. 287–307.

Huang, C.F. & Hunter, J.A. (1984) The tube-wave method of estimating in-situ rock fracture permeability in fluid-filled boreholes. *Geoexploration*, 22 (3 & 4), 245–259.

Ito, K., Takeno, N., Seki, Y., Naito, K. & Watanabe, Y. (2006) Numerical simulation and parameter estimation using borehole flowmeter logging in low permeability rocks. In: *Proceedings of The TOUGH Symposium, Lawrence Berkeley National Laboratory Berkeley, California, 15–17 May 2006*.

Keys, W.S. (1967) Well logging in ground-water hydrology. Presented at: *The Eighth Annual Logging Symposium, Trans., SPWLA*, pp. 10–18.

Keys, W.S. (1979) Borehole geophysics in igneous and metamorphic rocks. Presented at: *The SPWLA twentieth Annual Logging Symposium, The Log Analyst*, 20 (4), 14–28.

Keys, W.S. & Brown, R.F. (1978) The use of temperature logs to trace the movement of injected water. *Ground Water*, 16 (1), 32–48.

Keys, W.S. (1988) *Borehole geophysics applied to ground-water hydrology*. U.S. Geological Survey Open-File Report 87-539. U.S.G.S.

Keys, W.S. & MacCary, L.M. (1971) *Application of borehole geophysics to water-resources investigations, Techniques of Water-Resources Investigations*, U.S. Geological Survey, Chapter E1, Book 2. U.S.G.S.

Klepikova, M.V., Le Borgne, T., Bour, O. & Davy, P. (2011) A methodology for using borehole temperature-depth profiles under ambient, single and cross-borehole pumping conditions to estimate fracture hydraulic properties. *Journal of Hydrology*, 407 (1–4), 145–152.

Lane Jr., J.W., Haeni, F.P. & Day-Lewis, F.D. (1998a) Use of time-lapsed attenuation-difference radar tomography methods to monitor saline tracer transport in fractured crystalline rock. *Seventh Int. Conf. on Ground Penetrating Radar, 27-30 May 1998, University of Kansas, KS, Lawrence, USA*. pp. 26–33.

Lane Jr., J.W., Haeni, F.P. & Versteeg, R. (1998b) Use of a multi-offset borehole-radar reflection method in fractured crystalline bedrock at Mirror Lake, Grafton County, New Hampshire. In: *Proceedings of Symposium on the Application of Geophysics to Engineering and Environmental Problems, 22-26 March 1998, Chicago, Illinois, U.S.A*. Wheat Ridge, Colorado, Environmental and Engineering Geophysical Society. pp. 359–368.

Li, Y.D., Rabbel, W. & Wang, R. (1994) Investigation of permeable fracture zones by tube-wave analysis. *Geophysical Journal International*, 116 (3), 739–753.

Liu, L. & Li, Y. (2001) Identification of liquefaction and deformation features using ground penetrating radar in the New Madrid seismic zone, USA. *Journal of Applied Geophysics*, 47 (3 & 4), 199–215.

Liu, S. & Sato, M. (2006) Subsurface water-filled fracture detection by borehole radar: a case history. *Journal of Environmental & Engineering Geophysics*, 11 (2), 95–101.

Mack, T.J. (1993) Detection of contaminant plumes by borehole geophysical logging. *Ground Water Monitoring and Remediation*, 13 (1), 107–114.

McNeill, J.D. (1980) *Electromagnetic terrain conductivity measurement at low induction numbers*. Technical Note- TN 6, Geonics Limited, Ontario, Canada.

McNeill, J.D., Bosnar, M. & Snelgrove, F.B. (1990) *Resolution of an electromagnetic borehole conductivity logger for geotechnical and ground water applications*. Technical Note TN-25, Geonics Limited, Ontario, Canada.

Metzger, L.F. & Izbicki, J.A. (2013) Electromagnetic-induction logging to monitor changing chloride concentrations. *Ground Water*, 51 (1), 108–121.

Michalski, A. (1989) Application of temperature and electrical conductivity logging in ground water monitoring. *Ground Water Monitoring and Remediation*, 9 (3), 112–118.

Molz, F.J., Morin, R.H., Hess, A.E., Melville, J.G. & Guven, O. (1989) The impeller meter for measuring aquifer permeability variations: Evaluation and comparison with other tests. *Water Resources Research*, 25 (7), 1677–1683.

Morin, R.H., Carleton, G.B. & Poirier, S. (1997) Fractured-aquifer hydrogeology from geophysical logs; the Passaic Formation, New Jersey. *Ground Water*, 35 (2), 328–338.

Mwenifumbo, C.J., Barrash, W. & Knoll, M.D. (2009) Capacitive conductivity logging and electrical stratigraphy in a high-resistivity aquifer, Boise Hydrogeophysical Research Site. *Geophysics*, 74 (3), E125–E133.

Nakashima, Y. & Kikuchi, T. (2007) Estimation of the apertures of water-saturated fractures by nuclear magnetic resonance well logging. *Geophysical Prospecting*, 55 (2), 235–254.

National Academy of Sciences (1996) *Rock fractures and fluid flow: contemporary understanding and applications*. Committee on 'Fracture Characterization and Fluid Flow', National Research Council, Washington D.C., USA.

Nelson, P.H., Magnusson, K.A. & Rachiele, R. (1982) Application of borehole geophysics at an experimental waste storage site. *Geophysical Prospecting*, 30 (6), 910–934.

Olsson, O., Anderson, P., Carlsten, S., Falk, L., Niva, B. & Sandberg, E. (1992a) Fracture characterization in crystalline rock by borehole radar. In: Pilon, J. A. (ed.) *Ground Penetrating Radar, Geological Survey of Canada, Paper 90–4*, pp. 139–150.

Olsson, O., Duran, O., Jamtlid, A. & Stenberg, L. (1984) Geophysical investigations in Sweden for the characterization of a site for radioactive waste disposal – an overview. *Geoexploration*, 22 (3 & 4), 187–201.

Olsson, O., Falk, L., Forslund, O., Lundmark, L. & Sandberg, E. (1992b) Borehole radar applied to the characterization of hydraulically conductive fracture zones in crystalline rock. *Geophysical Prospecting*, 40 (2), 109–142.

Paillet, F. (1994) *Application of borehole geophysics in the characterization of flow in fractured rocks*. U.S. Geological Survey Water Resources Investigation Report, 93-4214, Denver, Colorado. U.S.G.S.

Paillet, F. (2004) Borehole flowmeter applications in irregular and large-diameter boreholes. *Journal of Applied Geophysics*, 55 (1 & 2), 39–59.

Paillet, F.L., Cheng, C.H. & Hsieh, P. (1987a) Experimental verification of acoustic waveform and VSP seismic tube wave measurements of fracture permeability. *Trans. SPWLA 28th Ann. Logging Symp.*, pp. 191–211.

Paillet, F.L. & Ellefsen, K.J. (2005) Downhole applications of geophysics. In: Butler, D.K. (ed.) *Near-Surface Geophysics, Part 1, Chapter 12, Series: Investigations in Geophysics: No. 33*, Tulsa, Oklahoma, U.S.A., Society of Exploration Geophysicists. pp. 439–471.

Paillet, F.L., Hess, A.E., Cheng, C.H. & Hardin, E. (1987b) Characterization of fracture permeability with high-resolution vertical flow measurements during borehole pumping. *Ground Water*, 25 (1), 28–40.

Pflug, K.A., Mwenifumbo, C.J. & Killeen, P.G. (1997) Full waveform acoustic logging applications in mineral exploration and mining. In: Gubins, A.G. (ed.) *Proceedings of Exploration 97: Fourth Decennial International Conference on Mineral Exploration, 14-18 September 1997, Toronto, Canada*, GEO F/X Div. of AG Inf. Sys. Ltd. Canada, pp. 477–480.

Radhakrishna, I., Rao, T.G., Athavale, R.N., Singh, V.S., Subrahmanyam, K., Anjaneyulu, G.R., Venkateswarlu, T. & Deshmukh, S.D. (1979) *Evaluation of shallow aquifers in Lower Maner Basin*, Indo-German Project, NGRI, Hyderabad, India, Technical Report No. GH 10-HG3.

Ramirez, A.L. & Lytle, R.J. (1986) Investigation of fracture flow paths using alterant geophysical tomography. *Tech. Note, Int. J. Rock Mech. Min. Sci. & Geomech.*, Abstract, 23 (2), 165–169.

Repsold, H. (1989) *Well logging in groundwater development*. International contribution to Hydrogeology Series, Vol. 9, UNESCO IHP, A A Balkema Publishers.

Schmitt, D.R., Mwenifumbo, C.J., Pflug, K.A. & Meglis, I.L. (2003) *Geophysical logging for elastic properties in hardrock: A tutorial*. In: Eaton D.W., Milkereit, B. & Salisbury, M.H. (eds.) Hardrock Seismic Exploration, Geophysical Developments Series No. 10. Tulsa, Oklahoma, U.S.A., Society of Exploration Geophysicists.

Serra, O. (1984) *Fundamentals of Well-Log Interpretation*. Vol. 1 & 2, Series: Development in Petroleum Science, 15A & 15 B. Amsterdam, The Netherlands. Elsevier.

Silliman, S. & Robinson, R. (1989) Identifying fracture interconnections between boreholes using natural temperature profiling: I. Conceptual basis. *Ground Water,* 27 (3), 393–402.

Singha, K., Kimball, K. & Lane Jr., J.W. (2000) *Borehole-radar methods: tools for characterization of fractured rock*. U.S. Geological Survey Fact Sheet: 054-00, U.S.G.S.

Soonawala, N.M. (1983) Geophysical logging in granites. *Geoexploration*, 21 (3), 221–230.

Sorey, M.L. (1971) Measurement of vertical groundwater velocity from temperature profiles in wells. *Water Resources Research*, 7 (4), 963–970.

Stumm, F., Chu, A. & Lange, A.D. (2001) *Use of advanced borehole geophysical techniques to delineate fractured-rock ground-water flow, faults, foliation, and fractures along the western part of Manhattan, New York*. U.S. Geological Survey Open File Report 01-196, U.S.G.S.

Tang, X.M. & Cheng, C.H. (1993) Borehole Stoneley wave propagation across permeable structures. *Geophysical Prospecting*, 41 (2), 165–187.

Tang, X.M., Cheng, C.H. & Paillet, F.L. (1991) Modeling borehole Stoneley wave propagation across permeable in-situ fractures. *Trans. SPWLA 32nd Ann. Logging Symp. Paper GG*.

Telford, W.M., Geldart, L.P. & Sheriff, R.R. (1990) *Applied Geophysics*, Cambridge University Press.

Tsang, C.-F., Hufschmied, P.& Hale, F.V. (1990) Determination of fracture inflow parameters with a borehole fluid conductivity logging method. *Water Resources Research*, 26 (4), 561–578.

Tsang, C.-F. & Doughty, C. (2003) Multirate flowing fluid electrical conductivity logging method. *Water Resources Research*, 39 (12), 1354 doi:10.1029/2003WR002308.

Vasvari, V. (2011) On the applicability of dual laterolog for the determination of fracture parameters in hard rock aquifers. *Austrian Journal of Earth Sciences*, 104 (2), 80–89.

West, F.G., Kintzinger, P.R. & Laughlin, A.W. (1975) *Geophysical Logging in Los Alamos Scientific Laboratory Geothermal Test Hole No. 2*. US ERDA Informal Report LA-6112-MS, Los Alamos Sci. Lab., University of California.

Williams, J.H. & Johnson, C.D. (2004) Acoustic and optical borehole-wall imaging for fractured-rock aquifer studies. *Journal Applied Geophysics*, 55 (1 & 2), 151–159.

Williams, J.H., Lane, Jr., J.W., Singha, K. & Haeni, F.P. (2002) *Application of advanced geophysical logging methods in the characterization of a fractured-sedimentary bedrock aquifer, Ventura County, California*. U.S. Geological Survey Water Resources Investigations Report 00-4083, U.S.G.S.

Williams, J.H., Lapham, W.W. & Barringer, T.H. (1993) Application of electromagnetic logging to contamination investigations in glacial sand-and-gravel aquifers. *Ground Water Monitoring & Remediation*, 13 (3), 129–138.

Young, S.C., Julian, H.E., Pearson, H.S., Molz, F.J. & Boman, G.K. (1998) *Application of electromagnetic borehole flowmeter*. US EPA, Report EPA/600/R-98/058.

Young, S.C. & Pearson, H.S. (1995) The electromagnetic borehole flowmeter: description and application. *Ground Water Monitoring and Remediation*. 15 (1), 138–147.

Zhou, C., Liu, L. & Lane Jr., J.W. (2001) Nonlinear inversion of borehole-radar tomography data to reconstruct velocity and attenuation distribution in earth materials. *Journal Applied Geophysics*, 47 (3 & 4), 271–284.

Chapter 13

Integrated geophysical survey

13.1 INTRODUCTION

In hard rock groundwater exploration, the aim of subsurface characterization is to delineate the weathered zone and its saturated thickness, saprolite, basic intrusives, structures and saturated fractured zones occurring up to ≈200 m depth. Also, it is to decipher fracture orientation and connectivity and the water content, flow and quality. It is likely that information on these aspects will be obtained through geophysics which basically provides subsurface physical property distribution or variation. The delineation of the weathered zone, mostly within 30 m depth, generally does not pose any problem. The conventional DC resistivity sounding or imaging is adequate for the purpose. The problem starts with the investigation of deeper horizons to which one must look for a better yield and potable water; the delineation of underlying productive saprolite whose thickness and character vary and that of deeper fractured zones which are generally thin (a few centimetres to a few metres). As far as resistivity of the saprolite is concerned, under normal situation with fresh water saturation, it is mostly associated with intermediate values compared to the weathered zone and underlying compact formations and represents a layer with transitional increase in resistivity with depth. Detection and characterization of this layer by the resistivity method alone is quite difficult if it is 'thin' (thickness compared to depth of occurrence), because of transitional resistivity, non-uniqueness (equivalence) and poor resolution leading to a range of equivalent models. Similarly, it is difficult to quantify the parameters of thin conductive localized fractured zones in highly resistive compact bedrock due to suppression and decreased resolution even with reduced scale of measurement. Like DC resistivity, the commonly used EM methods also suffer from similar limitations and are generally used in combination. Integrating the DC and EM results with the inferences drawn from remote sensing, geological and borehole information, hydrogeochemical data, other geophysical methods and techniques and combining the interpretational procedures sequentially or through joint inversion can improve resolution and minimize (not eliminate) non-uniqueness or uncertainty in target characterization and depth estimations. The integration of methods or a multi-method approach has almost become a necessity to solve critical hydrogeological complexities holistically (Chandra *et al.*, 2002a).

Prior to selecting the geophysical methods for an integrated approach it is necessary to perceive the hydrogeophysical objective of the project vis-a-vis hydrogeological conditions, geophysical targets, hydrogeophysical model, capability of available geophysical methods, detectability of target, attainable lateral and vertical resolutions vs. required resolutions and depth of investigation. It is followed by acquiring details of terrain condition and clearance, cultural interferences, requirement of appropriate method, parameters and scale of measurement. The data collected through different methods for integration should be at a compatible scale. Currently a variety of high resolution geophysical methods are available giving flexibility in selection for near-surface investigations that can minimize time-consuming labour-intensive conventional surveys, increase success rate in investigations and thus make the overall programme cost effective. Integration of modern techniques with traditional ones not only requires infrastructural development but also the availability of expertise. Nevertheless, the conventional DC resistivity method even with its inherent limitations remains the favourable cost-effective basic method with minimal infrastructure and expertise for low-budget groundwater exploration in hard rock.

It is well known that inversion of geophysical data suffers from non-uniqueness. A variety of equivalent geophysical models can be speculated and derived from the same data. Besides, all the methods whether traditional or modern have one or the other inherent limitations in providing the required hydrogeological information. Though a number of optimization techniques are in vogue to minimize it and increase the confidence level in parameter estimation, to overcome this and make the derived geophysical model of subsurface hydrogeological conditions reliable, it is essential to design approaches through alternative or integrated geophysical methods and techniques which are sensitive to different physical properties of the subsurface. For example, electrical and/or electromagnetic methods sensitive to conductivity characteristics of the fractured zone, signifying the quantity and quality of groundwater in it, could be merged with seismic, dealing with its mechanical property to convincingly extract the depth of interface and fractured zone. The essence of an integrated approach is to enhance the number of evidences derived from various methods and combine them to constrain the interpretation, refine the model and reduce the ambiguity. Such integration conventionally comprises economically available methods of magnetic and DC resistivity or electromagnetic and the seismic. It is appropriate to use low resolution cost-effective methods for fast reconnaissance to identify the area of interest and combine high resolution or specific methods for the detailed study. Depending on hydrogeological conditions and the availability of equipment a variety of integrations is possible. However, prior to deploying any method or technique its capability under given hydrogeological condition vis-a-vis the objective should be essentially realized through forward modeling. Also, the simplified forward model considered is to be judged against expected complexities and sensitivity and robustness of the model to input perturbations.

The reconnaissance on the ground can be done by a combination of conventional magnetic and VLF or single high frequency FEM profiling and detailing by a combination from VES or ERT, GRP, multi-frequency and multi-spacing FEM or dual moment TEM and the high resolution reflection seismic (HRS) whichever is found suitable and available. For reconnaissance carrying out multi-electrode spacing DC resistivity profiling or ERT is tedious, time-consuming and expensive. The reconnoitory magnetic survey in a grid form or parallel profiles is quite common in hard rock to delineate basic intrusives and structures, demarcate fault zones and lithological contact and also approximate the bed rock topography. The distant source electromagnetic (VLF) profiling can be used to obtain resistivity and thickness variations of weathered zone and locate shallow litho-contacts. In crystalline rocks of Nigeria Beeson and Jones (1988) used FEM profiling and VES to locate about 85% successful well sites. Bernard and Valla (1991) combined Schlumberger resistivity profiling with VLF while Sharma and Baranwal (2005) combined Schlumberger VES with VLF for well site pinpointing. The gravity method can also be combined for reconnaissance to map lateral variations in density/lithology, valley fill, bed rock topography (Ibrahim and Hinze, 1972; Hansen, 1984) and structure. Murty and Raghavan (2002) used gravity surveys in granitic terrain to delineate a thick weathered zone. The integration of gravity and seismic refraction could be useful in delineating buried valleys. Van Overmeeren (1975 and 1980) combined the results of gravity and seismic refraction surveys. Though there are a good number of references on gravity survey, it is not a commonly used method for groundwater in hard rock terrain. It is mostly used as a micro-gravity measurement to detect cavities in limestone. The emerging practice, particularly for large area coverage is, to carry out heliborne geophysical surveys and follow-up by high resolution ground geophysics to validate the results and pin-point the sites.

Prior to geophysical surveys, geomorphological, geological, hydrogeological and hydrogeochemical mapping is done to understand the groundwater conditions and assess favourable locations vis-a-vis groundwater demand. It helps select areas where geophysical investigations are essential. The sequential integrated approach to investigation from 'regional to local' i.e., satellite imagery and airborne survey data to detailed ground geophysical survey is discussed below.

13.2 MAPPING OF LINEAMENTS FROM SATELLITE IMAGERY

The exploration of discontinuous aquifers in hard rock starts with remote sensing study involving analysis of satellite imagery data and derived lineament map which gives a synoptic superficial view of structural features in the area. A systematic mapping of geological lineaments helps optimize the area of interest for ground geophysical surveys. Generally the lineaments are associated with geological structures like faulted and fractured zones, discontinuities, foliation and other linear features such as dykes and hence present a regional overview of the structural framework. However, there may not be any correlation between the orientation of the lineaments and that of the fractures. Since these structural features are favourable for groundwater occurrences, the mapping of lineaments is quite useful. The regionality, higher density and criss-crossing of lineaments revealing a greater density of fracturing are indicative of better groundwater prospect and yield. There are numerous examples confirming the location of high yielding wells on prominent lineaments and lineament intersections, showing the utility of a lineament map (Fig. 13.1). A positive correlation is generally obtained between the yield of a well and its proximity to the lineament. However, there could always remain a possibility of the fractured zone not being picked up as a lineament and high yielding wells being located even in areas without any lineament. For a detailed study, imageries at 1:50,000 scale are required.

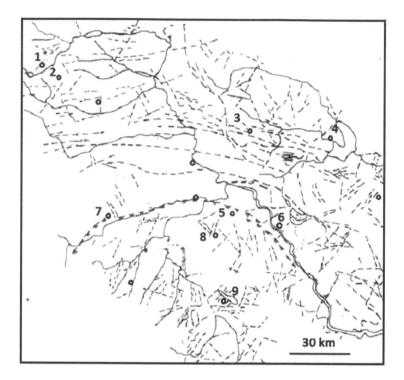

Figure 13.1 Lineament map prepared from satellite imageries for a hard rock area in the eastern part of India. Some of the moderate to high discharge (2.5–25 lps) wells drilled on the basis of integrated surveys are shown by circles (1...9) (Source CGWB, 2012).

13.3 AIRBORNE GEOPHYSICAL SURVEYS

In addition to satellite imageries, the aeromagnetic maps which exist over a large part and are being added to rapidly can be studied and interpreted for delineating the structures. Aeromagnetic data have been found to be an extremely valuable support in locating potential aquifers in hard rocks. The combined study of satellite imagery and aeromagnetic data could be a preparatory effort to help demarcate the zones of interest for surface geophysical surveys on a regional scale. Such attempts have been made for hard rock terrains in several parts of the world including eastern parts of India. The aeromagnetic maps of two areas from India around well locations 4 and 6 shown in figure 13.1 are presented in figures 5.9 and 5.10. Paterson (1989) suggests that for proper interpretation, aeromagnetic map over a minimum 10 km × 10 km area at 1:50,000 scale should be considered and magnetic modeling is done where dykes are suspected. Nowadays a heliborne geophysical survey is preferred for regional groundwater exploration. For the purpose, high resolution heliborne magnetic and TEM data are collected together, yielding very high data density cost effectively and rapidly which is not at all possible by ground survey. The combination is quite useful in hard rock providing subsurface structural image through heliMAG and conductivity image through heliTEM, thereby ascertaining the role of structures in forming the ground-water repositories. Interpreting heliMAG data for magnetic linears and anomalies revealing subsurface structures and then analyzing heliTEM data along and across these structures as depth of investigation (DOI) sections with 1-d Occam's inversion of TEM along the sections could be an effective approach to locate promising sites. Lineaments derived from heliborne magnetic data and satellite imagery may not match at places and also may not exist at sites where groundwater exploration is to be carried out. Nevertheless, locating the nearest magnetic lineament/structure is advantageous in placing ERT and shifting or accommodating the drilling site. Besides these, the heliborne survey results make it possible to investigate the regionality of hydrogeological features and boundary conditions which are quite significant in groundwater modeling.

13.4   GEOLOGICAL AND BOREHOLE INFORMATION

Prior geological and borehole information are required at two stages. At first this information is required to plan a geophysical programme and to select techniques for the best or desired output. The scale of geological information varies according to the type of geophysical survey. For example, reconnaissance magnetic survey requires major structural information such as presence and orientation of dykes and faults etc. while the detailed electrical resistivity survey would require information on subsurface lithological variations and the trend of the target geometry. In this, a forward modeling exercise is carried out to see the target response and efficacy of the technique in picking the target. Secondly, borehole information is required at the time of interpretation involving inversion where a guess model is generated through existing information. Also, the geophysical parameters obtained from ground surveys are validated through borehole information, required to standardize the parameters and check how sensible or realistic they are in terms of hydrogeological conditions.

13.5   SELECTED SURFACE GEOPHYSICAL METHODS AND TECHNIQUES FOR INTEGRATION

Prior to discussing the integration of geophysical surveys a few selected methods which help characterize the subsurface conditions in hard rock and constrain the interpretations are described here briefly. The basic objective is to merge conventional ones with modern methods and techniques whichever is available or required specifically for the objective, minimize non-uniqueness and get the best and maximum out of geophysical investigations.

13.5.1   Seismic surveys

The use of seismic surveys was initiated through the refraction method to assess the depth to shallow compact rock or thickness of the weathered zone, delineate the fractured zone underlying the weathered material and variations in lithology. It has been quite successful as the weathered zone and saprolite have a distinctly lower seismic velocity (P wave velocity ≈ 1500 m/s) compared to the underlying compact rock

(P wave velocity >4000 m/s) and varies with degree of weathering. Recent seismic refraction studies for weathered zones are by Befus et al. (2011), Clarke and Burbank (2011) and Holbrook et al. (2014). Based on reduction in seismic velocity due to fracturing Clarke and Burbank suggest a method of numerical modeling of seismic refraction data to estimate apparent fracture-density in shallow bedrock. Significantly, the seismic method has also been used to estimate the depth to groundwater level and saturation by Wallace (1970), Bradley et al. (1987) and Bachrach and Nur (1998). It is quite an important study showing the effect of saturation and partial saturation on seismic velocity but not being economical for the purpose of water table identification, the applications have been limited. Such approaches may be effective in identifying the thickness of a saturated weathered zone. The USGS publication on refraction method by Haeni (1988) and the papers by Lankston (1990), Steeples and Miller (1990), Steeples (2005) and Pelton (2005) present the details of seismic refraction and reflection methods and their applications in groundwater and near-surface investigations. Eaton et al. (2003) present various aspects of seismic surveys in hard rock. Applications for fracture and fault detection are outlined by Ernstson and Kirsch (2009). The fracture orientation can be detected by azimuthal refraction survey (Hansen and Lane, 1995). The vertical fractured zones can be detected by shear wave splitting analysis (Sun and Jones, 1993; Hardage, 2011). In refraction surveys velocity inversion due to the presence of a low velocity layer underlying a high velocity layer is a serious limitation (Whiteley and Greenhalgh, 1979). The reflection seismic is significantly useful to overcome this limitation, but at the same time it needs specific survey design and parameters and processing to remove the effect of near-surface heterogeneities (Meekes et al., 1990 and Gruber and Rieger, 2003). The high resolution seismic is precise, with high vertical and horizontal resolutions but being expensive, requiring better logistics and infrastructure its foremost application in groundwater exploration is limited to integrating with other geophysical methods for minimizing or removing the non-uniqueness and ambiguities in interpretations and enhancing the resolution wherever required. HRS can be useful in basalt flow areas where DC resistivity or EM may not yield definitive information on deeper aquifers, in thin vesicular and fractured zones which are also with low seismic velocity, occurring between compact highly resistive basalt flows. The cost is however reduced by using non-destructive seismic sources. The major development in application of the seismic method in hard rock, relevant to groundwater, is brought in by its use in mineral exploration and the characterization of a nuclear waste disposal site for delineating fracture free rock volumes and depth zones. Juhlin and Palm (2003) present in detail the procedure and parameters for seismic data acquisition, processing and interpretations for hard rock.

Hunter et al. (1984) and Miller et al. (1989) conducted a shallow seismic reflection survey to map the overburden-bedrock interface. Juhlin and Palm (2003) conducted HRS surveys in the granites of Sweden and brought out several structural features like dolerite-granite contact and fractured zones. According to them a shallow seismic reflection survey could prove to be an ideal tool for identifying sub-horizontal fractured zones. They further indicated that a dolerite-granite interface and a saturated fractured zone are significant sources of seismic reflection as the former is a high impedance boundary and the latter is a low impedance boundary and these can be distinguished by studying the polarity of reflected waves. The deeper fractured zones (up to 300–400 m depth) can be traced by HRS as shown by Mair and Green (1981) who conducted

HRS in 'homogeneous' granite batholith and identified the fractured zones at about 300 m depth which were confirmed by geophysical logs of boreholes drilled.

To detect thin layers within 50–60 m depth high resolution S-wave refraction and reflection surveys are being used. Johnson and Clark (1992) used S wave reflection technique and showed that by S wave a better resolution can be obtained in detecting thin layers. The S-wave with a lower velocity compared to P-wave, has a shorter wave length and therefore may improve the resolution. Dasios et al. (1999) brought out the advantages of shear wave surveys in near surface investigations. Guy (2006) indicated that the S-wave resolution is more in saturated overburden than in dry overburden. According to Johnson and Clark (1992) and Widess (1973) theoretically the minimum resolvable bed thickness is 1/4 to 1/8 the wavelength of seismic reflection and it varies with the S/N ratio. For example with a seismic velocity 5000 ft/s ($\approx$1500 m/s) and predominant wavelet frequency 100 Hz, the wavelength is 50 ft (or 15.24 m) and the minimum resolvable bed thickness would be 6.25 to 12 ft (or 1.9 to 3.8 m). The relation between resolvable minimum bed thickness and predominant wavelet frequency and seismic wave velocity as computed by Johnson and Clark (1992) is shown in figure 13.2. Some of the examples of high-resolution S-wave refraction and reflection surveys for shallow investigations are Ellefsen et al. (2005) and Inazaki (2006). Guy (2006) however suggests combined analysis of P and S-waves complementing each other in imaging the near-surface sequence. Using S and P wave reflection technique, delineation of a shallow sub-horizontal thin fractured zone with improved resolution can be attempted.

13.5.2 Passive seismic

Seismic methods discussed above use an active artificial seismic energy source. The method of passive seismic uses ambient micro-tremors as a seismic source. The goal is to obtain a subsurface structural image analogous to that obtained from conventional reflection seismic by recording ambient noise-field of the Earth (Artman, 2006). The seismic tremors from subsurface sources at low frequency, about 1–6 Hz, are recorded over a time period (1–3 months) through an array of sensors spread in a regular grid (1 km × 1 km or more) fashion and analysis of these can provide information on subsurface structural features, lithologic heterogeneities and bedrock topography. Measurements can also be made for shorter duration with a single sensor. The 3d model of P and S wave travel times yields information on structures and the ratio of P to S wave velocity (V_p/V_s) yields lithologic information (Martakis et al., 2011). The basic difference between conventional seismic and passive seismic is that while the former involves two-way travel of a seismic wave the latter involves only one-way travel of a seismic wave to the surface. It is an emerging technology, having relatively low-cost, with no environmental hazard as no active source is used and effective in areas difficult or inaccessible for conventional surveys. Martakis et al. (2006) and Tselentis et al. (2007) present the details of data acquisition, modeling and inversion. They point out that areas with high seismicity are favourable targets to exploit the potential of passive seismic tomography. Passive seismic measurements have been made at the ground surface as well as in boreholes for source location and monitoring fracture growth process due to hydraulic fracturing which induces micro-tremors (Li, 1996; Maxwell et al., 2002).

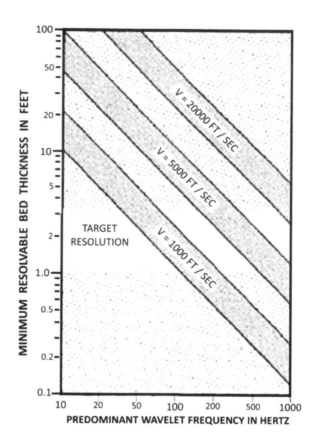

Figure 13.2 The minimum resolvable bed thickness is a function of predominant reflection frequency and velocity. The upper and lower bounds for each velocity assume that the minimum resolvable bed thickness ranges from λ/4 to λ/8, where λ is the length of the seismic wavelet in the ground (Source: Johnson and Clark, 1992).

To estimate the depth to bedrock the H/V method proposed by Nakamura in 1989 is used. The method can be employed to estimate the thickness of sediments overlying the bed rock, map bed rock topography and infer the structures. In this method the vertical (V) and two horizontal (H) components of ambient seismic noise are measured for determining the fundamental seismic resonance frequency of a site. It is obtained by analyzing the spectral ratio of horizontal and vertical components. Using the inverse proportion relationship between the main resonance frequency of a soil and its thickness (Delgado *et al.*, 2000) the sediment thickness is determined. Lane *et al.* (2008) used this method effectively in estimating the depth to bedrock and indicated its limitation where litho-contact is transitional such as gradational weathering and the contrast in acoustic impedance of bedrock and overburden is less. Other than the research papers referred to here details of the H/V method can be obtained from

Nakamura (1989), Ibs-von Seht and Wohlenberg (1999), Parolai *et al.* (2002), Haefner *et al.* (2010) and Cakir (2011).

13.5.3 Ground penetrating radar

Ground penetrating radar (GPR) is also known as ground probing radar or georadar. It is a non-invasive high resolution surface geophysical technique. It is used in bore-holes also as discussed in Chapter 12. GPR is used for shallow investigations mostly within a few tens of metres depth. It is in wide use for a diverse range of near surface investigations and applications are abundant (Davis and Annan, 1989). Even the tree roots can be mapped (van Schoor and Colvin, 2009). According to Bristow and Jol (2003) …. "GPR provides high-resolution images of the shallow subsurface that cannot be derived by any other non-destructive method". In hard rock groundwater exploration it can be used to study the heterogeneities in the top weathered zone which is generally 20–30 m thick and delineate the underlying compact bedrock and could also be used to identify shallow fractures in it. Under favourable subsurface conditions, with the development in acquisition technology, it can yield information up to 80 to 100 m depth. GPR uses a high frequency electromagnetic field. The frequency generally ranges from 10 to 1000 MHz. Use of such frequencies gives significantly high resolution to detect features as thin as a few tens of millimetres (Davis and Annan, 1989) but the field attenuates faster with increasing frequency and deeper information cannot be obtained. The attenuation, besides frequency depends on conductivity of the medium in which the EM field propagates. Because of this GPR cannot be used in a conductive subsurface environment such as saturated clay and saline water saturated formations. GPR works well in a resistive environment.

In frequency domain electromagnetic (FEM) and very low frequency (VLF) surveys the EM field used is in the range of 100 Hz to 30 kHz where induction takes place but when a very high frequency is used the EM field behaves as wave propagation. So, GPR behaves as seismic wave propagation with refractions and reflections, however the reflected phase is mostly considered. The basic difference is that unlike seismic, high frequency GPR wave propagation and reflection are controlled by electrical conductivity and a contrast in the dielectric constant of the medium. The principle of GPR is discussed in detail by Davis and Annan (1989). There are issues of journals dealing with GPR exclusively, e.g. Journal of Applied Geophysics (Vol. 33, No. 1–3, 1995). Also, the readers can consult the review papers by Annan (2005 a and b).

In GPR operation, the radiated wave front of a high frequency electromagnetic wave penetrates into the ground from a transmitting antenna placed on the ground. The wave travels through the subsurface and hits interfaces separating materials of different dielectric properties. These interfaces could be between soil and weathered material, heterogeneities within the weathered material, weathered overburden and compact rock, and fractures in the compact rock. The wave is reflected back from the interfaces and recorded by the receiver antenna. The depth to the reflector is obtained from electromagnetic wave velocity and reflection time to reach the receiver. At high frequency, the velocity of EM wave propagation depends on the dielectric permittivity of the medium. The relative dielectric constant (ε_r) is expressed as $\varepsilon_r = \varepsilon/\varepsilon_0$ where ε_0 is permittivity of free space and ε is permittivity of the medium. The value of

ε_0 is 8.854×10^{-12} Farad/m. The attenuation (α) and skin depth (δ) are given by the simplified relations

$$\alpha \approx \sigma/2\sqrt{\mu/\varepsilon} \quad \text{and} \quad \delta \approx 2/\sigma\sqrt{\varepsilon/\mu}$$

where μ is assumed to be equal to $\mu_0 = 4\pi \times 10^{-7}$ H/m. The dielectric constant of water is 80. The dielectric constant for some common materials is presented in Table 13.1.

The GPR has been used to delineate bed rock topography (Davis and Annan, 1989). It has also been used to detect fractures in hard rocks in several parts of the world. Travassos and Menezes (2004) delineated a shallow saturated fractured zone in granitic terrain of Brazil. Porsani *et al.* (2006) showed its superiority in delineating shallow sub-horizontal fractures over vertical fractures. Stevens *et al.* (1995) used GPR to delineate dipping fractures in parts of the Canadian Shield and fractures up to a depth of about 60 m were inferred. Hansen and Lane (1995) carried out GPR surveys for delineating fractures in crystalline bedrock of Massachusetts. For delineating deeper fractures antennas of lesser frequency are used. By reducing the frequency the attenuation is reduced and the S/N ratio is enhanced which helps in detecting reflections from deeper saturated fractures. Tsoflias and Becker (2008) suggest the use of signal amplitude and phase to image bed rock fractures and characterize their aperture and saturating fluid conductivity (up to 1 S/m). Nascimento da Silva *et al.* (2004) used GPR in combination with DC resistivity in a resistive terrain. While DC resistivity yielded the spatial pattern of the fractured zones, GPR characterized the individual fractures. Seol *et al.* (2001) gave a scheme for finding the strike direction of a fracture combining the data obtained along a profile using three different configurations of antenna, viz., perpendicular-broadside, parallel-broadside, and cross-polarization. Gendzwill *et al.* (1994) compared the results of HRS and 190 MHz GPR surveys from granitic terrain. According to them though the resolution of HRS was 5m and that of GPR was 0.6 m the GPR features had corresponding seismic expression. While HRS showed clear reflections up to 180 m depth the GPR penetration was maximum 50 m. However, the shallowest HRS reflections were beyond 30 m depth only. From GPR fractures and fractured zones in the depth range of 5 to 50 m were recognized. It indicates that for fractured zone investigations up to about 30–50 m depth GPR (if there is no salinity) and beyond that HRS can be used. The GPR is effective in characterizing the vadose zone, identifying areas for artificial recharge and shallow groundwater contaminations. Knight (2001) presents the details of GPR and its applicability in contaminant study by identifying the presence of liquid organic contaminant. The possibility of long-term monitoring of contamination due to oil leakages is reported by Dolgiy *et al.* (2006).

13.5.4 Nuclear magnetic resonance measurement

The surface geophysical methods presented so far are used to delineate aquifers indirectly. The Nuclear Magnetic Resonance (NMR) method is the only surface method which identifies the subsurface water content directly – a remarkable and fascinating breakthrough in hydrogeophysics. It opens up a unique geophysical perspective of groundwater resource quantification within ≈ 100 m depth and generation of hydrogeological models. By varying the field parameters, MRS method can investigate a large subsurface volume approximated by a cube ranging about 2.5×10^4 to 10×10^6 m^3

Table 13.1 Typical dielectric constant, electrical conductivity/resistivity, velocity and attenuation observed in common geological materials at 100 MHz (Source: Davis and Annan, 1989).

Material	K	σ (mS/m)	V (m/ns)	α (dB/m)
Air	1	0	0.30	0
Distilled water	80	0.01	0.033	2×10^{-3}
Fresh water	80	0.5	0.033	0.1
Sea water	80	3×10^4	0.01	10^3
Dry sand	3–5	0.01	0.15	0.01
Saturated sand	20–30	0.1–1.0	0.06	0.03–0.3
Limestone	4–8	0.5–2	0.12	0.4–1
Shales	5–15	1–100	0.09	1–100
Silts	5–30	1–100	0.07	1–100
Clays	5–40	2–1000	0.06	1–300
Granite	4–6	0.01–1	0.13	0.01–1
Dry salt	5–6	0.01–1	0.13	0.01–1
Ice	3–4	0.01	0.16	0.01

K: dielectric constant, σ: electrical conductivity,
V: velocity of EM waves, α: attenuation

in size (Legchenko *et al.*, 2006). It is a major advantage of this method. It could also be a disadvantage of the method in hard rock because of the low volume of water in fractures compared to the large volume investigated (Roy, personal communication) being unable to create enough response above the noise level. Being sensitive only to water, it exploits the magnetic moment of hydrogen atom nuclei abundant in water molecules in pore spaces. In a simple language, water molecule contains a hydrogen atom whose nucleus holds one proton. If the protons can be energized and their overall response is measured, the total number of protons can be estimated and hence the aggregate population of hydrogen atoms or water content can be assessed. Schirov *et al.* (1991), Yaramanci (2000) and Lubczynski and Roy (2004) present a brief history of the development of surface NMR sounding. It is a non-invasive method and in use for groundwater since the late 1970s. The method has been mainly used for sedimentary formations to estimate the vertical distribution of water content or fluid saturation, porosity, free water content or permeability and transmissivity. In hard rock the method has been experimented with to characterize the water content in weathered and fractured zones and holds the potential for effective utilization under favourable condition. A brief outline of surface NMR or Surface Magnetic Resonance Sounding (SMRS) measurements is presented here and the details can be obtained in the papers referred.

The principle of Nuclear Magnetic Resonance (NMR) or Proton Magnetic Resonance (Raghunathan and Jagannathan, 1996) and its application in groundwater has been described in detail in several research papers viz., Yaramanci (2000), Legchenko and Valla (2002), Roy and Lubczynski (2003) and Roy (2007). The NMR measurement is based on the inherent spin property of the proton carrying a positive charge. The proton spins to generate a magnetic field and is considered as a spinning magnetic dipole with its axis parallel to the axis of spin. In the presence of an ambient static magnetic field the magnetic dipoles are aligned with the static field and precess at a frequency which depends on the magnitude of the static field at that point. This frequency is known as the Larmor frequency (ω_0) and relates to the static field by gyromagnetic ratio (γ_p)

$$\omega_0 = 2\pi f_0 = \gamma_P |B_0|$$

where B_0 denotes a static magnetic field and γ_P denotes the gyromagnetic ratio of a proton. For a hydrogen proton in the water molecule $\gamma_P = 0.2675$ rad s^{-1} nT^{-1} or 0.04258 Hz nT^{-1}. For Earth's magnetic field varying from 20,000 to 60,000 nT, the Larmor frequency (f_0) correspondingly ranges between 800 and 2800 Hz. Since γ_p has an explicit value for a type of nucleus, the Larmor frequency can be selected to investigate specific nuclei (Legchenko et al., 2002). Here, the hydrogen protons are studied.

The precession axis of a proton is aligned either parallel or antiparallel to the static magnetic field direction. As per the energy configuration, the low energy state relates to parallel alignment and the high energy state relates to antiparallel. The slight excess proton population with low energy state over the high energy state population yields a net magnetization in the direction of the ambient static magnetic field. This magnetization is very small and therefore to generate a measurable signal an external field is applied. The proton in the static magnetic field can absorb external electromagnetic energy created by an oscillating magnetic field at a specific frequency – the Larmor frequency of the static magnetic field at the point of measurement. When the oscillating magnetic field is applied, its component perpendicular to the static field creates a torque and changes the spin axis of the proton. On turning-off the oscillating field, the protons generate a relaxation magnetic field at the Larmor frequency and come to equilibrium realigning with the static field. The various attributes of this relaxation magnetic field is measured to know the amount of protons present.

IRIS (1998, 2004), Lubczynski and Roy (2004), Baltassat et al. (2006) and Roy (2007) present the field procedures for data acquisition and processing and inversion. The SMRS or MRS measurement uses earth's magnetic field as static field. The data acquisition is initiated through measurement of earth's total magnetic field intensity at the MRS site using a magnetometer. The homogeneity or uniformity in the magnetic field is also checked (Roy, 2007). Conventionally a VES or TEM sounding is conducted at the MRS site to know the subsurface depth-wise resistivity variation and assess its effect on phase and amplitude of the MRS signal. Goldman et al. (1994) show the usefulness of combining TEM with MRS in differentiating fresh and saline water. Further, they mention that the combined inversion of these may also solve the equivalence problem causing overestimation of conductivity and underestimation of thickness of the intermediate conductive layer. Aquifers under a thick conductive layer do not give a measurable signal (Lubczynski and Roy, 2004). According to Trushkin et al. (1994) resistivity around 50 ohm.m has no effect on the MRS signal. Baltassat et al. (2006) show the effect of resistivity of the medium on depth of detection of a 1-m thick layer of free water (Fig. 13.3).

The MRS measurement is carried out by laying a copper wire transmitter/receiver loop on the ground. The loop can be a circle or a square. Its diameter or side can vary in size from 10 to 150 m. The larger size gives a larger volume of investigation. The investigated volume is approximated by a cube of $1.5 \times a$ where 'a' is the side of a square loop (Legchenko et al., 2004). To improve the signal, the loop can be laid as

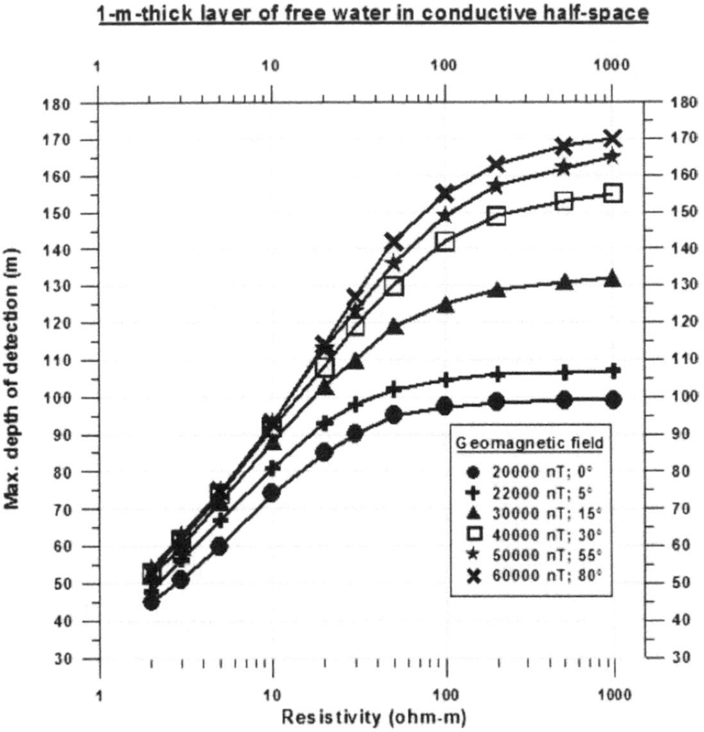

1-m-thick layer of free water in conductive half-space

Figure 13.3 The maximum depth of detection calculated for a 1-m thick layer of free water (w = 100%) vs. the half-space resistivity (Source: Baltassat *et al.*, 2006).

a figure of eight as proposed by Trushkin *et al.* (1994) shown in figure 13.4, but the depth of investigation is reduced to almost half. Lubczynski and Roy (2004) indicate the superiority of an eight shaped-loop configuration over square loop in identifying shallow fractured zones in noisy area and suggest that the longest axis of an eight shaped loop should be parallel to the strike of fault/fractured zone. According to Roy (personal communication), MRS measurement with an eight shaped loop focused over a lineament could be useful in estimating the water content of the fractured zone underneath. By this configuration the signal received from water in a narrow and shallow volume of the fractured zone is enough to make measurement above noise level and detection and characterization possible.

Once the local magnetic field is known at a site, the instrument is tuned to the Larmor frequency (f_0) for the local magnetic field, which is proportional to the field intensity. Its value is obtained using the formula given above. By the transmitter, an excitation current at the Larmor frequency determined is passed for a few tens of milliseconds (≈ 40 ms) through the square or eight shaped (in high noise areas) large loop laid on the ground. Since in addition to variation in the size of loop, variation in pulse moment also gives depth discrimination (Roy, 2007), measurement is carried out varying the excitation pulse moment which is the product of the intensity of loop

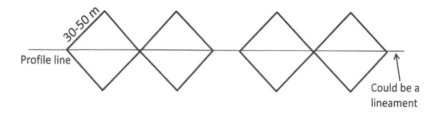

Plan of Eight shaped loop

Figure 13.4 The eight-shaped loop to enhance signal to noise ratio. Effective in shallow detection as depth of investigation is reduced

current at Larmor frequency and pulse duration. A large loop with high pulse moment gives deeper information (Fig. 13.5). So, a large number of measurements are made and stacked with increased excitation pulse moments for information from different depths and enhanced signal to noise ratio.

After the excitation current is abruptly turned off, protons in the hydrogen atoms of groundwater molecules produce a relaxation magnetic field, i.e., they dissipate energy over a certain length of time and return to equilibrium. Using the same loop as the receiver, the amplitude of the relaxation field is measured which induces a voltage in the loop and is directly linked to the number of protons and the water content (IRIS, 1998). The decay of the relaxation field is given by the relation

$$E(t) = E_0 e^{\left(\frac{-t}{T_2^*}\right)} \sin\left(2\pi f_0 t + \phi_0\right)$$

where T_2^* is the time constant of decay which describes how magnetization returns to its equilibrium state (it is the time required for decay of field to $1/e$ or 37% of its maximum value after turn off) and ϕ_0 is phase shift between excitation current and induced relaxation voltage measured in the loop as receiver (Fig. 13.6). The signal can be measured only after the 'dead time' (30 to 40 ms) so the initial amplitude E_0 is obtained by extrapolation (Baltassat *et al.*, 2006). The T_2^* is of the order of a few tens of milliseconds for small pores with clay bound water and of the order of a few hundred milliseconds in free water in the pores which can flow. Hence, pore size distribution and permeability can be measured from the relaxation time. According to Legchenko *et al.* (2004) MRS is more reliable to estimate transmissivity than hydraulic conductivity.

The electromagnetic field is subject to attenuation depending on the frequency used and the conductivity of the medium. Therefore, at low latitudes where earth's magnetic field is low the corresponding Larmor frequency is low and attenuation is less but at the same time the resonance response being proportional to the square of the magnetic field, at low latitude the response will be relatively poor and prone to effects of noise compared to high latitudes with higher values of magnetic field. At low latitudes, Lubczynski and Roy (2004) based on their experience from southern Africa suggest conducting MRS measurement during the night or early morning.

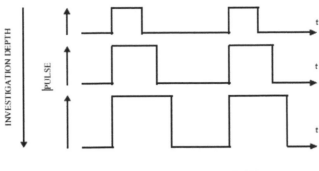

PULSE AND DEPTH INVESTIGATION

Figure 13.5 Increase in depth of investigation with increase in pulse moment which is the product of intensity of excitation current at Larmor frequency in the loop and the pulse duration (Source: IRIS, 1998).

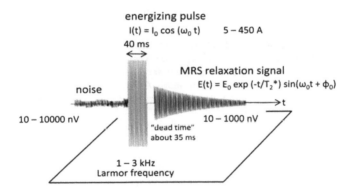

Figure 13.6 The proton excitation and relaxation signal envelop (Source: Turu-Michels, 1997).

Legchenko *et al.* (1998) carried out MRS measurements along with horizontal loop electromagnetic (HLEM) in metasediments and crystallines of Saudi Arabia and found a good correlation with drilling results. Since HLEM cannot discriminate water content from other conductive material in the fractures, MRS found better applicability. However, MRS measurements get perturbed by variations in magnetic susceptibility causing magnetic field inhomogeneities which is not uncommon in hard rock. Legchenko *et al.* (2004) carried out MRS with ERT in a granitic terrain in France and concluded that MRS cannot resolve aquifers at depths greater than the half loop size and support from ERT or a borehole is required. Further, Legchenko *et al.* (2006) carried out MRS measurements in the hard rock of south India where the weathered zone held more water than the underlying fissured or fractured zones. To assess the threshold value for the instrument using a 50 m × 50 m square loop of two turns and to ascertain the accuracy of data inversion, numerical modeling was done considering the weathered

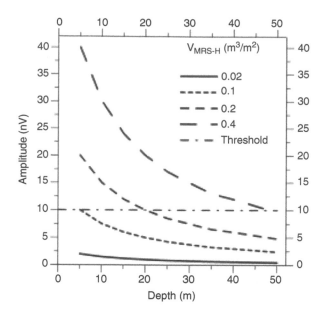

Figure 13.7 The maximum amplitude of MRS signal vs. depth of aquifer, for different volumes of water in the subsurface (50 m × 50-m loop, two turns). The threshold signal is 10 nV. A 20 m thick aquifer with 1% water content can be detected up to 20 m depth and with water content of 2% can be detected up to 50 m depth. Assuming horizontally stratified earth V_{MRS-H} provides an estimate of the volume of water per surface unit in a layer of thickness Δz. (Source: Legchenko *et al.*, 2006).

and fractured zone as a single unit with low effective porosity. They found that a 20-m thick aquifer up to 20 m depth with 1% water content and up to 50 m depth with 2% water content can be detected and inferred that underlying fractured zones with hardly 0.1 to 0.2% effective porosity cannot be detected or their water content is underestimated (Fig. 13.7). The weathered zone aquifer with 2% water content can be detected if its thickness is more than 10 m. It was indicated that MRS resolution decreases with depth and a deeper aquifer can be resolved only if there is no water in the layers overlying it. That is, the MRS signal from the fractured zone can be obtained if the overlying weathered zone is dry. It was further pointed out that aquifers with dimensions smaller than the loop configuration, which may be observed in the weathered zone as pockets, behave as 2d structures and error is introduced in inversion of MRS data obtained over such structures. The MRS results obtained along a profile by Legchenko *et al.* (2006) are shown in figure 13.8.

Baltassat *et al.* (2006) conducted MRS in the hard rock of south India. They also concluded that with a detection threshold of 10 nV a 10-m thick aquifer of less than 1% water content cannot be detected if it is deeper than 20 m and a 20 m thick aquifer with 1.5% water content can be detected up to 50 m depth by increasing the loop size appropriately. Vouillamoz *et al.* (2005) used MRS for characterizing weathered and fractured zone aquifers in granites and associated basement rock of

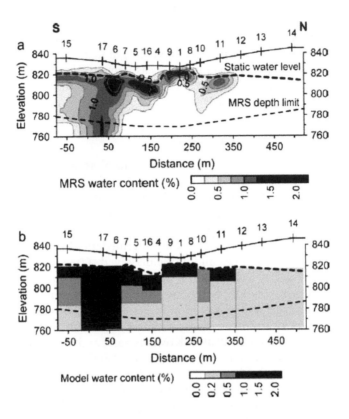

Figure 13.8 MRS results from hard rock aquifer in southern India obtained by Legchenko *et al.* (2006). The MRS cross-section is across the outlet of the watershed. (a) cross section of water content distribution derived from inversion of field measurements; (b) a 2D block model of the MRS water content. The thick and thin dashed lines show the static water level measured in boreholes and depth of investigation of MRS respectively.

Burkina Faso. A square loop of 125 m side length was used and the depth up to 80 m was investigated. According to them thin fractures occurring at depths more than half the loop size cannot be detected. All these studies reveal that by varying the loop size and pulse moment MRS can be usefully deployed in determining water content and the hydraulic parameters of the weathered zone aquifer and shallow fractured zones. Besides, it is essential that MRS measurements be combined with at least one of the surface methods, viz., VES, ERT, HLEM or TEM. In this regard it would be appropriate to attempt joint inversion of hydraulic test data with MRS and TEM data as suggested by Behroozmand *et al.* (2012) and Vilhelmsen *et al.* (2014).

The problems in MRS measurement are 50 or 60 Hz power line noise, the lateral inhomogeneities deviating from 1D assumption for inversion and the heterogeneities in magnetic field intensity caused by basic dyke and crystalline bed rock. To minimize the effect of powerline noise Larsen *et al.* (2014) use the post processing method of modeling the noise and removing it from MRS data and then applying multichannel

Wiener filtering for further removal of noise. They further indicate that by this the use of eight-shaped loop required in noisy area may not be necessary as that hampers the depth of investigation. Roy (personal communication) indicates that magnetite-bearing aquifers pose a problem by creating large magnetic gradient at pore scale thereby reducing T_2^* often below the aperture window of measurement. Vouillmoz *et al.* (2011) used spin-echo methodology to reduce the effect of magnetic field heterogeneities causing underestimation of aquifer parameters.

13.5.5   Radon gas measurement

The radon gas measurement involves the geochemical aspect of groundwater exploration in hard rocks. Its integration with a geophysical survey can be effective in delineating fractured zones and also identifying the hazardous areas. Radon (^{222}Rn) is a radioactive isotope and has a half life of 3.825 days. It occurs as a colourless and odourless noble gas. It is non-polar, monatomic and chemically inert. The ^{222}Rn is derived from disintegration of radium, an intermediate decay product of uranium (^{238}U). It occurs in soil as soil-gas and in the dissolved state in groundwater (Tanner, 1992; Alexander and Devocelle, 1997). As indicated by Hammond *et al.* (1988), radium and radon enter solution, but in low salinity environments, radium is strongly adsorbed on the rock surfaces while radon is dissolved and moves with the fluid. The radium (^{226}Ra) in rocks and minerals is, however, the major factor in regulating the supply of ^{222}Rn in air and groundwater (Krishnaswami and Seidemann, 1988). Radon present in groundwater gets released into the air. While decaying, radon produces harmful alpha radiation and is a health hazard from air (indoor) and groundwater. The radon isotopes are present almost everywhere in very minute concentrations. The unit of measuring its concentration is picoCurie per litre (pCi/l) or Becquerel per cubic metre (Bq/m^3); 1 pCi/l is equivalent to 37 Bq/m^3. On an average the outdoor air radon concentration is 0.4 pCi/l. The US EPA recommended upper limit of indoor air radon concentrations is 4 pCi/l and above this requires remediation (LeGrand, 1987). A radon concentration of 10,000 pCi/l in water can contribute about 1 pCi/l to indoor air. The US EPA limit of radon in water is 300 pCi/l which can contribute about 0.03 pCi/l in air (US EPA, 2012). Because of its short half-life the distance radon can move before decay is limited (Alexander and Devocelle, 1997).

Its diffusion from soil and rocks can be used for hydrogeological characterization in igneous and metamorphic rocks, particularly in the granitic terrain, that holds more than the normal amount of source material of uranium having a half-life long enough to persist in its original stage (LeGrand, 1987). The helium, radon and carbon dioxide – the sub-terrestrial gases diffusing and getting exhausted upward in the atmosphere through faults and an interconnected network of vertical and horizontal fractures as conduits, according to Ciotoli *et al.* (1998), are the most reliable geochemical signature of fluid circulation and can be used as faults and fractured zone tracers. The ^{222}Rn gas comes out with these gases as traces. However, the geological condition has a great control on radon concentration (Ranger, 1993). Its concentration and movement in the subsurface is controlled by composition, texture, fracture connectivity, permeability and moisture content. Radon moves faster through dry coarse material than wet fine material (LeGrand, 1987). The maximum concentration of radon is in granitic terrain followed by high grade metamorphic rocks like granite gneisses and schists and areas with pegmatites (Brutsaert *et al.*, 1981; Ranger, 1993; Frishman *et al.*, 1993). Srilatha *et al.* (2014) reported a higher concentration of radon in pink granite compared to grey granite in parts of Karnataka, India and could be linked with variations in mineral composition, texture and alteration.

Measurement of radon has been done for various purposes, like health hazard, study of active fault zones and precursor to earthquakes (Fu *et al.*, 2008; Mahajan, *et al.*, 2010), as natural tracer, submarine groundwater discharge (Burnett & Dulaiova, 2003; Peterson *et al.*, 2008), groundwater flow dynamics (Kafri, 2001), flow rates (Cook *et al.*, 1999), rock-water interaction (Senior and Vogel, 1989), neotectonic activities (Choubey *et al.*, 2001, Ramola *et al.*, 2008) and fractured zone detection on surface (Pointet 1989; Wright and Carruthers, 1993) and in boreholes (Hammond *et al.*, 1988). Banwell and Parizek (1988) preferred groundwater ^{4}He profile across lineaments to identify or confirm the fractured zones rather than the groundwater ^{222}Rn profile as the latter is more influenced by geochemical and hydrogeological factors than fracture concentration. For a regional study on this, Wattananikorn *et al.* (2008) suggested the use of airborne data collected for studying variations in potassium, uranium and thorium in the near-surface formations.

In hard rock the localized concentration of radon can be related to the presence of faults, vertical and horizontal fracture and shear zones. As mentioned in Reddy *et al.* (1996), Pointet (1989) shows the usefulness of shallow soil-gas radon concentration measurement in identifying high yielding wells on fractured zones in the granitic terrain of southern India. Wright and Carruthers (1993) also carry out electromagnetic and radon measurements across lineaments in the crystalline area of Zimbabwe and show a positive correlation. The EM anomalies on fractured zones are associated with radon maxima. Reddy *et al.* (1996) based on their field studies measuring radon concentrations at high and low yielding wells reveal that observations made by Pointet (1989) cannot be generalized. In this regard Lachassagne *et al.* (2001) clarify that higher radon concentration can be observed only where hydraulically active fractures are present. Also, they indicate the effect of topography on radon concentration. However, further studies by Reddy *et al.* (2006) support the view of Pointet and Wright and Carruthers. The radon profile only indicates the surface manifestation of a lineament or fractured zone. However, to locate a borehole site, it is necessary to know its subsurface attitude through geophysical measurements. It is likely that wells tapping deeper fractures show a higher concentration of radon compared to shallow wells in the weathered zone but it cannot be a guiding factor and specific study is required as Hammond *et al.* (1988) indicate low radon concentrations in deep fractured zones also. In fact, radon measurement could be used as the most sensitive 'permeability indicator' for thin deeply connected fractured zones where other methods and even MRS may fail to generate adequate response (Roy, personal communication). Using radon measurement to locate lineament is a cost-effective complimentary geochemical approach. In this connection the method of sampling is very important. Byegard *et al.* (2002) analyze ^{222}Rn measurements on wells by collecting water from fractures using packers. They use high pressured stainless steel vessels for sample collection. The radon concentration of samples collected in these vessels vary from 400 to 600 Bq/l whereas the samples collected in plastic beakers show lower concentrations.

The majority of studies are on its hazardous nature and risk assessment. In granites radon concentration can go as high as 100,000 pCi/l (Brutsaert *et al.*, 1981). Veeger and

Ruderman (1998) indicate higher concentration of dissolved radon in bed rock aquifer compared to that in near surface aquifers and hence, in general, lesser risk in using groundwater from shallow aquifers. Therefore, radon measurement is very important and it should form a part of groundwater exploration in hard rocks. Radon profiles, particularly across lineaments should be conducted to identify the radon hotspots. The wells or lineaments showing very high concentration of radon may be avoided or necessary arrangements may be made for the use of the water. It is even suggested that while hydrofracturing radon measurements should be made to ensure the concentration of radon is within limits (Burkhart et al., 2013).

13.6   INTEGRATION OF ELECTRICAL AND ELECTROMAGNETIC METHODS

Almost all the surface geophysical methods have been used in groundwater exploration. However, all these methods are not equally responsive to the groundwater targets. Out of these, DC resistivity and EM methods are the most responsive to variations in groundwater conditions – quantity as well as quality and widely used in hard rock. The methods can be used for monitoring as well. DC resistivity and EM methods are cost-effective and interpretation software is easily available. In hard rock though the basic assumptions for a vertical electrical resistivity sounding (VES) are seldom met with, the technique works satisfactorily. To pinpoint a site on a thick weathered zone, saprolite and fractured zones, even in an area selected through hydrogeological and geomorphological studies, conducting VES at random may not yield the location of the best site. For this, it is desired that gradient resistivity profiling (GRP) and ERT are conducted in combination. To demarcate fractured zones GRP can be combined with FEM profiling also. To delineate the contact zone of the exposed linear ridge of basic dyke or quartz reef with the country rock a series of short-spaced, small-loop TEM soundings can be conducted on either side of it along profiles orthogonal to the trend of the ridge. It may not be possible to conduct GRP in this direction due to the topographic variations, the difficulty in placing large-spaced current electrodes and the errors in measurements. The VES or ERT can be combined with a profile of TEM soundings in case a highly conductive zone or layer at depth is suspected. Because, for a conductive layer at depth overlying the bedrock, e.g., in a coastal hard rock area (Chandra et al. 1983), the VES curve is of QH or KQH type and in such cases the depth to the bed rock is generally overestimated (Chandra, 1984) and the support from TEM can help resolve the ambiguity in depth estimation. A cost effective practice is to carry out parallel profile multi-spacing (AB) GRP (Chandra et al., 2002b) across a lineament to confirm the presence and linear extension of conductive fractured zone followed by VES along and ERT along and across the linear conductive zone. The ERT gives a 2d picture and has almost replaced VES and resistivity profiling but by using it the data acquisition cost goes up. The ERT with conventional configuration and 10 m spaced 120 electrodes gives information up to about 240 m depth. An important point to be considered while combining ERT or VES with TEM is that once a high resistivity thick layer is encountered at depth underlying the conductive to very conductive (less than 10 ohm.m) shallow layers, by TEM it is very difficult to pick up still deeper thin conductive layers immediately underlying the thick high resistive layer, which may be an aquifer. While in VES or ERT the conductive layer below a thick resistive layer at shallow depth may be picked up but only with under- or overestimation of resistivity and thickness. In such cases a support from HRS may be useful. Practically in TEM, the top of the high resistivity layer at depth underlying highly conductive overburden becomes the depth of investigation.

The DC resistivity as well as EM suffer from equivalence and suppression and bring in non-uniqueness and because of this the precise estimation of depth to resistive substratum and the parameters of the layer immediately overlying may be difficult at times, particularly, when resistivity increases transitionally with depth. However, TEM being more sensitive to the conductive layer may help minimize the ambiguities and equivalence to some extent by inverting it jointly with VES. A support from geological and borehole information can minimize this. Also, a support from other geophysical methods is essential, of which seismic is the most significant. The depth to compact substratum obtained from seismic is used in VES interpretation. Once the depth is fixed the resistivity of the layer immediately overlying the substratum is precisely obtained and its suitability as aquifer is characterized (Chandra and Ramakrishna, 1979). Van Overmeeren (1981) makes a similar attempt in removing equivalence from VES interpretation and correctly estimating the layer resistivity. Shtivelman and Goldman (2000) use shallow seismic reflection results to improve the inversion of TEM sounding. Meju (2005) combines resistivity and TEM with seismic refraction profile to delineate the fractured zone. The combination of seismic refraction or reflection with DC resistivity or EM is quite effective in hard rock.

13.7 PROCEDURE FOR INTEGRATED FIELD SURVEYS

The geophysical surveys for groundwater are carried out for two purposes, either to assess the suitability of a site already selected as per the demand and habitation or to select suitable sites in a large area. For the latter a systematic and detailed approach can be made. The geological map, subsurface hydrogeological conditions, lineament maps and also the aeromagnetic, heliborne TEM and MAG maps, if available, are studied and areas of interest are defined. Otherwise, for integrated geophysical studies at regional scale over a large area, it is always better to initiate with heliborne TEM and MAG surveys. As already mentioned, to assess the groundwater suitability, the combination of heliborne magnetic and TEM can yield information on subsurface structures and depth-wise conductivity variations along and across the structures. The DOI maps mentioned in Chapter 11 (Section 11.8.1) can be prepared to approximate the suitable fractured zones in granitic terrain. It is followed by VES or ERT at selected sites to pinpoint the locations. The interpretation of heliTEM can be constrained by ERT or the lithological and geophysical logs of existing boreholes. The TEM interpretation can be constrained by seismic survey also. For this either the passive seismic measurements are carried out to decipher the depth to the bedrock or information from conventional HRS can be incorporated in the inversion of TEM data as suggested by Burschil *et al.* (2012). Hoyer *et al.* (2011) combines heliTEM with HRS to obtain the structural information. It helps constrain as well as conceptualize the inverted TEM results and tailor the inversion scheme for the entire surveyed area. An attempt is made in pilot aquifer mapping studies reported by Ahmed (2014) to integrate HRS with ERT and HeliTEM to distinguish the basalt flows and their hydrogeological character. Integration of geophysical techniques is more pertinent in basalt-flow terrains because of the depth-wise repetition of high and low resistivity layers representing massive basalt, vesicular basalt and interflows which results in underestimations of true layer-resistivities and hence poor identifications of aquifers.

To have a preliminary idea about quantitative variations in resistivity with depth, ground surveys are initiated by simply observing VES at a few locations selected randomly or geologically within the areas of interest. Out of these a few VES are observed near existing boreholes, if any, to standardize the resistivity ranges. If the lithologs or geophysical logs of existing boreholes are available forward modeling can be attempted. The sample DC resistivity measurement and forward modeling exercises determine the applicability or need of other geophysical methods in the area. The VES are generally observed with Schlumberger configuration. Since in granitic terrain, compaction of formation increases with depth, generally an increase in resistivity with depth is observed and the VES curves are of A, H, AA or HA types. The curve shows 45° slope at larger current electrode separations indicating the presence of compact formation at depth. Generally, the last 45° slope segment should have 3 measurement points (AB/2). In the case of a deeper thick layer with saline water K and Q type curves are obtained.

For reconnaissance ground magnetic and Very Low Frequency (VLF) surveys, which are quite fast can be conducted. Qualitatively the magnetic survey helps locate basic dykes and also lithological contacts and thickening of weathered zone. It can be interpreted quantitatively for source depth and bed rock topography. The magnetic profiles are kept long enough to provide a regional trend. Similarly, the VLF profiles are also kept large so that current-density pseudo-sections for a considerable depth can be prepared. The VLF profiles may also yield information on thickening or thinning of the weathered zone and shallow saturated fractures. Basically VLF is useful in areas with a thin resistive weathered zone. It may not be useful in areas with a thick weathered zone of varying resistivity as it will produce too many anomalies and it is difficult to select one for necessary detailing by other techniques. These reconnaissance surveys yield information for planning detailed survey.

In general, a thick weathered zone and saprolite yield moderately. So to estimate the saturated thickness of the weathered zone and saprolite, their characterization can be attempted simply by conventional VES or ERT. Locating a thick saturated weathered zone and saprolite near aeromagnetic or photo-lineaments is almost certain to provide successful results. The VES or ERT can be conducted either along a profile to generate a cross-section or in a grid pattern. Since, lateral variations in hydrogeological conditions are quite high, preparation of cross-sections based on VES in a hard rock area is not suggested unless spot measurements are made at a very close intervals. Better would be to carry out roll-on ERT along selected profiles, if required, for continuous lateral information for a defined depth.

For a better yield, efforts need to be made through detailed ground geophysical surveys to delineate the saturated fractured zones. The detailed surveys at a site should comprise at least 3 profile lines each 500 m long and with 50 m profile-interval. This is essential to determine the strike of the target to be drilled and correlate it with the strike of geological structures and lineaments. The longer the strike length of the target anomaly the better would be the yield of a well drilled on it. Also, it renders flexibility in selecting a drilling site, particularly when placing a drilling rig exactly on the anomaly is difficult and also the point of demand may lie on or near the extension of the target-strike.

The suitable areas or zones identified by VLF or magnetic survey can be detailed by DC resistivity and multi-frequency, multi-spacing electromagnetic profiling also. Resistivity profiling can be conducted by gradient array (GRP) of electrodes on 3 parallel lines using only one pair of current electrodes set-up collinear with one of the lines. The electromagnetic profiling can be done along 3 parallel profiles or at least along one profile across the selected GRP anomaly to possibly discriminate between the weathered and saprolite zone, the effect of a conductive weathered zone, underlying fractured zone and its attitude. The interpretation of GRP is mainly qualitative. So, the points selected are covered by VES or ERT to confirm the subsurface resistivity distribution and choose the best local anomaly. The depth to fractured zones occurring within 100 m may be empirically estimated from VES using curve break and factor methods. For this, the VES are conducted up to about 200–250 m current electrode spacing (AB/2) with sequential smaller increments in AB/2 and least change in potential electrode spacing (MN). The kinks in the curve with increased current are picked up to get the fracture depth. It can also be done through computation and plotting of apparent conductance (C_A) for each current electrode separation as suggested by Sharma and Biswas (2013). The apparent conductance is defined as $C_A = I_D + I_R/2V_{app}$, where I_D and I_R are current flows in the direct and reverse directions generally obtained using simple resistivity meters and V_{app} is the applied voltage. Higher C_A at a particular AB/2 indicates the presence of a conductive fractured zone at equivalent depth. It can also be attempted through resistance values generally measured by constant-current or constant-voltage resistivity meters.

The first stage in DC resistivity interpretation is to interpret the smoothed VES curve obtained near existing boreholes. The layer model inferred solely from VES, yielding geoelectrical characteristics is identified with the lithological boundaries encountered in the borehole or layer boundaries obtained from its geophysical logs. If the match is not satisfactory, VES is reinterpreted by fixing the depth or resistivity of geoelectrical layers in conformity with litho-interfaces or log derived resistivities and interfaces. It yields an improved geoelectrical model acceptable in terms of lithology and hydrogeological condition and forms the base for inversion of other VES data from the area. Through borehole information, the validation and standardization exercises in geophysical and hydrogeological domains respectively are essential steps for any inversion of surface geophysical measurement. In case more than one geophysical method is used several approaches in data inversion can be adopted. The second stage is to empirically interpret the unsmoothed VES curves for curve-breaks and fractures.

Meekes and van Will (1991) use seismic reflection, VES and TEM and compare the effectiveness of a combined method with the individual methods through a traditional approach of interpretation. They generate theoretical TEM response through layer model obtained from VES data and the model is modified iteratively. It is a sequential approach of inversion in which the layer parameters obtained from one method is used as input to generate the theoretical response of another method. Then the layer model is iteratively improved to match with observed data from both the methods. This is straightforward and can be attempted through software supplied with the equipment. However, the joint inversion of VES and TEM (Raiche et al., 1985) is better than sequential inversion. Meekes and van Will further point out that while comparing VES and TEM responses over a multi-thin layer sequence with contrasting resistivities to get the best fit, respectively the mean and longitudinal resistivities should be considered as the nature of current flow differs. A similar phenomenon is observed while comparing VES deduced resistivities with the borehole resistivity data having numerous thin layers which are combined to match with a VES generated layer model (Chandra, 1984). In DC resistivity as well as electromagnetic the resolution decreases with depth and combining VES with TEM may reduce the range of equivalence depending on depth-wise resistivity distribution but cannot eliminate it. The deeper the equivalent layer, the wider is the range of equivalence and hence the seismic with better resolution at depth is brought in. Fitterman et al. (1988) and Sharma and Kaikkonen (1999) analyse the scope of minimizing the range of equivalence by combining VES and TEM sounding interpretations and conclude that even with integration the most difficult to resolve equivalence is the case where resistivity increases with depth. While integrating the interpretation by VES and TEM it is better to use first layer parameters obtained from VES.

The analysis of anisotropy or fracture induced anisotropy in resistivity and identification of fractured zone orientation can be carried out by complementing radial Schlumberger and square array resistivity soundings. Once a borehole is drilled and a shallow yielding fractured zone is tapped within 50 m depth, its orientation can be assessed by MAM and the direction of flow through SP measurements done before, during and immediately after pumping the well and therefore these measurements may be carried out in combination. Since the variations in resistivity in general and near-surface variations in particular affect MAM and SP measurements, it is necessary that these measurements are carried out in combination with ERT to make quantitative interpretations. Even without the borehole drilled and pumped, the SP survey being quite economic in terms of time and cost, it may be added as parallel profiles to ERT to supplement a dimension of possible groundwater flow dynamics related to the structures. In a case where the MAM response is feeble, time-lapsed MAM measurements can be made before and after injecting high-concentration salt water slug into the fractured zone to ascertain the fracture orientation. White (1988, 1994) suggests resistivity-rectangle measurements between two distant current electrodes placed approximately in the direction of flow, which in the present case is the direction of fracture. The current electrode separation can be fixed based on test VES carried out near the well and the depth of the fracture. SP and even MAM potentials may vary with time and therefore time-lapsed repeat measurements can be used to evaluate the temporal changes caused by recharging water, groundwater flow variations and movement of contaminant plumes. The time-lapse ERT measurements can also be made with an injecting salt water slug to assess the fracture orientation and groundwater movement.

The exact depth of the fractured zone and its contribution to total discharge from a borehole are inferred through borehole logging using a minimum combination of resistivity, temperature-fluid conductivity, caliper and flow measurement. For the purpose of fracture characterization acoustic and gamma-gamma logging can be included as P wave velocity and electron density are affected by the presence of fracture. Neutron and NMR logging can be used to assess the hydraulic parameters of weathered and fractured zones. Borehole radar is used to define the extent of fractures encountered in a borehole and BHTV to assess borehole wall conditions and the fractures.

13.8   CASE STUDIES

Geophysical investigation in granites and metasediments in the eastern part of India (Chandra *et al.*, 1994) was initiated through the analysis of lineament maps and existing aeromagnetic data wherever available and systematic integrated field surveys through magnetic, followed by, VLF, multi-frequency horizontal loop coplanar (HLEM), Schlumberger, Wenner and gradient resistivity profiling (GRP), Schlumberger sounding (VES), self potential (SP) and mise-a-la-masse (MAM) profiling and borehole logging. The electrical and electromagnetic results from this area are discussed in the relevant chapters.

The results of integrated surveys in two areas, in granites of West Bengal and metasediments of Jharkhand, India are shown in figures 13.9 and 13.10. The lineament and aeromagnetic maps of these areas are shown in figures 5.9 and 5.10. The ground geophysical survey profiles were placed across the lineaments and the aeromagnetic linears. The reconnaissance ground magnetic profiling at the site (1) in granites (Fig. 13.9 a) revealed an anomaly expected to be associated with a northwest dipping dyke-like body. It was followed by HLEM profiling with 100 m T-R spacing at four frequencies (112.5, 337.5, 1012.5 and 3037.5 Hz). The anomalies for 1012.5 and 3037.5 Hz frequencies are prominent. The presence of relatively conductive overburden, as revealed by VES, enhanced the anomaly. The shape of the anomaly is indicative of the presence of northwest dipping parallel conductors underlying the overburden. It was further confirmed by GRP with AB: 500 m and MN: 20 m. The GRP manifested the presence of three prominent resistivity lows of the order of 400 to 500 ohm.m in the background resistivity of about 1000 ohm.m. The GRP anomalies confirmed the HLEM anomalies. The VES observed at GRP low revealed the presence of 24 m thick weathered zone and saprolite of 148 to 212 ohm.m resistivity. All these led to the selection of a drilling site (1) at the 200 m station. The borehole drilled at site (1) up to a depth of 198.76 m encountered weathered zone and saprolite up to 25 m and saturated fractures F1 at 48.36–49.36 m (thickness: 1 m) and F2 at 110.32–110.53 m (thickness: 0.21 m). The discharge from the weathered zone (9.5–13.6 m) was 2.5 lps and that from F1 and F2 was 2.0 and 3.0 lps. Two dry fractures were encountered at 63 and 99 m depth. The electrical resistivity log of the borehole reveals the saturated fractured zones with low resistivity, dry fractures with relatively high resistivity and compact formations with very high resistivity of the order of 10,000 ohm.m.

Near this site, magnetic, HLEM and GRP surveys were also conducted along a NE-SW profile to select another borehole site (Fig. 13.9 c). The parallel magnetic profiles indicate the presence of a SW dipping magnetic body (dyke). The GRP low towards SW of the dyke is associated with HLEM anomaly. The inphase and quadrature responses are enhanced at 1012.5 Hz and 3037.5 Hz frequencies indicating thickening of the conductive overburden and the possibility of a saturated fractured zone immediately underlying it. The borehole (2) drilled up to 200.82 m on GRP/HLEM anomaly encountered about 17 m thick weathered zone and saprolite with a discharge of 1.41 lps and a fracture at 44.8 m depth with a discharge of 1.91 lps. Another test borehole (3) was drilled on the GRP high up to a depth of 300.76 m. It was totally dry with only a meagre discharge of 0.21 lps from the top 25.47 m thick weathered and saprolite zone.

A similar approach was made to locate drilling sites (EW, OW1 and OW2, Fig. 13.10) in metasediments of Jharkhand, India. The metasediments comprise schists,

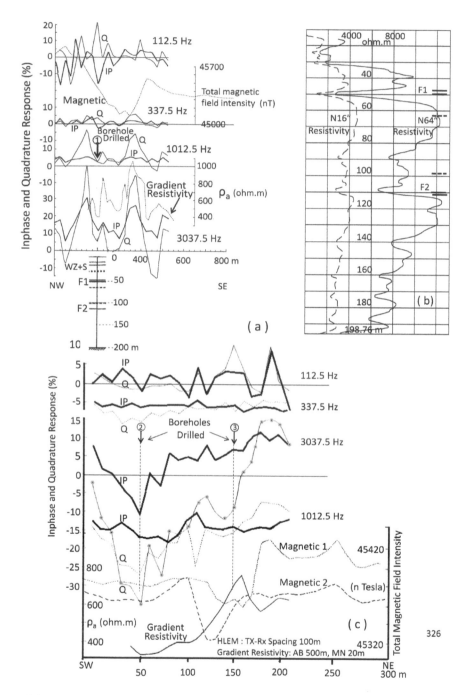

Figure 13.9 Integrated magnetic, gradient resistivity and multi-frequency horizontal loop EM profiling in granites, West Bengal, India for locating borehole drilling sites; (a) for borehole 1 and (b) N16″ and N64″ resistivity log of the borehole 1 drilled; dry fractures at 63 and 99 m depth are shown by dotted lines and saturated fractures with continuous lines; log records more than 12,000 ohm.m resistivity for the compact formation; discharge from weathered zone and saprolite (WZ + S) was 2.5 lps and (c) profiling for borehole 2 and 3; borehole 2 encountered fracture at 44.8 m and borehole 3 on GRP high was dry. Location on the aeromagnetic map is shown in figure 5.9.

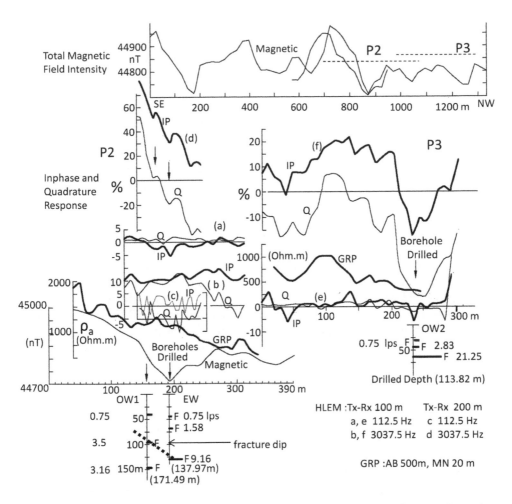

Figure 13.10 Integrated magnetic, gradient resistivity and multi-frequency horizontal loop EM profiling in metasediments, Jharkhand, India for locating borehole drilling site. Location on the aeromagnetic map is shown in figure 5.10.

quartzites and phyllites. These are highly folded with steeply dipping flanks. A long ground magnetic profile was placed across the NNE-SSW trending aeromagnetic linear and the lineament shown in figure 5.10. The model for the magnetic anomaly at 875 m station of Profiles P2 indicates a shallow wide tabular body dipping towards northwest. On the basis of the magnetic anomaly HLEM profiles and GRP were placed on P2. At this site only two frequencies, viz., 112.5 Hz and 3037.5 Hz were used. In addition to 100 m T-R spacing 200 m spacing was also used. On P2 the inphase and quadrature responses for 100 m and 200 m coil spacing at low frequency are considerable. The high frequency response for 200 m spacing showed a step like anomaly with kinks and can be attributed to thickening of overburden and parallel dipping conductors. The GRP also showed a step like decrease in resistivity from 1500 ohm.m to 600 ohm.m towards northwest. The VES at this site revealed 7 m thick weathered zone of 128 ohm.m resistivity underlain by 22 m thick saprolite of 198 ohm.m resistivity. On P2 two sites (EW and OW1) were selected on the basis of HLEM anomalies and drilling rig accessibility. The boreholes encountered high yielding fractures up to 147 m depth and the deeper fractures showed higher yield (Fig. 13.10). The correlation of the natural gamma radioactivity logs of these two boreholes (Fig. 12.5 a and b) indicates dip of beds towards northwest. Since the GRP showed a decreasing trend towards northwest, another GRP HLEM combination parallel profile (P3) was placed over 300m stretch. The GRP showed lesser resistivities in the range of 750 to 300 ohm.m. The high frequency HLEM showed a prominent anomaly towards the north eastern end of the profile indicative of conductive overburden. The VES at this site revealed the presence of 6–8 m thick top layer of 22 ohm.m resistivity followed by 35–40 m thick layer of 60–70 ohm.m resistivity. Compared to VES on P2 the VES on P3 revealed relatively conductive layers and helped understand the cause of high frequency HLEM anomaly. The low frequency anomaly was not significant. Borehole OW2 was located towards northwest on the basis of GRP and HLEM anomalies. The borehole encountered a very high yielding fracture in the depth range 57.86 to 68.10 m and further drilling could not be done due to back pressure of water. A comparison of the results of P2 with P3 reveals enhancement of HLEM response by conductive overburden and presence of conductive fractures at depths around 50–70 m. The reduction in GRP resistivity in P3 compared to P2 is revealed by the reduction in layer resistivities obtained from VES. Overall the results substantiate the effectiveness of integrated geophysical surveys in delineating the shallow fracture (F1) within 100 m depth. The HLEM anomalies are to be confirmed by GRP and GRP highs definitely are negatives for groundwater.

A magnetic low anomaly across a lineament associated with wide and contrasting GRP low and a high frequency FEM anomaly (as low frequency anomalies are generally subdued) can be considered for locating a water well drilling site. The minimum requirement is of lineament mapping, gradient resistivity profiling across the lineament supported by ERT or VES. Paterson (1989) made an attempt to analyse the successful wells in this area vis-a-vis geophysical anomalies which reveals the minimum combination of methods required for groundwater exploration in hard rock (Table 13.2). However, the effectiveness of geophysical methods cannot be judged either on a statistical basis or on the basis of well discharge alone because the wells are generally located on the 'best' amongst the local anomalies which may not always be a good anomaly.

13.9   RESEARCH STUDIES AND FIELD EXPERIMENTS

Application of geophysics to a large extent has yielded information on aquifer availability and it continues to grow with modern techniques bringing in precision. It is to primarily aim at precise characterization of saturated fractures at all scales and functional depths. Also, it may have to focus on infiltration through the vadose zone, groundwater movement through fractures and the quality aspects. The research and field activities in hard rock hydrogeophysics can be grouped into two major domains. The first is in the geophysical domain with sub-domains of heliborne, surface and

Table 13.2 Comparison of success of wells drilled on various geophysical anomalies (Paterson, 1989) in hard rock of Jharkhand State, India. The discharge of boreholes drilled ranged from 1 to about 25 litres per second.

Anomaly	Total no. of wells drilled	No. of wells with yield more than 2 lps (%)	No. of wells with yield less than 2 lps (%)	No. of wells with yield more than 2 lps in the absence of an anomaly
Photo-lineament	15	10 (67)	5 (33)	5
Aeromagnetic lineament	7	6 (86)	1 (14)	1
Very Low Frequency (VLF)	12	11 (92)	1 (8)	0
Gradient Resistivity Profiling (GRP)	20	17 (85)	3 (15)	0
Frequency domain electromagnetic (HLEM)	14	11 (79)	3 (11)	0
GRP + HLEM	13	11 (85)	2 (15)	0
GRP + VLF	9	8 (89)	1 (11)	0
VLF + HLEM	9	8 (89)	1 (11)	0
GRP + HLEM + VLF	8	7 (88)	1 (12)	0
No geophysical anomaly	5	0 (0)	5 (100)	–

borehole geophysics and their integrations and the second is in the hydrogeophysical domain focussed on hydrogeological transformation of geophysical derivatives and their applicabilities in groundwater flow and quality modeling. The primary and vital objective is how to remove uncertainties in geophysical interpretations and hydrogeophysical models so that applications are definitive and convincing. Hubbard *et al.* (1999), Linde *et al.* (2004), Auken *et al.* (2006), Kowalsky *et al.* (2006), Robinson *et al.* (2008), Ferre *et al.* (2009), Moller *et al.* (2009), Binley *et al.* (2010), Auken and Christiansen (2012), and Linde (2014) discuss some of the relevant aspects of R&D studies, joint inversion of hydrogeological and geophysical data and hydrogeophysical data base. Krasny and Sharp (2007) present a detailed list of significant challenges in fractured rock hydrogeology which remain largely unresolved. Readers may refer to these publications. A few key aspects are discussed below.

Groundwater geophysics in hard rock is becoming more challenging while investigating hydrogeologically significant deeper thin fractured zone targets, more so when the fractured zone characterization demands its 3-dimensional disposition. Geophysical methods, in general, have reduced resolution with depth. A strong need is of improving resolution and minimizing non-uniqueness. The electrical resistivity method which is most appropriate in characterizing the aquifer in terms of quantity and quality has well known inherent limitations of transition, equivalence, suppression, top soil effect, inhomogeneity and anisotropy. The use of passive seismic as a regular method to have a control on depth is to be explored further (Lane *et al.* 2008). The efficacy of combination methods in detecting and characterizing thin relatively conductive fractured zones is to be established with standardization of area-specific optimum combination methods and approach in data acquisition and inversion. Though there is a need of combining other geophysical method having better resolving power, but unlike oil and mineral, groundwater exploration has to be cost effective which constrains the multi-method approach. It is therefore imperative to develop groundwater geophysics in a multi-disciplinary environment.

An emerging modern technology is high resolution heliborne EM and MAG surveys. It is cost effective. Its applicability in fracture detection is to be ascertained. For fracture detection multi-component heliTEM measurements are required. It is necessary to measure the X component at early time gates as accurately as possible which is mostly with smaller signal strength and the noise. Research activities on instrumentation may be focused on improving the X component signal strength and scope of jointly inverting X component data with MAG data and ERT. Also to be explored is the combination of horizontal and vertical coils heliborne FEM and MAG. Besides, research activities may also be focused on joint inversion of multi component different loop configurations ground TEM measurements with gradient resistivity profiling and/or multi-parametric borehole data to identify and characterize the conductive, yielding fractured zones. The basic difference is that in groundwater exploration the conductivity contrasts are not prominent as in mineral exploration and such activities are to be carried out with this in mind.

Another area where research activities on hard rock heliborne and surface geophysics may be focused is the delineation of thin aquifers at depth in flood basalt terrain where electrical and electromagnetic methods suffer limitations. While attempting integration of high resolution surface methods, it is vital to include high data density- high resolution heliborne geophysics and establish the efficacy of surface as well as heliborne surveys.

Though the integrated use of modern high resolution methods and techniques is, in general essential, the major constrain being cost of equipment, they remain mostly underutilized. For using a combination method approach it may be an obligatory incentive to develop 'economic-versions' of modern instruments with tailored specifications such that they can be expansively used on a large-scale for groundwater investigations up to minimum 200m depth.

The wealth of hydrogeological information that can be deduced from geophysical data is almost boundless. It has the potential to offer hydrogeophysical information required for exploration, development, management, monitoring and rehabilitation. Hence, the second domain starts with precise hydrogeological transformation of geophysical observations and information and generation of dependable hydrogeological models including dynamic process and heterogeneities at different scales. It may require further studies on relations between hydrogeological and geophysical properties, their scales of variation in space and time, dimensionality, scale of measurement vis-a-vis resolution and heterogeneities and the limitations and uncertainties. It warrants studies on petrophysics, hydrogeological complexities, related transfer functions, joint inversion of hydrogeological and geophysical data and standardizations. Based on these only the geophysical information and models can be used conceivably and effectively in varied hydrogeological studies, predictive modeling and resource management.

Another essential aspect is the generation of a comprehensive national hydrogeophysical data and information bank which will manage all the surface, airborne and borehole data related to groundwater at different levels created by various agencies getting enriched time to time with the inflow of data. It will help generate variable grid-size aquifer-grid as per requirement. With the acquisition of high density geophysical data through heliborne surveys, aquifer-gridding can also be attempted for hard rock terrain. Besides, it is necessary to compile case studies on past success and failures and prepare guidelines for best practices in different hard rock terrains.

REFERENCES

Ahmed, S. (2014) A new chapter in groundwater geophysics in India: 3D aquifer mapping through heliborne transient electromagnetic investigations. *Journal of Geological Society of India*, 84 (4), 501–503.

Alexander, W.G. & Devocelle, L.L. (1997) Mapping indoor radon potential using geology and soil permeability. *The 1997 International Radon Symposium, 2–5 November 1997, Cincinnati, Ohio, U.S.A.* pp. I.4.1 – I.4.16.

Annan, A.P. (2005a) GPR methods for hydrogeological studies. In: Rubin, Y. & Hubbard, S.S. (eds.) *Hydrogeophysics, Chapter 7, Water Science and Technology Library, Vol. 50*. The Netherlands, Springer. pp. 185–213.

Annan, A.P. (2005b) Ground-penetrating radar. In: Butler, D.K. (ed.) *Near-Surface Geophysics Part 1, Chapter 11, SEG Investigations in Geophysics Series: No. 13*, Tulsa, Oklahoma, USA, Society of Exploration Geophysicists, *SEG Publication, Tulsa, Oklahoma, USA*. pp. 357–438.

Artman, B. (2006) Imaging passive seismic data. *Geophysics*, 71 (4), SI 177–SI 187.

Auken, E. & Christiansen, A. V. (2012) Integrating, geophysics, geology and hydrology for enhanced hydrogeological modelling. Presented at: *The American Geophysical Union Meeting, 3–7 December 2012, San Francisco, California, USA* (Abstract).

Auken, E., Pellerin, L., Christensen, N.B., & Sorensen, K. (2006) A survey of current trends in near-surface electrical and electromagnetic methods. *Geophysics*, 71 (5), G249–G260.

Bachrach, R. & Nur, A. (1998) High-resolution shallow-seismic experiments in sand, Part I: water table, fluid flow and saturation. *Geophysics*, 63 (4), 1225–1233.

Baltassat, J.M., Krishnamurthy, N.S., Girard, J.F., Datta, S., Dewandel, B., Chandra, S., Descloitres, M., Legchenko, A., Robain, H., Ananda Rao, V., & Ahmed, S. (2006) *Proton magnetic resonance technique in weathered-fractured aquifers*. NGRI-BRGM Report, IFCPAR Project 2700-W1, BRGM 2003-ARN-11.

Banwell, G.M. & Parizek, R.R. (1988) Helium 4 and Radon 222 concentrations in groundwater and soil gas as indicators of zones of fracture concentration in unexposed rock. *Journal of Geophysical Research*, 93 (B1), 355–366.

Beeson, S. & Jones, C.R.C. (1988) The combined EMT/VES geophysical method for siting boreholes. *Ground Water*, 26 (1), 54–63.

Befus, K.M., Sheehan, A.F., Leopold, M. Anderson, S.P. & Anderson, R.S. (2011) Seismic constraints on critical zone architecture, Boulder Creek Watershed, Front Range, Colorado. *Vadose Zone Journal*, 10 (3) 915–927.

Behroozmand, A.A., Auken, E., Fiandaca, G. & Christiansen A.V. (2012) Improvement in MRS parameter estimation by joint and laterally constrained inversion of MRS and TEM data. *Geophysics*, 77 (4), WB191–WB200.

Bernard, J. & Valla, P. (1991) Groundwater exploration in fissured media with electrical and VLF methods. *Geoexploration*, 27 (1–2), 81–91.

Binley, A., Cassiani, G. & Deiana, R. (2010) Hydrogeophysics: opportunities and challenges. *Bollettino di Geofisica Teorica ed Applicata*, 51 (4), pp. 267–284.

Bradley, A.B., Steeples, D.W. Miller, R.D. & Sophocleous, M. (1987) Seismic reflection study of a shallow aquifer during a pumping test. *Ground Water*, 25 (6), 703–709.

Bristow, C.S. & Jol, H.M. (2003) An introduction to ground penetrating radar (GPR) in sediments. In: Bristow, C.S. & Jol, H.M. (eds.) *Ground Penetrating Radar in Sediments*, The Geological Society, London. Special Publication No. 211, pp. 1–7.

Brutsaert, W.F., Norton, S.A., Hess, C.T. & Williams, J.S. (1981) Geologic and hydrologic factors controlling radon-222 in ground water of Maine. *Ground Water,* 19 (4), 407–417.

Burkhart, J., Huber, T. & Bolling, G. (2013) Potential radon release during fracking in Colorado. In: The 2013 *International Radon Symposium, 22–25 September 2013, Springfield, Illinois, USA.* pp 20–27.

Burnett, W.C. & Dulaiova, H. (2003) Estimating the dynamics of groundwater input into the coastal zone via continuous radon-222 measurements. *Journal of Environmental Radioactivity,* 69 (1 & 2), pp. 21–35.

Burschil, T., Wiederhold, H. & Auken, E. (2012) Seismic results as a-priori knowledge for airborne TEM data inversion – a case study. *Journal of Applied Geophysics,* 80, 121–128.

Byegard, J., Rameback, H. & Widestrand, H. (2002) *Use of radon concentrations for estimation of fracture apertures – Part I: Some method developments, preliminary measurements and laboratory experiments.* Aspo Hard Rock Laboratory, International Progress Report IPR-02-68.

Cakir, R. & Walsh, T. J. (2011) *Shallow seismic site characterizations at 23 strong-motion station sites in and near Washington State.* Technical report submitted to U.S. Geological Survey, Contract Award No. G10AP00027.

Central Ground Water Board (2012) *Groundwater studies in Kasai-Subarnarekha river basins.* CGWB, Min. of Water Resources, Govt. of India, Published Report.

Chandra, P.C. (1984) *Correlation of resistivity sounding and electric logs.* [Lecture] delivered at the training course on 'Geophysical Logging of Water Wells', organized by Central Ground Water Board, Min. of Water Resources, Govt. of India, Hyderabad, August 1984.

Chandra, P.C. & Ramakrishna, A. (1979) *Resistivity surveys in parts of Pudukkottai district, Tamil Nadu with special reference to exploration of deeper aquifers.* Central Ground Water Board, Min. of Water Resources, Govt. of India, Tech. Report.

Chandra, P.C., Ramakrishna, A. & Singh H. (1983) Results of surveys for groundwater in parts of Cannanore district, Kerala. *Jour. Association of Exploration Geophysicists,* 3 (4), 11–15.

Chandra, P.C., Ramakrishna, A., Singh, S.C. & Reddy, P.H.P. (2002a) Geophysical approach to delineate saturated fracture zones in hard rocks. In: Thangarajan, M., Rai, S.N. and Singh, V.S. (eds.) *IGC 2002: Int. Conf. on Sust. Dev. & Management GW Resources in Semi-Arid Region with Sp. Ref. hard Rock. February 20–22,2002, Dindigul, Tamil Nadu, India,* A.A. Balkema. pp. 75–84.

Chandra, P.C., Reddy, P.H.P. & Singh, S.C. (1994) *Geophysical studies for groundwater exploration in Kasai and Subarnarekha River basins.* (UNDP Project) Central Ground Water Board, Min. of Water Resources, Govt. of India, Tech. Report.

Chandra, P.C., Srivastava, M.M., Adil, M., Bhowmic, M.K., Pandey, K.S., Haq, S. & Singh, U.B. (2002b) Geoelectrical investigations for groundwater in quartzitic sandstones and granites of Sonbhadra district, U.P. In: Proceedings of the Int. Conf. on Hydrology and Watershed Management, 18–20 December 2002, JNTU Hyderabad.

Choubey, V.M., Bartarya, S.K. Saini, N.K. & Ramola, R.C. (2001) Impact of geohydrology and neotectonic activity on radon concentration in groundwater of intermontane Doon Valley, Outer Himalaya, India. *Environmental Geology,* 40 (3), 257–266.

Ciotoli, G., Etiope, G., Guerra, M., & Lombardi, S. (1999) The detection of concealed faults in the Ofanto Basin using the correlation between soil-gas fracture surveys. *Tectonophysics,* 301 (3–4), 321–332.

Clarke, B.A. & Burbank, D.W. (2011) Quantifying bedrock-fracture patterns within the shallow subsurface: Implications for rock mass strength, bedrock landslides and erodibility. *Journal of Geophysical Research,* 116, F04009, doi: 10.1029/2011JF001987.

Cook, P.G., Love, A.J., & Dighton, J.C. (1999) Inferring ground water flow in fractured rock from dissolved radon. *Ground Water,* 37 (4), 606–610.

Davis J.L. & Annan, P. (1989) Ground-penetrating radar for high-resolution mapping of soil and rock stratigraphy. *Geophysical Prospecting,* 37 (5), 531–551.

Dasios, A., McCann, C., Astin, T.R., McCann, D.M. & Fenning, P. (1999) Seismic imaging of the shallow subsurface: shear-wave case histories. *Geophysical Prospecting,* 47 (4), pp. 565–591.

Dolgiy, A., Dolgiy A.A., Markulis, V. & Zolotarev, V. (2006) Long-term application of GPR technique for estimation of ground contamination degree. In: *Near Surface Geoscience-2006, Proceedings of the 12th European Meeting of Environmental & Engineering Geophysics, 4–6 September 2006, Helsinki, Finland.* The Netherlands, EAGE.

Eaton, D.W., Milkereit, B. & Salisbury, M.H. (eds.) (2003) *Hardrock seismic exploration. Geophysical Developments Series 10.* Tulsa, Oklahoma, U.S.A. Society of Exploration Geophysicists.

Ellefsen, K.J., Tuttle, G.J., Williams, J.M. & Lucius, J.E. (2005) *S-wave refraction survey of alluvial aggregate.* U.S. Geological Survey Scientific Investigations Report 2005–5012. U.S.G.S.

Ernstson, K. & Kirsch, R. (2009) Aquifer structures: fracture zones and caves. In: Kirsch, R. (ed.) *Groundwater Geophysics: A Tool for Hydrogeology, Chapter 15.* 2nd edition, Berlin Heidelberg, Springer-Verlag Publication, pp. 447–474.

Ferre, Ty., Bentley, L., Binley, A., Linde, N., Kemna A., Singha, K., Holliger, K., Husman, J.A. & Minsley, B. (2009) Critical steps for the continuing advancement of hydrogeophysics. *Eos Transactions of American Geophysical Union,* 90 (23), 200–201.

Fitterman, D.V., Meekes, J.A.C. & Ritsema, I.L. (1988) Equivalence behavior of three electrical sounding methods as applied to hydrogeological problems. Presented at: *The 50th Annual Meeting and Technical Exhibition of the European Association of Exploration Geophysicists, 6–8 June 1988, The Hague, The Netherlands.*

Frishman, D., Day, W.C., Folger, P.F., Wanty, R.B., Briggs, P.H. & Poeter, E. (1993) Bedrock geologic controls on radon abundance in domestic well water, Conifer, Colorado. Presented at: *The 1993 International Radon Conference, September 1993, Denver, Colorado, USA.* pp. IV 20 – IV 30.

Fu, C.-C., Yang, T.F., Du, J., Walia, V., Chen, Y.-G., Liu, T.-K. & Chen, C.-H. (2008) Variations of helium and radon concentrations in soil gases from an active fault zone in southern Taiwan. *Radiation Measurements,* 43 (Supplement 1), S 348–S 352.

Gendzwill, D.J., Serzu, M.H. & Lodha, G.S. (1994) High resolution seismic reflection surveys to detect fracture zones at the AECL underground research laboratory. *Canadian Journal Exploration Geophysics,* 30 (1), 28–38.

Goldman, M., Rabinovich, B., Rabinovich, M., Gilad, D., Gev, I. & Schirov, M. (1994) Application of the integrated NMR-TDEM method in groundwater exploration in Israel. *Journal of Applied Geophysics,* 31 (1–4), 27–52.

Gruber, W. & Rieger, R. (2003) High resolution seismic reflection-constraints and pitfalls in groundwater exploration. *RMZ-Materials and Geoenvironment,* 50 (1), 133–136.

Guy, E.D. (2006) High-resolution P- and S-wave seismic reflection investigation of a shallow stratigraphic sequence. *The Electronic Journal of Geotechnical Engineering (EJGE),* 11 (B).

Haefner, R.J., Sheets, R.A. & Andrews, R.E. (2010) Evaluation of the horizontal-to-vertical spectral ratio (HVSR) seismic method to determine sediment thickness in the vicinity of the South Well Field, Franklin County, Ohio. *Ohio Jour. of Science,* 110 (4), 77–85.

Haeni, F.P. (1988) *Application of seismic-refraction techniques to hydrologic studies.* U.S. Geological Survey Techniques of Water-Resources Investigations Chapter D2, Book 2. U.S.G.S.

Hammond, D.E., Leslie, B.W. & Ku, T.-L. (1988) ^{222}Rn concentrations in deep formation waters and the geohydrology of the Cajon Pass borehole. *Geophysical Research Letters*, 15 (9), 1045–1048.

Hansen, D.S. (1984) Gravity delineation of a buried valley in quartzite. *Ground Water*, 22 (6), 773–779.

Hansen, B.P. & Lane, J.W. (1995) *Use of surface and borehole geophysical surveys to determine fracture orientation and other site characteristics in crystalline bedrock terrain, Millville and Uxbridge, Massachusetts.* U.S. Geological Survey Water-Resources Investigations Report 95-4121, U.S.G.S.

Hardage, B. (2011) Fracture identification and evaluation using S waves. *Search and Discovery Article #40792 (2011), Geophysical Corner Column, AAPG April-August Explorers, 2011.*

Holbrook, W., Riebe, C. S., Elwaseif, M., Hayes, J. L., Basler-Reeder, K., Harry, D. L., Malazian, A., Dosseto, A., Hartsough, P. C. & Hapmans, J. W. (2014) Geophysical constraints on deep weathering and water storage potential in the Southern Sierra Critical Zone Observatory. *Earth Surface Processes and Landforms*, 39 (3), 366–380.

Hoyer, A.-S., Lykke-Andersen, H., Jorgensen, F. & Auken, E. (2011) Combined interpretation of SkyTEM and high-resolution seismic data. *Physics and Chemistry of the Earth*, Parts A/B/C 36 (16), 1386–1397.

Hubbard, S.S., Rubin, Y. & Majer, E. (1999) Spatial correlation structure estimation using geophysical and hydrogeological data. *Water Resources Research*, 35 (6), 1809–1825.

Hunter, J.A., Pullan, S.E., Burns, R.A., Gagne, R.M. & Good, R.L. (1984) Shallow seismic reflection mapping of the overburden-bedrock interface with the engineering seismograph-some simple techniques. *Geophysics*, 49 (8), 1381–1385.

Ibrahim, A. & Hinze, W.J. (1972) Mapping buried topography with gravity. *Ground Water*, 10 (3), 18–23.

Ibs-von Seht, M. & Wohlenberg, J. (1999) Microtremors measurements used to map thickness of soft soil sediments. *Bulletin of the Seismological Society of America*, 89 (1), 250–259.

Inazaki, T. (2006) High-resolution S-wave reflection survey in urban areas using a woven belt type land streamer. In: *Near Surface Geoscience-2006, Proceedings of the 12th European Meeting of Environmental & Engineering Geophysics, 4–6 September 2006, Helsinki, Finland.* The Netherlands, EAGE.

IRIS (1998) *The proton magnetic resonance method for groundwater investigations.* IRIS Instruments, France.

IRIS (2004) *Magnetic resonance sounding: step-by-step operation of NUMIS system.* IRIS Instruments, France.

Johnson, W.J. & Clark, J.C. (1992) High resolution shear wave reflection surveying for hydrogeological investigations. *DOE/CH-9211, U.S. Dept. of Energy.*

Juhlin, C. & Palm, H. (2003) Experiences from shallow reflection seismic over granitic rocks in Sweden. In: Eaton, D.W., Milkereit, B. & Salisbury, M.H. (eds.) *Hardrock Seismic Exploration, Chapter 6, Geophysical Development Series No. 10.* Tulsa, Oklahoma, U.S.A. Society of Exploration Geophysicists. pp. 93–109.

Kafri, U. (2001) Radon in groundwater as a tracer to assess flow velocities: two test cases from Israel. *Environmental Geology*, 40 (3), 392–398.

Knight, R. (2001) Ground penetrating radar for environmental applications. *Annual Review of Earth and Planetary Sciences*, 29, 229–255.

Kowalsky, M.B., Chen, J. & Hubbard, S.S. (2006) Joint inversion of geophysical and hydrological data for improved subsurface characterization. *The Leading Edge*, 26 (5), 730–734.

Krasny, J. & Sharp Jr., J.M. (2007) Hydrogeology of fractured rocks from particular fractures to regional approaches: State-of-the-art and future challenges. In: Krasny, J. & Sharp Jr., J.M. (eds.), *Groundwater in Fractured Rocks: Selected Papers from the Groundwater in Fractured*

Rocks International Conference, 15–19 September, 2003, Prague, IAH Selected Paper Series, Volume 9. The Netherlands, Taylor & Francis/Balkema. pp. 1–30.

Krishnaswami, S. & Seidemann, D.E. (1988) Comparative study of ^{222}Rn, ^{40}Ar, ^{39}Ar and ^{37}Ar leakage from rocks and minerals: implications for the role of nanopores in gas transport through natural silicates. *Geochimica et Cosmochimica Acta*, 52 (3), 655–658.

Lachassagne, P., Pinault, J.-L. & Laporte, P. (2001) Radon 222 emanometry: A relevant methodology for water well siting in hard rock aquifers. *Water Resources Research*, 37 (12), 3131–3148.

Lane Jr., J.W., White, E.A., Steele, G.V. & Cannia, J.C. (2008) Estimation of bedrock depth using the horizontal-to-vertical (H/V) ambient-noise seismic method. Presented at: *The Symposium on the Application of Geophysics to Engineering and Environmental Problems, 6–10 April 2008, Philadelphia, Pennsylvania, Proceedings: Denver, Colorado, Environmental & Engineering Geophysical Society*. 13 p.

Lankston, R.W. (1990) High-resolution refraction seismic data acquisition and interpretation. In: Ward, S.H. (ed.) *Geotechnical and Environmental Geophysics, Vol. I: Review and Tutorial*. Tulsa, Oklahoma, USA, Society of Exploration Geophysicists, pp. 45–73.

Larsen, J.J., Dalgaard, E. & Auken, E. (2014) Noise cancelling of MRS signals combining model-based removal of powerline harmonics and multichannel Wiener filtering. *Geophysical Journal International*, 196 (2), 828–836.

Legchenko, A.V., Baltassat, J.M., Beauce, A., Makki, M.A. & Al-Gaydi, B.A. (1998) Application of the surface proton magnetic resonance method for the detection of fractured granite aquifers. In: *Proceedings of the IV Meeting of the Environmental and Engineering Geophysical Society, 14–17 September 1998, Barcelona*, pp. 163–166, (Abstract).

Legchenko, A., Baltassat, J.-M., Beauce, A. & Bernard, J. (2002) Nuclear magnetic resonance as a geophysical tool for hydrogeologists. *Journal of Applied Geophysics*, 50 (1–2), 21–46.

Legchenko, A., Baltassat, J.-M., Bobachev, A., Martin, C., Robain, H. & Vouillamoz, J.-M. (2004) Magnetic resonance sounding applied to aquifer characterization. *Ground Water*, 42 (3), 363–373.

Legchenko, A. & Valla, P. (2002) A review of the basic principles for proton magnetic resonance sounding measurements. *Journal of Applied Geophysics*, 50 (1–2), 3–19.

Legchenko, A., Descloitres, M., Bost, A., Ruiz, L., Reddy, M., Girard, J.-F., Sekhar, M., Mohan Kumar, M.S. & Braun, J.-J. (2006) Resolution of MRS applied to the characterization of hard-rock aquifers. *Ground Water*, 44 (4), 547–554.

LeGrand, H.E. (1987) Radon and radium emanations from fractured crystalline rocks-a conceptual hydrogeological model. *Ground Water*, 25 (10), 59–69.

Linde, N. (2014) Falsification and corroboration of conceptual hydrological models using geophysical data. *Wiley Interdisciplinary Reviews: Water*, 1 (2), 151–171 (Abstract).

Linde N., Chen, J., Kowalsky, M.B. & Hubbard, S. (2004) Hydrogeophysical parameter estimation approaches for field scale characterization. In: Vereecken, H., Binley A., Cassiani, G., Revil, A. and Titov, K. (eds.) *Applied Hydrogeophysics, Chapter 2, NATO Science Series IV: Earth and Environmental Sciences-Vol. 71*. Dordrecht, The Netherlands, Springer. pp. 9–44.

Li, Y.-P. (1996) Microearthquake analysis for hydraulic fracturing process. *Acta Seismologica Sinica*, 9 (3), 377–387.

Lubczynski, M. & Roy, J. (2004) Magnetic resonance sounding: new method for ground water assessment. *Ground Water*, 42 (2), 291–303.

Mahajan, S., Walia, V., Bajwa, B.S., Kumar, A., Singh, S., Seth, N., Dhar, S., Gill, G.S. & Yang, T. F. (2010) Soil-gas radon/helium surveys in some neotectonic areas of NW Himalayan foothills, India. *Natural Hazards & Earth System Sciences*, 10 (6), 1221–1227.

Mair, J.A. & Green, A.G. (1981) High-resolution seismic reflection profiles reveal fracture zones within a 'homogeneous' granite batholiths. *Nature*, 294 (5840), 439–442.

Martakis, N., Kapotas, S. & Tselentis, G.-A. (2006) Integrated passive seismic acquisition and methodology, Case Studies. *Geophysical Prospecting,* 54 (6), 829–847.

Martakis, N., Tselentis, A. & Paraskevopoulos, P. (2011) High resolution passive seismic tomography – A new exploration tool for hydrocarbon investigation, recent results from a successful case history in Albania. Presented at: The *GEO-India, 12–14 January 2011, Greater Noida, New Delhi, India.* [extended abstract].

Maxwell, S.C., Urbancic, T.I. & Prince, M. (2002) Passive seismic imaging of hydraulic fractures. Presented at: The *Canadian Society of Exploration Geophysicist, National Convention: 'Taking Exploration to the Edge', 2002.*

Meekes, J.A.C., Scheffers, B.C. & Ridder, J. (1990) Optimization of high-resolution seismic reflection parameters for hydrogeological investigations in the Netherlands. *First Break,* 8 (7), 263–270.

Meekes, J.A.C. & van Will, M.F.P. (1991) Comparison of seismic reflection and combined TEM/VES methods for hydrogeological mapping. *First Break,* 9 (12), 543–551.

Meju, M. A. (2005) *Non-invasive characterization of fractured crystalline rocks using a combined multicomponent transient electromagnetic, resistivity and seismic approach.* In: Harvey P.K., Brewer, T.S., Pezard, P.A. & Petrov V.A. (eds.) Petrophysical Properties of Crystalline Rocks, Special Publications 240, London, The Geological Society, pp. 195–206.

Miller, R.D., Steeples, D.W. & Brannan, M. (1989) Mapping a bedrock surface under dry alluvium with shallow seismic reflections. *Geophysics,* 54 (12), 1528–1534.

Moller, I., Søndergaard, V.H., Jørgensen, F., Auken, E. & Christiansen, A.V. (2009) Integrated management and utilization of hydrogeophysical data on a national scale. *Near Surface Geophysics,* 7 (5 & 6), pp. 647–659.

Murty, B.V.S. & Raghavan, V.K. (2002) The gravity method in groundwater exploration in crystalline rocks: a study in the peninsular granitic region of Hyderabad, India. *Hydrogeology Journal,* 10 (2), 307–321.

Nakamura, Y. (1989) *A method for dynamic characteristics estimations of subsurface using microtremors on the ground surface.* Quarterly Report, Railway Technical Research Institute, Japan, 30, 25–33 (Abstract).

Nascimento da Silva, C.C., de Medeiros, W.E., de Sa, E.F.J. & Neto, P.X. (2004) Resistivity and ground-penetrating radar images of fractures in a crystalline aquifer: a case study in Caicara farm-NE Brazil. *Journal of Applied Geophysics,* 56 (4), 295–307.

Parolai, S., Bormann, P. & Milkert, C. (2002) New relationships between V_s, thickness of sediments, and resonance frequency calculated by the H/V ratio of seismic noise for Cologne Area (Germany). *Bulletin of the Seismological Society of America,* 92 (6), 2521–2527.

Paterson, N.R. (1989) *Final report on consultancy in groundwater geophysics, Central Ground Water Board, Kasai-Subarnarekha Project Jamshedpur, Bihar, India. March 30–April 6, 1989.* UNDP Project IND/84/011 for UN/DTCD, Paterson, Grant & Watson Ltd. Canada

Pelton, J.R. (2005) Near-surface seismology: surface based methods. In: Butler, D.K. (ed.) *Near-Surface Geophysics Part 1, Chapter 8, SEG Investigations in Geophysics Series: No. 13,* Tulsa, Oklahoma, USA, Society of Exploration Geophysicists, *SEG Publication, Tulsa, Oklahoma, USA,* pp. 219–263.

Peterson, R.N., Burnett, W.C., Taniguchi, M., Chen, J., Santos, I.R. & Ishitobi, T. (2008) Radon and radium isotope assessment of submarine groundwater discharge in the Yellow River delta, China. *Journal of Geophysical Research,* 113, C09021, doi:10.1029/2008JC004776, 2008.

Pointet, T. (1989) Exploration of fractured zones by radon determination in the soil. Presented at: *The International Workshop on Appropriate Methodologies for Development and Management of Groundwater Resources in Developing Countries,* Vol. 3, NGRI, Hyderabad, pp. 37–47.

Porsani, J.L., Sauck, W.A. & Junior, A.O.S. (2006) GPR for mapping fractures and as a guide for the extraction of ornamental granite from a quarry: A case study from southern Brazil. *Journal of Applied Geophysics*, 58 (3), 177–187.

Raghunathan, P. & Jagannathan, N.R. (1996) Magnetic resonance imaging: basic concepts and applications. *Current Science*, 70 (8), 698–708.

Raiche, A.P., Jupp, D.L.B., Rutter, H. & Vozoff, K. (1985) The joint use of coincident loop transient electromagnetic, and Schlumberger sounding to resolve layered structures. *Geophysics*, 50 (10), 1618–1627.

Ramola, R.C., Choubey, V.M., Negi, M.S., Prasad, Y. & Prasad, G. (2008) Radon occurrence in soil–gas and groundwater around an active landslide. *Radiation Measurements*, 43 (1), 98–101.

Ranger, L.S. (1993) Geologic control of radon in the Greater Atlanta region, Georgia. Presented at: *The 1993 International Radon Conference, 20–22 September 1993, Denver, Colorado, USA*. pp. IVP 5 – IVP11.

Reddy, D.V., Sukhija, B.S., Nagabhushanam, P., Reddy, G.K., Kumar, D. & Lachassagne, P. (2006) Soil gas radon emanometry: A tool for delineation of fractures for groundwater in granitic terrains. *Journal of Hydrology*, 329 (1–2), 186–195.

Reddy, D.V., Sukhija, B.S. & Rama (1996) Search for correlation between radon and high-yield borewells in granitic terrain. *Journal of Applied Geophysics*, 34 (3), 221–228.

Robinson D.A., Binley, A., Crook, N., Day-Lewis, F.D., Ferré, T.P.A., Grauch, V.J.S., Knight, R., Knoll, M., Lakshmi, V., Miller, R., Nyquist, J., Pellerin, L., Singha, K. & Slater, L. (2008) Advancing process-based watershed hydrological researchusing near-surface geophysics: a vision for, and review of, electrical and magnetic geophysical methods. *Hydrological Processes*, 22 (18), 3604–3635.

Roy, J. (2007) MRS: New GW geophysical technique. In: Milkereit, B. (ed.) *Exploration 07: Proceedings of the Fifth Decennial Int. Conf. on Mineral Exploration, Sept 9 to 12 2007, Toronto, Canada*. pp. 1125–1129.

Roy, J. Expert, Ground Water Geophysics, IGP, Montreal, Canada (Personal Communication, 10th January, 2015).

Roy, J. & Lubczynski, M. (2003) The magnetic resonance sounding technique and its use for groundwater investigations. *Hydrogeology Journal*, 11 (4), 455–465.

Schirov, M., Legchenko, A. & Créer, G. (1991) A new direct non-invasive groundwater detection technology for Australia. *Exploration Geophysics*, 22 (2), 333–338.

Senior, L.A. & Vogel, K.L. (1989) Geochemistry of radium-226 and radium-228 and radon-222 in ground water in the Chickies quartzite, Southeastern Pennsylvania. In: Daniel, C.C. III, White, R.K. & Stone, P.A., (eds.): *Ground Water in the Piedmont: Proceedings of a Conference on Ground Water in the Piedmont of the Eastern United States, 16–18 October 1989, Charlotte, North Carolina*. South Carolina, Clemson University. pp. 547–565.

Seol, J.S., Kim, J.-H., Song, Y. & Chung, S.-H. (2001) Finding the strike direction of fractures using GPR. *Geophysical Prospecting*, 49 (3), 300–308.

Sharma, S.P. & Baranwal, V.C. (2005) Delineation of groundwater-bearing fracture zones in a hardrock area integrating Very Low Frequency electromagnetic and resistivity data. *Journal of Applied Geophysics*, 57 (2), 155–166.

Sharma, S.P. & Biswas, A. (2013) A practical solution in delineating thin conducting structures and suppression problem in direct current resistivity sounding. *Journal of Earth System Science*, 122 (4), 1065–1080.

Sharma, S.P. & Kaikkonen, P. (1999) Appraisal of equivalence and suppression problems in 1D EM and DC measurements using global optimization and joint inversion. *Geophysical Prospecting*, 47 (2), 219–249.

Shtivelman, V. & Goldman, M. (2000) Integration of shallow reflection seismics and time domain electromagnetics for detailed study of the coastal aquifer in the Nitzanim area of Israel. *Journal of Applied Geophysics*, 44 (2–3), 197–215.

Srilatha, M.C., Rangaswamy, D. R. & Sannappa, J. (2014) Studies on concentration of Radon and Physicochemical parameters in ground water around Ramanagara and Tumkur districts, Karnataka, India. *International Journal of Advanced Scientific & Technical Research*, 2 (4), 641–660.

Steeples, D.W. (2005) Shallow seismic methods. Rubin, Y. & Hubbard, S.S. (eds.) *Hydrogeophysics: Chapter 8, Water Science and Technology Library, Vol. 50*, The Netherlands, Springer. pp. 215–251,

Steeples, D.W. & Miller, R.D. (1990) Seismic reflection methods applied to engineering, environmental and groundwater problems. In: Ward, S.H. (ed.) *Geotechnical and Environmental Geophysics, Vol. I: Review and Tutorial*. Tulsa, Oklahoma, USA, Society of Exploration Geophysicists. pp. 1–30.

Stevens, K.M., Lodha, G.S., Holloway, A.L. & Soonawala, N.M. (1995) The application of ground penetrating radar for mapping fractures in plutonic rocks within the Whiteshell Research Area, Pinawa, Manitoba, Canada. *Journal of Applied Geophysics*, 33 (1–3), 125–141

Sun, Z. & Jones, M.J. (1993) *VSP multi-algorithm shear-wave anisotropy study*. Consortium for *Research* in Elastic Wave Exploration Seismology (CREWES) Research Report, 5, 6.1–6.22.

Tanner, A.B. (1992) *Bibliography of radon in the outdoor environment and selected references on gas mobility in the ground*. U.S. Geological Survey Open-file Report 92-351-A. U.S.G.S.

Travassos, J. de M. & Menezes, P. de T. L. (2004), GPR exploration for groundwater in a crystalline rock terrain. *Journal of Applied Geophysics*, 55 (3 & 4), 239–248.

Trushkin, D.V., Shushakov, O.A. & Legchenko, A.V. (1994) The potential of a noise-reducing antenna for surface NMR groundwater surveys in the magnetic field. *Geophysical Prospecting*, 42 (8), 855–862.

Tselentis, G.-A., Serpetsidaki, A., Martakis, N., Sokos, E., Paraskevopoulos, P. & Kapotas, S. (2007) Local high-resolution passive seismic tomography and Kohonen neural networks – Application at the Rio-Antirrio Strait, Central Greece. *Geophysics*, 72 (4), B93–B106.

Tsoflias G.P. & Becker, M.W. (2008) Ground-penetrating-radar response to fracture-fluid salinity: Why lower frequencies are favorable for resolving salinity changes. *Geophysics*, 73 (5), J25–J30.

Turu-Michels, V. (1997) Some principles, methods and devices for surface nuclear magnetic resonance (SNMR) and logging SMR. Available from: *http://www.igeotest.fr.*

US EPA (2012) *Report to Congress: Radon in drinking water regulation*. Office of Water (4607 M) EPA 815-R-12-002.

Van Overmeeren, R.A. (1975) a combination of gravity and seismic refraction measurements, applied to groundwater explorations near Taltal, Province of Antofagasta, Chile. *Geophysical Prospecting*, 23 (2), 248–258.

Van Overmeeren, R.A. (1980) Tracing by gravity of a narrow buried Graben structure, detected by seismic refraction, for ground-water Investigations in North Chile. *Geophysical Prospecting*, 28 (3), 392–407.

Van Overmeeren, R.A. (1981) A combination of electrical resistivity, seismic refraction and gravity measurements for groundwater exploration in Sudan. *Geophysics*, 46 (9), 1304–1313.

Van Schoor, M. & Colvin, C. (2009) Tree root mapping with ground penetrating radar. In: *Proceedings of 11th SAGA Biennial Technical Meeting & Exhibition, Swaziland 16–18 September 2009*, pp. 771–374.

Veeger, A.I. & Ruderman, N.C. (1998) Hydrogeologic controls on Radon-222 in a buried valley-fractured bedrock aquifer system. *Ground Water,* 36 (4), 596–604.

Vilhelmsen, T.N., Behroozmand, A.A. Christensen, S. & Nielsen, T.H. (2014) Joint inversion of aquifer test, MRS, and TEM data. *Water Resources Research,* 50 (5) 3956–3975.

Vouillamoz, J.M., Descloitres, M., Toe, G. & Legchenko, A. (2005) Characterization of crystalline basement aquifers with MRS: comparison with boreholes and pumping tests data in Burkina Faso. *Near Surface Geophysics,* 3 (3), 205–213.

Vouillamoz, J.M., Legchenko, A. & Nandgiri, L. (2011) Characterizing aquifers when using magnetic resonance sounding in a heterogeneous geomagnetic field. *Near Surface Geophysics,* 9 (2), 135–144.

Wallace, D.E. (1970) Some limitations of seismic refraction methods in geohydrological surveys of deep alluvial basins. *Ground Water,* 8 (6), 8–13.

Wattananikorn, K., Emharuthai, S. & Wanaphongse, P. (2008) A feasibility study of geogenic indoor radon mapping from airborne radiometric survey in northern Thailand. *Radiation Measurements,* 43 (1), 85–90.

White, P.A. (1988) Measurement of ground–water parameters using salt-water injection and surface resistivity. *Ground Water,* 26 (2), 179–186.

White, P.A. (1994) Electrode array for measuring groundwater flow direction and velocity. *Geophysics,* 59 (2), 192–201.

Whiteley, R.J. & Greenhalgh, S.A. (1979) Velocity inversion and the shallow seismic refraction method. *Geoexploration,* 17 (2), 125–141.

Widess, M.B. (1973) How thin is a thin bed? *Geophysics,* 38 (6), 1176–1180.

Wright, E.P. & Carruthers, R.M. (1993) Radon gas in soil air as a potential exploration tool in crystalline basement aquifers. In: Banks, D and Banks Sheila (eds.) *Hydrogeology of Hard Rocks: Memoires of the XXIV Congress of IAH, 28 June–2 July, As, Oslo, Norway,* IAH Press. pp. 672–683.

Yaramanci, U. (2000) Surface nuclear magnetic resonance (SMNR)-A new method for exploration of groundwater and aquifer properties. *Annali Di Geofisica,* 43 (6), 1159–1175.

Chapter 14

Geophysical methods in management of aquifer recharge & groundwater contamination study

14.1 INTRODUCTION

In some of the tropical hard rock areas dependability on groundwater is high and also increasing continuously. Earlier groundwater was abstracted in these areas from weathered zone aquifer through hand dug wells. The growing demand prompted for the sinking of public as well as private shallow bore wells. These wells were at the most affected by seasonal water level decline and some of them running dry in summer months indicating a temporal variation in saturated weathered zone thickness and consequent alteration in aquifer dimension and storage. The excessive withdrawal of groundwater more than its annual natural replenishment resulted in a wide-spread common phenomenon of declining groundwater level. The shallow weathered zone aquifers, distributed discontinuously as pockets with limited storage capacity got dewatered or dried-up at many places rendering a large number of wells dry through-out. It triggered drilling of deeper wells and also successive deepening of wells to exploit the saturated fractured zones for a better yield. With increased depth to groundwater level and reduced availability the stress on physical as well as economic access to groundwater also increased heavily. Overall, it resulted in a huge draft of groundwater inconsistent with its annual natural recharge. Consequentially the water level continued to decline, eventually forcing the most important resource to the present near-irreversible precarious situation in some of the tropical hard rock areas. Besides, in urban areas while draft has increased enormously, the decrease in permeable surface area has increased the surface runoff and reduced the natural recharge, causing localized lowering of groundwater levels. In coastal tracts, depleting the level of groundwater has caused sea water intrusion, resulting in a deterioration in groundwater quality.

The major issue emerged is sustainability – controlling the declining water level and restoring or improving water quality for which management of aquifer recharge, also known as managed aquifer recharge (MAR) or artificial recharge (AR) of depleting or drying up and poor water quality aquifer is essential. The term MAR appears a better substitute for AR. Also, it has been named as 'water banking' (Murray and Harris, 2010). It is essential for aquifer management in such areas. A large volume of dewatered weathered zones is available where water can be stored by MAR if recharging water of required quality and quantity is consistently available. MAR has several merits. It can improve the quality of water in aquifers, like dilution of fluoride (Muralidharan *et al.*, 2002; Andrade *et al.*, 2012), reduce sea water intrusion in coastal tracts by flushing as well as pushing the interface back (Chandra, 1993), reduce land subsidence (Chandra and Trivedi, 2000) and augment fresh water pockets in inland saline areas (Trivedi *et al.*, 2000). According to Bouwer (2002) it enhances the use of surface water as groundwater by improving its quality through geopurification – the 'soil-aquifer treatment' (SAT) and helps store surplus surface water in wet periods for use in water shortages during dry periods and reuse of water. MAR structure is constructed in hydrogeologically suitable and socially viable areas. Mostly it is cost effective. The aquifer modification in hard rock through hydrofracturing also comes under MAR.

14.2 MANAGED AQUIFER RECHARGE

The MAR or AR of groundwater refers to the transmission of surface water through human intervention into the subsurface storage space available, at a rate higher than the natural replenishment. The process is to accelerate the percolation of water at the surface, which may not otherwise percolate. Bouwer (2002) defines artificial recharge as *"engineered systems, where surface water is put on or in the ground for infiltration and subsequent movement to aquifers to augment the groundwater resources"*. This requires some man-made structures or changes at the surface compatible with the subsurface condition. The design and management involves geological, geochemical, hydrological, biological and engineering aspects (Bouwer, 2002). The main aim is to store water in the dewatered aquifer, transmit water to the saturated zone and enhance replenishment of groundwater.

MAR is a very old concept and has been in practice over centuries in several parts of the world. The practice of MAR can be traced to 600 AD (Sakthivadivel, 2007) in India to 475 BC in China (Wang *et al.*, 2014). In India, CGWB – a central government agency and other state agencies and NGOs carry out MAR studies on a regular basis using different techniques. MAR activities in the US started since the beginning of 20th century as reported by Weeks (2002) in an historical overview of artificial recharge in the US. Early references on MAR are from Illinois, USA (Suter and Harmeson, 1960) and basalt aquifers of Washington state (Price, 1961) which were adopted more than sixty years back to reduce groundwater shortages. In Australia MAR studies were initiated in mid 1960s, in Buderkin Delta, Queensland (Dillon *et al.*, 2009). In Latin American and African countries also practice of traditional MAR is quite old and the recent extensive MAR applications started since late 1970s. One of the successful artificial recharge schemes is the Atlantis Water Resource Management Scheme in South Africa (DWA, 2010).

There are several methods of MAR. Their applications depend on surface and subsurface lithologic conditions, terrain, groundwater conditions, availability of clean or treated recharging water and the objective. The details of MAR techniques are given in various publications. Some of the references are Bower (2002), CGWB (1994), Dillon (2005), Gale and Dillon (2005) and Dillon *et al.* (2009). The common type of MAR structures are basins for surface spreading if a large land area is available to increase the wetted area and residence time, furrows, ditches, trenches, percolation tanks, check dams, recharge shafts, subsurface dykes and injection wells. All these structures can be used to recharge an unconfined aquifer – a weathered zone aquifer in hard rock. However, the injection well is mainly used to recharge semi-confined and confined aquifers or specific fractured zones in granites or vesicular zones in basaltic terrain encountered at depth. MAR also includes transfer of water from one aquifer to another (Murray and Harris, 2010). Bhattacharya (2010) reports the use of injection wells for recharging deeper aquifers in the Deccan basalt. Most of the MAR techniques and structures are simple, low-cost and can be easily constructed with local materials, methods and people. In addition to these the dug well recharge method initiated by C.G.W.B. in India can also be adopted, wherein the excess storm water of agricultural fields being collected and filtered through a sand-gravel filter pit constructed near the dug well is put into it through a connector pipe.

As mentioned, the selection of MAR method varies with the hydrogeological conditions, availability of surface water for recharge and its quality. Forming a part of the local water resource management programme, MAR should be environmentally viable and less-expensive (Tizro *et al.*, 2010). Muralidharan *et al.* (2007) studied the efficiency of check dams in a small catchment in granitic terrain by continuous monitoring of the aquifer response- the water table. According to them recharge through the check dam is much higher than natural recharge and repeated filling of check dam enhances the groundwater recharge. Renganayaki and Elango (2013) reviewed the benefits of check dams in hard rock terrain in terms of the increase in groundwater storage through a rise in water level and also improvement in water quality.

As discussed in chapter 2, hard rock terrain is mostly heterogeneous in character with rapid changes in hydrogeological conditions. Groundwater in hard rock occurs mainly in the weathered zone. The underlying isolated fractures and joints also store groundwater but mostly act as conduits to transport groundwater locally to regionally depending on their dimensions and interconnections. The recharging water infiltrates through the top unsaturated permeable part, gets stored in the saturated weathered zone and channelized into the fracture network connected to the weathered zone. The best suited areas for MAR are those where weathered zone aquifers can store a large quantity of recharged water and do not allow it to move out quickly. The viability of areas/sites and depth zones for artificial groundwater recharge and its success requires adequate information on the wells available in the area (upstream and downstream), depth to water table, water level fluctuation, direction of groundwater movement, permeability of the near-surface material, spatial variation in hydrologic properties of unsaturated and saturated weathered zone, depth and extent of low permeability layers within the weathered zone, depth to compact bed rock, presence and orientation of fractures, their near-surface connections with the weathered zone, faults and dykes, depth of aquifer zones in basaltic terrain, the quality of water in weathered-zone aquifer and deeper aquifers including fresh/saline interface, if it exists, quality of recharging water and parameters like rainfall-its frequency and duration, evaporation and infiltration rate etc. Important consideration in MAR being infiltration of water through the top unsaturated part of the weathered material, information on hydraulic characteristics of the unsaturated zone and the surface soil is essentially required. All the above information can be obtained through hydrogeological, geophysical, hydrometeorological, hydrological and hydrochemical studies of the area. Besides, it is necessary to collect information on irrigation practices in the area, possible extent of land submergence due to MAR structure and other local information related to social viability. Areas with a shallow water table within 3 m below ground level are not considered for MAR as it may cause water logging and/or stagnation of water in the root zone subsequently.

While a MAR scheme in granitic terrain is to store water in weathered zone, maintain the water table in it through constrained movement of groundwater and possibly recharge the fractured zones at depth, in basaltic terrain it is a bit different. The groundwater situation in basaltic terrain is typical, where a number of near horizontal lava flows are stacked one over the other. They present alternate layering of hard and compact basalts and vesicular basalts with thin interflow beds of marine or fluvial sediments or weathered material from the parent rock. The weathered zone, joints and fractured zones in the compact basalt and the vesicular zones having considerable permeability are the groundwater repositories, presenting a multi-aquifer system in places. Therefore, unlike granitic terrain, in basaltic terrain besides recharging the permeable weathered zone, the deeper aquifer system comprising a sequence of separated vesicular and fractured zones, that can receive and store water, are also recharged.

In urban areas, where surface runoff is high, MAR of groundwater can be done by a convenient structure of roof top rain water harvesting by intercepting the rainfall and storing it in the subsurface. However, it depends on the subsurface storage space available. Urban areas with deeper water level having declining trend, near surface permeable layer and vadose zone devoid of clay lenses are best suited for rain water harvesting.

14.2.1 Geophysical investigations

Hard rock areas with complex hydrogeological condition, in addition to necessary hydrogeological and hydrochemical studies, warrant detailed subsurface information through geophysical investigations prior to construction of a MAR structure. Application of geophysics in MAR studies does not differ much from that in exploration and there is an array of geophysical methods and techniques which can be applied. Each geophysical method, whether conventional or modern high-resolution, provides valuable information that can be used appropriately or subjectively for MAR and contamination studies. Only, re-orientation in approach through judicious and economic integration and innovative shallow application is required to see how best the method could be utilized for the near-surface investigation. For the desired information, data are subjectively interpreted with hydrogeological inputs and the inferences are drawn on a contextual perspective. The main deliverables from geophysical investigation are quantitative map on feasibility of MAR structures for an area, suitable location and depth zone for maximum percolation and storage of recharged water, preferential flow path of recharged water, demarcation of favourable subsurface sections for construction of subsurface dykes to constrain the groundwater movement, precise location of monitoring wells and the impact assessment. The involvement of geophysical investigations for site and zone selection for MAR in hard rock is essential. It definitely increases the cost, but only marginally. Nevertheless it helps understand the local subsurface conditions, makes the MAR structure effective and helps assess the extent of areas which can get maximum benefit from the structure.

The surface resistivity method, viz., conventional resistivity sounding and profiling or resistivity imaging (ERT) is generally used for the purpose to assess depth-wise and lateral changes in resistivity that can be interpreted in terms of hydrogeological characteristics and MAR site suitability (Andrade, 2011; Ahmed *et al.*, 2014), viz., top soil conditions, weathered zone, macro-fractures in granites, shallow vesicular and fractured basalt layers and cavities in limestones. Obtaining a recognizable geophysical response from a deeper crystalline basement fractured zone or vesicular/fractured basalt zone to be recharged only through the resistivity method may not be always possible. It depends on its size, shape, depth of occurrence and also whether it is dry, saturated or clay filled or filled with secondary minerals, its physical property contrast with the surrounding and the character of the overburden. Integration with other geophysical methods is necessary. In this regard, the knowledge of geological conditions and causes responsible for the development of fractured zones, joints and faults is essential to understand the geophysical response, deduce relevant parameters, assess the degree of uncertainty in estimation and select other geophysical methods. The self potential (SP) and mise-a-la-masse (MAM) or charged body are the other techniques of electrical methods that can be used to decipher the direction of movement of recharged water in the subsurface.

The EM methods – the Very Low Frequency (VLF) and controlled source frequency domain EM (FEM) in frequency domain and the Time Domain EM (TEM) can also be used for the purpose. Cook *et al.* (1992) used FEM, TEM and DC resistivity to define the recharge areas and found FEM suitable as DC needs electrode contact and TEM requires measurement as early as 6–10 micro seconds for near surface assessment. The FEM surveys are conducted at a number of frequencies and transmitter-receiver coil separations to locate the fractured zones at depth. The EM soil-conductivity mapping is quite fast in field operation and helps derive a comprehensive soil-conductivity/resistivity map of an area. As in DC resistivity, the conductive overburden (surface layer and the weathered zone) of varying thickness induces masking anomalies and the ambiguity. Using VLF, the fractured zone behaving as a sheet like conductor can be detected if oriented in line with the VLF transmitter and the overburden resistivity is sufficiently high. Because of the high frequency (15–30 kHz) used, VLF has a better resolving power to demarcate the lateral disposition of the macro-fractures, at the same time the response gets attenuated quickly with depth, more so if the overburden is conductive. The VLF-R mode in which electrodes are just placed on the ground for resistivity measurements is quite fast for weathered zone resistivity mapping.

The magnetic and seismic are other methods employed depending on the demand of the problem. The weathered and fractured zone, dykes and faults show a variation in magnetic susceptibility as compared to the compact rock. In general, the weathered/fractured/fault zone is associated with comparatively low magnetic susceptibility and the basic dykes with higher susceptibility. The ground magnetic surveys provide first hand information on the presence of prominent faults and fractured zones, concealed dykes and the bed rock topography and thus can help in locating a MAR structure.

For precise subsurface characterization combination techniques that are sensitive to different physical properties are required. The varied degree of weathering and presence of fracture cause variation in seismic wave velocity and therefore it is possible to delineate these zones by shallow refraction and reflection seismic methods. Also, these methods provide a quantitative support in interpretation of electrical and electromagnetic anomalies. The techniques such as ERT and GPR are gaining importance because of their relatively easy field operations, interpretations and continuous site coverage for shallow/near surface investigations and are useful in MAR site selection.

The role of borehole logging is important in defining the depth zone suitable for recharge. It becomes significant in multi-aquifer system in vesicular and compact zones in basaltic terrain having different aquifer piezometric heads. The purpose is to precisely identify the aquifers. The minimum combination logging could be electrical, natural gamma radioactivity, temperature and fluid conductivity and flowmeter. Some of the MAR structures where geophysical investigations are helpful are detailed below.

14.2.1.1   Some managed aquifer recharge structures

Percolation tank

Percolation tanks are constructed for conservation as well as percolation of excess surface runoff for recharging the near surface aquifer downstream. The site for construction is so selected that the area under tank submergence has uniform and considerable thickness of weathered zone, saprolite and fractured zone with sufficient permeability and the downstream area as well has sufficiently thick weathered zone aquifer and covers a considerably large area to store the recharging water. It is an effective way of recharging groundwater in granitic and basaltic terrains. Percolation tanks are constructed by putting earthen bunds across small streams. Geophysical surveys through VES or ERT and GRP are quite useful in assessing the suitability of weathered zone and saprolite in terms of lithology and percolation capacity, fractured zones immediately underlying it and subsurface volume (storage capacity) which can accommodate the percolating water. Geophysical surveys are also conducted over a larger area downstream of the percolation tank to assess the aquifer getting recharged.

Check dam

A check dam for MAR is quite common in a hard rock area characterized by undulating topography. It is constructed across small streams with gentle slope for MAR through surplus surface runoff. The check dam is constructed in areas where a weathered zone is considerably thick so that sufficient water gets stored by recharging and the near surface layer in the ponded area is adequately permeable (CGWB, 1994). Presence of fractured zones is not desirable. Prior to construction of check dam, topographical survey, groundwater level, average annual rainfall, soil type, land use pattern in the area and the chemical and bacteriological data of the recharging water are collected. After a tentative section is fixed, a detailed geophysical survey comprising either close grid VES or ERT is conducted to know the weathered zone thickness and the presence of fractured zones. For locating the monitoring well also geophysical surveys are conducted.

Recharge pit and shaft

It is quite possible that a near surface layer of low permeability acts as a barrier to infiltration of surface water to the weathered zone and immediately underlying vesicular basalt aquifer with much better permeability. In such cases a deep recharge pit or shaft is excavated through the impermeable layer for easy access of the surface water to the aquifer. A recharge shaft is a pit of smaller cross-section. In basaltic terrain the recharge pit can be constructed by removing the near surface massive basalt blocks till the underlying vesicular basalt or interflow zone is reached (CGWB, 1994). The objective is that the pit should end in a permeable layer. So, prior to digging it is necessary that the hydrogeological characteristics of the layer underlying the top impermeable layer is known through ERT or possibly through EM-conductivity mapping which is fast and cost-effective for near surface investigation and may yield adequate information for the purpose.

Injection well

Injection well is a conventional borewell used for pumping-in treated water into the over-exploited unconfined as well as confined aquifer to augment groundwater storage in a multi-aquifer system. The stored water can be pumped out either by the same well or nearby well known as Aquifer Storage Recovery (ASR) or Aquifer Storage Transfer and Recovery (ASTR). In hard rock areas, individual or group of fractures encountered in a borehole can be recharged. Recharging of deeper aquifers through injection wells is in practice in Deccan Trap Basalts and Columbia River Basalts. For this, it is necessary that the geometry of the fractured zones in granites and vesicular and fractured zones in basalts, their hydraulic character and the hydrogeological boundaries are known through integrated geophysical surveys so that storage volume, quantity that can be pumped out and the movement of groundwater can be assessed.

Subsurface dyke

A subsurface dyke is a groundwater flow barrier constructed below the ground surface to arrest or retard the groundwater flow and increase storage of groundwater in a near surface aquifer for a longer time to make the shallow well sustainable. It is constructed in areas across ephemeral streams flowing through a valley with uneven topography and with low to moderate thickness of weathered zone. A fracture, if any, present within the bed rock is not favourable for subsurface dyke construction, as that would transmit groundwater in a particular direction downstream and attenuate the groundwater level build-up. It has a negative effect on downstream side and that is to be considered while constructing such dykes. For construction of a subsurface dyke, the depth to compact rock is assessed by shallow DC resistivity, EM or seismic survey. The survey reveals depth to bed rock in the valley as well as the characteristics of the underlying bedrock and weathered material In the case study by Singh *et al.* (1984) shown in figure 14.1, the thickness of the weathered zone estimated by DC resistivity and seismic refraction is more than 20 m along the stream course and therefore the site was not selected for construction of a subsurface dyke.

Rock fracturing and hydro-fracturing

Rock fracturing through borehole blasting is conducted to create artificial fractures which can help subsurface movement of groundwater and also possibly store it. Hydro-fracturing is basically a process of developing or revitalizing of a well in hard rock to improve its yield. It involves injection of high pressured water into the fractures and joints encountered in a well for cleaning them so that water flows out of these into the well at a better rate. Besides cleaning, the process helps fractures and joints connect to nearby other such structures so that groundwater moves towards the well. The pre and post-hydrofracturing geophysical logging of boreholes helps assess the development of fractures.

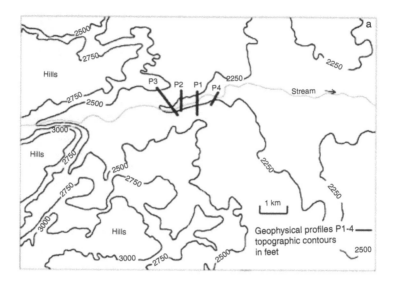

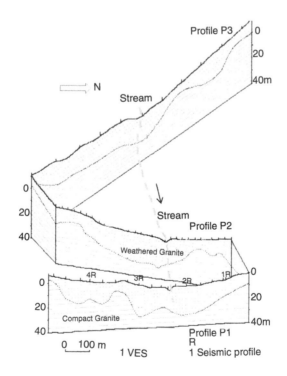

Figure 14.1 Elevation contour map and cross-sections based on electrical resistivity and refraction seismic survey across a valley in granitic terrain in south India to locate suitable section for construction of sub-surface dyke (Source: Singh et al., 1984).

14.2.1.2 Unsaturated zone characterization and monitoring recharge conditions

Calibrated DC resistivity sounding and profiling or ERT can be used to understand heterogeneity and estimate moisture variations in unsaturated zone. Depending on local hydrogeological conditions, DC resistivity or EM-conductivity measurements can be used to assess the hydraulic character also. Changes in physical properties due to dynamic processes can be monitored by collecting subsurface information as a function of time. It is an important aspect of impact assessment where surface and borehole geophysical measurements can contribute effectively. Surface and borehole geophysical measurements play a vital role in monitoring the abstraction and recharge conditions in an aquifer at shallow as well as deeper horizons. The variations in resistivity could be related to rate of recharge. Time-lapsed conventional surface resistivity measurements or ERT can record the variations at shallow level, e.g., resistivity variations due to moving recharge front or shifting of water quality interface. Also, time-lapsed EM imaging and GPR can be used for monitoring purposes from the surface. To get a near continuous, lateral, high resolution scanning of the unsaturated zone, its moisture content and also of the deeper zones, techniques like cross-borehole ERT (Daily *et al.*, 1992) and radar can be employed (Binley *et al.*, 2001). Time-lapsed surface ERT and cross-borehole ERT and GPR yield movement of infiltrating water, spatial variations in infiltration rate in an area and helps locate a zone suitable for the infiltration recharge system. The technique of neutron probing of an unsaturated zone is in practice to study the moisture content and its movement. The neutron logging and EM induction logging can also be used for monitoring. Besides, magnetic resonance sounding (MRS) can be useful to characterize the unsaturated zone (Costabel and Gunther, 2014). Time domain reflectometry (TDR) developed by Topp (1980) can be used for continuous monitoring of variations in soil moisture content and unsaturated zone changed by MAR (Verbist *et al.*, 2010, Schüth *et al.*, 2011).

14.3 GROUNDWATER CONTAMINATION STUDY

Concentration of any chemical constituent in groundwater beyond permissible limits, degrading the quality for the purpose of its use, is chemical contamination of ground-water. Its criteria depend on type of water use. Contamination could be of geogenic or anthropogenic origin. The contaminations of geogenic or geologic origin such as excess of fluoride, arsenic, iron and salinity are widespread and occur throughout the world. Similarly anthropogenic contamination of groundwater is also widespread. Anthropogenic contamination could be industrial, domestic and municipal, agricultural and environmental. Bacteriological contamination has not been considered here. Industrial contamination includes effluents carrying chemical compounds, trace elements and radioactive elements, effluents from tannery, acid mine drainage from mining, the domestic includes leakages from sanitary landfills and septic tanks to the unsaturated zone and unconfined aquifer, the agricultural includes return from irrigation water having fertilizers and pesticides and the environmental one includes changes in ground-water quality due to sea water intrusion in coastal tracts or salinity ingress from inland saline groundwater bodies etc. The contamination from landfills such as chlorides and acetic acid (Nobes, 1996) develops as a plume that moves in the subsurface outward and downward. The movement is controlled by the hydraulic property of the unsaturated zone and aquifer underlying it. The conventional investigation for contamination includes borehole drilling, monitoring and analysis for chemical constituents and their concentrations (Benson *et al.*, 1984).

In hydrogeologically heterogeneous hard rock, groundwater movement is dominated by flows through fractures and joints underlying the weathered zone. The network of fractures and joints can provide a pathway for movement of contaminants. The identification of fractured zones requires investigations at different scales and it is difficult to ascertain subsurface conditions precisely. The expensive drilling of boreholes for *in situ* sampling or monitoring at a location on a guess may result in failure and it becomes necessary to drill a number of boreholes to know subsurface conditions adequately at the contamination site. Even then the interpolation of findings between wells may not be satisfactory. Overall, this is a time-consuming and cost-prohibitive approach. Obviously, the geophysical measurements cannot replace micro-level information obtained through boreholes, but by generating a large density of data, help demarcate the area and target to a large extent as adequately and precisely as possible and also provide a more complete understanding of the site (Glaccum *et al.*, 1982) and a monitoring tool. Geophysics can be used as a first step in a systematic approach for contamination study. Applying non-intrusive surface geophysical methods in locating contaminated area and sites for optimal intrusive investigations through boreholes and their geophysical logging for monitoring is an economical approach. The geophysical applications are well documented in a number of publications. Some of the references on groundwater contamination are (Fried, 1975), Glaccum *et al.* (1982), National Academy Press (1984), Benson *et al.* (1984), EPA (1993), Paterson (1997), MacDougall *et al.* (2002) and Sara (2003).

14.3.1   Geophysical investigations

Geophysical methods are used to assess various physical properties of the subsurface and identify anomalies caused by contamination in groundwater at shallow levels. Compared to the use in selection of site for MAR, they have been applied more for contamination studies. Also, investigations are carried out in detail with better precision compared to MAR. The main aspects of geophysical studies are to locate where the contaminant is, how it is moving into the groundwater and where the contaminants can be dumped, i.e., to decipher the protective clay base layer that constrains the deeper movement of contaminants in groundwater. In this way, the area suitable for dumping solid municipal waste or liquid waste treatment pond that may not affect groundwater quality can be identified and also the best possible surface path for effluent transport can be located. The leakages from land-fill as well as from industrial effluents can be identified and movement of the plume can be detected, its dimensions mapped and time-lapse changes in dimensions and concentration monitored. Radioactive waste disposal site in hard rock without fractures that act as subsurface conduits can be identified. The protective low permeability layer for the fresh aquifer can be defined and site specific protection criteria can be fixed adequately with accuracy. The applications are multifarious-direct as well as indirect. A point is clear that chemical constituents of the contaminants (non-radioactive) cannot be identified by geophysical investigations except its concentration or variations in concentrations in terms of electrical conductivity (salinity) as this is the most common physical property altered considerably by a majority of chemical contaminants soluble in groundwater and the percent change in resistivities is more important than absolute resistivities. Because of this, DC resistivity and EM methods, whether surface or borehole, have been widely used and continue to be used in contaminant detection. There are numerous studies which have been conducted the world over. Some of the early references are of Cartwright and McComas (1968), Warner (1969), Merkel (1972) and Kelly (1976) on electrical resistivity surveys demonstrating the use of resistivity sounding in locating contaminant plume, Chandra *et al.* (1983) on the effect of effluents from tanneries on resistivity sounding, Greenhouse and Slaine (1983) and Greenhouse and Monier-Williams (1985) on EM and resistivity surveys, Karous *et al.* (1993) on repeat resistivity for detection of shallow groundwater contamination. There has been a significant development in the last three decades in geophysical characterization of contamination sites.

A single geophysical method may not always yield the desired result and achieve the objective. Subjective combination of primary and supplementary methods is preferred for subsurface characterization and monitoring as the methods are governed by different physical properties and can provide complementary hydrogeological information. By integrating selectively from the methods and techniques like resistivity and induced polarization tomography (ERT and IPT), EM conductivity mapping, FEM and TEM imaging, VLF-EM, VLF-R, GPR, high resolution shallow seismic survey (HRS) and the conventional techniques of electrical resistivity sounding and profiling, SP and MAM, it is possible to obtain a 3-d picture of the variations in the physical properties of the shallow subsurface affected by quality insurgence. The ERT, TEM, GPR and if required HRS can be conducted prior to the selection of municipal waste disposal site to determine the 3d continuity of overburden and heterogeneous clay lenses if present, that impedes infiltration and lateral movement of contaminated water. HRS having better resolution, it acts as a complimentary method yielding information regarding aquifer disposition and boundaries controlling the movement of contaminant plume. The solid waste landfills have density less than the natural surroundings and therefore a micro-gravity survey can be conducted for locating and demarcating the dimension of the landfill (Hinze *et al.*, 1990) and a follow-up by resistivity imaging (Forsberg *et al.*, 2006) or TEM will characterize lateral variations in the nature of landfill content. The magnetic method helps detect the buried metallic objects like drums, pipes and storage tanks. The leakages in sealing material or through impermeable bottom bed can be detected by the combination of SP and electrical resistivity survey. Besides surface geophysical methods and techniques, borehole geophysical methods are also used to characterize subsurface and obviously monitor the changes.

The approach of multiple methods and techniques is preferred to reduce non-uniqueness in inferences drawn from geophysical responses thus enhancing the level of confidence in characterization. Benson *et al.* (1984) used magnetic, electrical resistivity, electromagnetic and seismic refraction and reflection and GPR for hazardous waste site characterization. Glaccum *et al.* (1982) and Matias *et al.* (1994) combined DC resistivity and EM to map respectively the extent of leachate plume and contamination of groundwater from landfill (Fig. 14.2). Benson *et al.* (1997) used electrical resistivity and VLF and Nwankwo and Emujakporue (2012) used resistivity sounding and ERT to demarcate the contaminant plume as a high resistivity body generated by

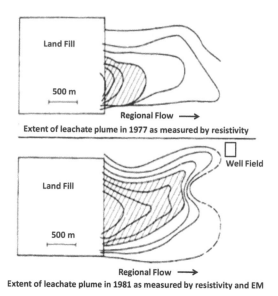

Figure 14.2 The growth and response of a leachate plume from landfill over a 4-year period in an expanding industrial area obtained from high density resistivity and EM data (Source: Glaccum *et al.*, 1982).

hydrocarbon leakages from shallow underground storage. Jin *et al.* (2008) used EM ground conductivity meter to locate diesel contamination in groundwater and Shevnin *et al.* (2003) used ERT to delineate the low resistivity zone associated with hydrocarbon spill developed after a lapse of 3–4 months. Merkel (1972) and Yuval and Oldenberg (1996) used resistivity and IP methods to detect acid mine drainage. For monitoring sea water intrusion, mapping the variations in electrical resistivity through a combination of time-lapsed ERT and TEM (Stewart and Gay, 1986 and Fitterman and Stewart, 1986) or FEM (Stewart, 1982) measurement is adequate of which TEM is time as well as cost effective but needs an equipment capable of early-time ($\sim$10 μs) measurements. Bauer *et al.* (2006) used surface and borehole to surface resistivity imaging to demarcate the brine occurrences. However, Frohlich *et al.* (1994) and Chandra (2005) used simple resistivity soundings to delineate the salinity interface.

The physical property generally affected is conductivity or resistivity of pore fluid and therefore the most common anomaly which can be detected is the variations in electrical conductivity due to contaminants. The application of these methods is effective only when the contaminant increases the total dissolved solid (TDS) and hence electrical conductivity (EC) of pore fluid to a level differentiable from the background. In the case where conductivity is not increased much higher or reduced much lower (by the presence of hydrocarbon) than the background or variations in conductivity due to lithology mask that due to the contaminant, it is difficult to detect the contaminant plume. The effect of lithological variation can however, be eliminated by repeat measurements and preparation of residual resistivity maps as shown by Hackbarth (1971).

The scale of changes in conductivity is important which makes these methods effective in delineating fresh and saline water aquifers in coastal areas where pore water salinity is high and electrical conductivity increases by several order. Where contaminants do not increase the EC, the DC resistivity and electromagnetic methods can be employed for indirect assessment, in demarcating the geometry of the aquifer which is expected to hold contaminated groundwater. Chandra *et al.* (2011) applied ERT and IPT and TEM soundings to delineate the clay barriers separating the arsenic free and arsenic contaminated aquifers. Adil *et al.* (1999) using resistivity sounding delineated fresh water aquifer in palaeo-channels in an inland saline area of India and defined the aquifer protection criteria based on lateral and depth-wise resistivity variations.

White (1988) demonstrated the scope of mapping the movement of a contaminant plume by injecting a small volume (1000–2000 litres) of highly concentrated salt water (20% by weight NaCl solution) in a well just below the water table and tracing its movement over time through surface resistivity measurements. The current electrode spacing is decided by conducting one or two resistivity soundings in the vicinity prior to injection and theoretically generating the layer models by introducing a thin salt water layer. The resistivity of the layer containing salt water (ρ_{sl}) is estimated using the formula $\frac{\rho_{sl}}{\rho_{sw}} = \frac{\rho_{ol}}{\rho_{ogw}}$, where ρ_{sw}, ρ_{ol} and ρ_{ogw} are the resistivity of the salt solution, original geoelectrical layer and original groundwater in the layer respectively. The resistivity profiling measurements with pre-defined current electrode spacing can be made at a definite station interval either along the lines radiating from the well or in a grid fashion. Grid measurement is preferred for easy movement of electrodes. The direction of the salt water slug is deciphered from the direction of maximum percentage of decrease in apparent resistivity. The velocity of slug movement can also be calculated by monitoring the apparent resistivity variations with time. Besides the construction of geoelectrical cross sections, resistivity contour maps and apparent resistivity pseudo-sections can help identify the conductive zones. Closing of contours and trend of contours are indicative of the boundaries of plume.

Application of the DC resistivity method has seen success as well as failure in contamination detection. It suffers limitations due to reasons like i) resistivity values obtained represent the combined effect of lithology and water quality (masking effect), ii) low resistivity topsoil at a site and lateral resistivity variations within it affect resistivity measurements, iii) poor resistivity contrast between contaminated and non-contaminated zones, iv) inadequate dimension of contaminated zone with respect to its depth of occurrence to produce an appreciable anomaly, v) averaging over a larger volume and vi) noisy data due to buried metallic objects.

Though the presence of water with higher EC reduces the resistivity, the correlation between electrical conductivity of contaminated pore fluid and measured apparent resistivity on the ground surface can be made for formations where changes in resistivity are only due to variations in electrical conductivity of saturating fluid. It may not be possible to correlate apparent resistivity values with EC of water if the formation is of mixed type e.g. clay mixing with sand in varying proportions. Even the maps for different electrode spacing may not yield much information. Cartwright and McComas (1968) and Klefstad *et al.* (1975) also describe these effects. In surface measurement a larger volume of subsurface is sampled as compared to the volume affected by contamination (Bevc and Morrison, 1991). Besides, if the contaminant is relatively deep buried and near surface layer is conductive or near surface layer is highly heterogeneous, surface resistivity measurements are ambiguous. An equivalent example of this limitation is the highs and lows in an apparent resistivity profile obtained only by varying the top soil resistivity, keeping other parameters fixed as shown in figure 6.7 and 6.42 of Chapter 6. In such cases potential response at the surface is enhanced by placing the current electrode in the contaminated depth zone or below the water table through the monitoring well or any existing borehole. In case of steel well casing, the casing itself can be used as the electrode. The other current electrode is placed on the surface at a distance about 5 to 10 times the expected lateral dimension of the plume. This is effective for inorganic contaminants which enhance the conductivity of pore fluid appreciably. The distribution of potential developed at the surface can be measured by a mobile potential electrode as gradient array along profile lines either radiating from the borehole or laid in a grid pattern. The technique of placing a current electrode in the contaminated zone thereby making it a source for current conduction along the preferred conductive plume is a form of mise-a-la-masse method, which has been used mostly in mineral prospecting with high conductivity contrasts (Osiensky and Donaldson, 1994). The shape or trend of the equipotential contour imitates the plume and gives an idea of the geometry of plume. Time lapse measurement would indicate the movement of the plume. The technique can be used for primary characterization to place additional monitoring wells. Before conducting such studies it is necessary that a few vertical electrical soundings on ERTs are conducted in the suspected area and a generalized subsurface geoelectrical layer model is conceived.

This technique can be further modified as indicated by Bevc and Morrison (1991). They conducted borehole-to-surface resistivity experimental measurements before and after injecting salt water into the fresh water aquifer and presented a method of enhancing the anomaly. For this, casing of the injection well was purposely fabricated with steel sections at defined depths to act as down hole-electrodes for current injection. The data acquisition can be through pole-pole and pole-dipole array of electrodes. In pole-pole array one current and one potential electrode is kept on the surface at a distance, while in pole-dipole array both the current electrodes are in the well and kept at a fixed spacing. The potential measurements are made on the surface in radial directions from the well. The details can be obtained from the paper by Bevc and Morrison (1991). The 25 to 40 percent change in resistivity due to the presence of conductive saline plume and its movement was interpreted through numerical modeling.

Where there are a number of steel cased wells around the contamination site, these wells can be used as linear electrodes to map the contaminant plume (Rucker et al., 2010). It is useful for detecting contaminant plumes below the unsaturated zone and where the top layer is highly conductive or constrains the application for noise free data collection. The data so collected can give better vertical resolution compared to the measurements made by an electrode array only at the surface. The modern techniques for the purpose are cross-well resistivity and radar tomography. Singha and Gorelick (2005) used the advanced technique of cross-well ERT to monitor the movement of injected saline water plume and Lane et al. (2004) used cross-well radar tomography to estimate the geometry of the contaminated area.

Electromagnetic surface methods (Greenhouse and Slaine, 1983, Ladwig, 1983) and borehole methods (McNeill et al., 1990) which also measure the electrical conductivity variations in the subsurface are used for groundwater contamination studies.

Out of these, the TEM method is becoming popular because of fast laying out time and operation and relatively smaller dimension of receiver-transmitter loop used for a relatively large depth of exploration. Also, it has capability of mapping very shallow targets with early time (6 to 10 μs) measurements. With a square loop size of 30 m × 30 m the depth up to 75 m could be explored (Hoekstra and Blohm, 1990). Because of these, it can be conveniently used even in urban areas in limited open spaces and parks. The loop size and transmitter moment are so selected that the signal is greater than the noise for the measurements at late delay times. The TEM is conducted at places with VES and jointly inverted to reduce the ambiguities due to equivalence. Also, a small spread VES can be conducted for better approximation of the top near surface layer resistivity and thickness which may be difficult to obtain from TEM at places (Buselli et al., 1990). Israil et al. (2011) used TEM, VES and ERT to delineate the extent of contamination in unconfined aquifer due to untreated sewage water irrigation. Pujari et al. (2011) used TEM to study the contamination effect of unlined ash pond of a thermal power plant on shallow groundwater within about 40 m depth. The terrain conductivity mapping using frequency domain electromagnetic is also useful in contamination studies. In areas with variations in topography the terrain conductivity shows a relation with topography. Monier-Williams et al. (1990) introduced topographic corrections using empirical relations derived from the cross-plot of measured apparent conductivity vs topography to assess the background conductivity in absence of any contamination. Barlow and Ryan (1985) used FEM with vertical- and horizontal-dipole Transmitter-Receiver (T-R) configuration and found vertical-dipole configuration useful in qualitatively differentiating the shallow and deeper contamination. Further, they attempted the usual linear as well as logarithmic contouring of EM data suggested by Greenhouse and Slaine (1983). The logarithmic contouring is in dimensionless units resulting from the ratio of the measured apparent conductivity to the background apparent conductivity. The contoured variable is

$$20\log_{10}\frac{\sigma_{\text{apparent measured at location X,Y}}}{\sigma_{\text{apparent background}}}$$

The selection of background conductivity, as pointed out by Barlow and Ryan (1985), is subjective and taken as the average low-apparent conductivity from the uncontaminated part of the area and also it may be necessary to select more than one background values for different T-R configurations. The advantage of plotting dimensionless logarithmic ratios is masking the background noise and clearly outlining the area of contamination.

14.3.1.1 Monitoring groundwater contamination

Besides time-lapsed high resolution surface and surface to borehole measurements including 3d ERT of the unsaturated zone (Nimmer et al., 2007), monitoring can also be done by borehole geophysical logging that detects groundwater contamination in situ. EPA (1993) presents a detailed list of borehole geophysical methods for the purpose. The electrical resistivity, temperature-fluid conductivity and EM logging are used to define the zones containing highly conductive fluid or down-hole groundwater quality variations. Obviously, the geophysical parameter measured would not give any indication of the chemical constituents. The increase in fluid conductivity could be

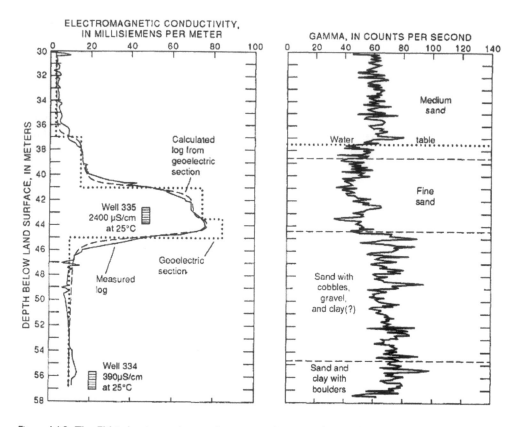

Figure 14.3 The EM induction and natural gamma radioactivity logs and litholog of a well at a landfill site; increase in conductivity is associated with the increase in EC of groundwater (Source: Williams *et al.*, 1993).

either due to soluble inorganic contaminants, the natural salinity or the present day sea water intrusion and is subjective. Repeat precision temperature logging of boreholes and their interpretation can yield movement of interface, and direction of flow in the borehole. The temperature variation with depth in the borehole is recorded also to correct the electrical conductivity values which changes with temperature. The logging can be repeated as time-lapsed measurements to evaluate the increase or decrease in conductivity with time. Repeat resistivity logging in water filled PVC cased borehole can be done using permanent multi-electrode arrangement against the expected or proven zone measuring the changes in resistivity or conductivity at regular time intervals. The EM induction, temperature-fluid conductivity and flow logging can also be used effectively for monitoring the movement of contaminated water below the water table. EM induction logging is effective for monitoring water quality in wells with PVC casing (McNeill *et al.*, 1990). There are several case studies on it. Mack (1993) and Williams *et al.* (1993) employed the combination of EM induction and natural gamma radioactivity logging of boreholes at landfill sites to identify the increase in conductivity due to groundwater contamination and the lithology respectively (Fig. 14.3). Metzger and

Izbicki (2012) used repeat EM induction logging measurements in PVC cased wells to estimate the change in chloride concentration. Chandra *et al.* (2001) used natural gamma radioactivity logs of boreholes along the coastline to estimate the hydraulic conductivity and depth zone vulnerable to sea water intrusion.

REFERENCES

Adil, M., Srivastava, M.M., Trivedi, B.B. & Chandra, P.C. (1999) Geoelectrical identification of fresh water aquifers in saline environ of Mathura district, Uttar Pradesh. In: Geological Survey of India *Proceedings of Golden Jubilee Seminar on Exploration Geophysics in India, Nov. 14–16 1995, Calcutta, India*, Calcutta, Geol. Survey of India, Special Publication 49, ISSN-0254-0436, pp. 461–470.

Ahmed, S., Arora, T., Sarah, S., Dar, F.A., Warsi, T., Gaur, T.K. & Raghuvender, P. (2014) Viewing Sub-Surface for an Effective Managed Aquifer Recharge from a Geophysical Acumen, In: Wintgens, T., Nättorp, A., Lakshmanan, E., & Asolekar, S.R. (eds.) *Natural Water Treatment Systems for Safe and Sustainable Water Supply in the Indian Context: Saph Pani, Chapter 19*. IWA Publishing Co. UK.

Andrade, R. (2011) Intervention of electrical resistance tomography (ERT) in resolving hydrological problems of a semi arid granite terrain of Southern India. *Journal of Geological Society of India*, 78 (4), pp. 337–344.

Andrade, R., Rangarajan, R. & Lakshmi Devi, D.N.V. (2012) Contaminated groundwater remediation – usage of water harvesting under favorable circumstances. *Earth Science India*, www.earthscienceindia.info, 5 (2), pp. 1–12.

Barlow, P.M. & Ryan, B.J. (1985) An electromagnetic method for delineating ground-water contamination, Wood River junction, Rhode Island. In: Subitzky, S. (ed.) *Selected Papers in the Hydrologic Sciences, 1985*, U.S. Geological Survey Water Supply Paper 2270, pp. 35–49.

Bauer, P., Supper, R., Zimmermann, S. & Kinzelbach, W. (2006) Geoelectrical imaging of groundwater salinization in the Okavango Delta, Botswana. *Journal of Applied Geophysics*, 60 (2), 126–141.

Benson, R.C., Glaccum, R.A. & Noel, M.R. (1984) *Geophysical techniques for sensing buried wastes and waste migration*. US EPA, EMSL, Las Vegas 600/7-84/064 (NTIS PB84-198449).

Benson, A.K., Payne K.L. & Stubben, M.A. (1997) Mapping groundwater contamination using dc resistivity and VLF geophysical methods – a case study. *Geophysics*, 62 (1), 80–86.

Bevc, D. & Morrison, H.F. (1991) Borehole-to-surface electrical resistivity monitoring of a salt water injection experiment. *Geophysics*, 56 (6), 769–777.

Bhattacharya, A.K. (2010) Artificial ground water recharge with a special reference to India. Int. Jour. of Res. & Reviews in Applied Sciences (IJRRAS), 4 (2), 214–221.

Binley, A., Winship, P. & Middleton, R. (2001) High-resolution characterization of vadose zone dynamics using cross-borehole radar. *Water Resources Research*, 37 (11), 2639–2652.

Bouwer, H. (2002) Artificial recharge of groundwater: hydrogeology and engineering. *Hydrogeology Journal*, 10 (1), 121–142.

Buselli, G., Barber, C., Davis, G.B. & Salama, R.B. (1990) Detection of groundwater contamination near waste disposal sites with transient electromagnetic and electrical methods. In: Ward, S.H. (ed.) *Geotechnical and Environmental Geophysics, Vol. II: Environmental and Groundwater*. Tulsa, Oklahoma, USA, Society of Exploration Geophysicists, pp. 27–40.

Cartwright, K. & McComas, M.R. (1968) Geophysical surveys in the vicinity of sanitary landfills in Northeastern Illinois. *Ground Water*, 6 (5), 23–30.

C.G.W.B. (1994) *Manual on artificial recharge of groundwater*. C.G.W.B., Min. of Water Resources, Govt. of India Publication, Technical Series- M, No. 3.

Chandra, P.C. (1993) Role of geophysical techniques in pinpointing artificial recharge sites in coastal areas. In: CGWB *Proceedings of Workshop on 'Artificial Recharge Studies in Coastal Areas, March 1993, Bhubaneswar, Odisha, India.*

Chandra, P.C. (2005) Geophysical investigations for groundwater in coastal tracts. *Jal Vigyan Sameeksha,* 20, 47–64.

Chandra, PC., Ramakrishna A., & Singh, Hira (1983) Tracing shear zones by geoelectrical techniques for groundwater in Dindigul area of Madurai district, Tamil Nadu. In: CGWB *Proceedings. Sem. Assess. Dev. & Mangment GW Res., New Delhi.*

Chandra P.C., Singh, S.C. & Reddy, P.H.P. (2001) Hydraulic conductivity estimation from natural gamma radioactivity borehole logs along Digha-Haldia coast line, West Bengal and its use in coastal groundwater development. *Geological Survey of India Spl. Pub. No. 65(ii),* pp. 237–239.

Chandra, P.C. & Trivedi, B.B. (2000) Land subsidence and long cracks in parts of Farrukhabad district, U.P.: an environmental impact of groundwater over-development. In: *Proc. All India Sem. on Ground Water Recharge: Strategy for Sustainable Water World, 15-16 July, 2000, Lucknow, U.P., India.* The Institution of Engineers (India). pp. 1–6.

Chandra, S., Ahmed, S., Nagaiah, E., Singh, S.K. & Chandra, P.C. (2011) Geophysical exploration for lithological control of arsenic contamination in groundwater in Middle Ganga Plains, India. *Physics and Chemistry of the Earth,* 36 (16), 1353–1362.

Cook, P.G., Walker, G.R., Busseli, G., Potts, I. & Dodds, A.R. (1992) The application of electromagnetic techniques to groundwater recharge investigations. *Journal of Hydrology,* 130 (1–4), 201–229.

Costabel S. & Gunther, T. (2014) Noninvasive estimation of water retention parameters by observing the capillary fringe with magnetic resonance sounding. Vadose Zone Journal, 13 (6), p. 14.

Daily, W., Ramirez, A. LaBrecque, D. & Nitao, J. (1992) Electrical resistivity tomography of vadose water movement. *Water Resources Research,* 28 (5), 1429–1442.

Department of Water Affairs (2010) Strategy and guideline development for national groundwater planning requirements. The Atlantis water Resource Management Scheme: 30 years of artificial groundwater recharge. *P RSA 000/00/11609/10-Activity 17 (AR5.1).*

Dillon, P.J. (2005) Future management of aquifer recharge, *Hydrogeology Journal,* 13 (1), 313–316.

Dillon, P., Pavelic, P., Page, D., Beringen, H. & Ward, J. (2009) *Managed aquifer recharge: an introduction.* Waterlines Report Series No. 13, National Water Commission, CSIRO, Australia.

EPA (1993) *Use of airborne, surface, and borehole geophysical techniques at contaminated sites: A reference guide.* Prepared by Eastern Research Group for US EPA: EPA/625/R-92/007.

Fitterman, D.V. & Stewart, M.T. (1986) Transient electromagnetic sounding for groundwater. *Geophysics,* 51 (4), 995–1005.

Fried, J.J. (1975) *Groundwater pollution:* Developments in Water Science Series 4. Amsterdam, The Netherlands, Elsevier Scientific Publishing Company.

Frohlich, R.K., Urish, D.W., Fuller, J. & Reilly, M. O. (1994) Use of geoelectrical methods in groundwater pollution surveys in a coastal environment. *Journal of Applied Geophysics,* 32 (2 & 3), 139–154.

Forsberg, K., Nilsson, A., Flyhammar, P. & Dahlin, L.T. (2006) Resistivity imaging for mapping of groundwater contamination at the municipal landfill La Chureca, Managua, Nicaragua. In: *Near Surface Geoscience 2006: Proceedings of the 12th European Meeting of Environmental & Engineering Geophysics, 4–6 September 2006, Helsinki, Finland.* The Netherlands, EAGE.

Gale, I. & Dillon, P. (2005) *Strategies for Managed Aquifer Recharge (MAR) in semi-arid areas.* Paris, UNESCO IHP Publication.

Glaccum, R.A., Benson, R.C. & Noel, M.R. (1982) Improving accuracy and cost-effectiveness of hazardous waste site investigations. *Ground Water Monitoring and Remediation*, 2 (3), 36–40.

Greenhouse, J.P. & Monier-Williams, M. (1985) Geophysical monitoring of groundwater contamination around waste disposal sites. *Ground Water Monitoring and Remediation*, 5 (4), 63–69.

Greenhouse, J.P. & Slaine, D.D. (1983) The use of reconnaissance electromagnetic methods to map contaminant migration. *Ground Water Monitoring and Remediation*, 3 (2), 47–59.

Hackbarth D.A. (1971) Field study of subsurface spent sulfite liquor movement using earth resistivity measurements. *Ground Water*, 9 (3), 11–16.

Hinze, W.J., Roberts, R.L. & Leap, D.I. (1990) Combined analysis of gravity and magnetic anomaly data in landfill investigations. In: Ward, S.H. (ed.) *Geotechnical and Environmental Geophysics, Vol. II: Environmental and Groundwater*. Tulsa, Oklahoma, USA, Society of Exploration Geophysicists, pp. 267–272.

Hoekstra, P. & Blohm, M.K. (1990) Case histories of time-domain electromagnetic soundings. In: Ward, S.H. (ed.) *Geotechnical and Environmental Geophysics, Vol. II: Environmental and Groundwater*. Tulsa, Oklahoma, USA, Society of Exploration Geophysicists, pp. 1–16.

Israil, M., Tezkan, B., Yogeshwar, P., von Papen, M., Sudha & Gupta, P.K. (2011) Mapping the groundwater contamination around Roorkee, India, using TEM and DC resistivity measurements: case study. Presented at: *The Workshop on Recent Advances in Ground & Airborne Electromagnetic Methods, 27-28 September 2011, AMD-DAE Hyderabad, India*. Abstract Volume pp. 57–59.

Jin, S., Fallgren, P., Cooper, J., Morris, J. & Urynowicz, M. (2008) Assessment of diesel contamination in groundwater using electromagnetic induction geophysical techniques. *Journal of Environmental Science & Health-Part A: Toxic/ Hazard Substance & Environmental Engineering*, 43 (6), 584–588.

Karous, M., Mares, S., Kelly, W.E. Anaton, J., &Havelka, J. & Stoje, V. (1993) Resistivity methods for monitoring spatial and temporal variations in groundwater contamination. In: Kovar, K. & Soveri, J. (eds.) *GQM 93: Proceedings of the Conference on Groundwater Quality Management, 6-9 September 1993, Tallinn, Estonia*. IAHS Press, 1994 Publication No. 220.

Kelly, W.E. (1976) Geoelectrical sounding for delineating groundwater contamination. *Ground Water*, 14 (1), 6–10.

Klefstad, G., Sendlein, L.V.A. & Palmquist, R.C. (1975) Limitations of the electrical resistivity method in landfill investigations. *Ground Water*, 13 (3), 418-427.

Ladwig, K.J. (1983) Electromagnetic induction methods for monitoring acid mine drainage. *Ground Water Monitoring and Remediation*, 3 (1), 46–51.

Lane Jr., J.W., Day-Lewis, F.D., Versteeg, R.J. & Casey, C.C. (2004) Object-based inversion of crosswell radar tomography data to monitor vegetable oil injection experiments. *Journal of Environmental & Engineering Geophysics*, 9 (2), 63–77.

MacDougall, K.A., Fenning, P., Cooke, D.A., Preston, H., Brown, A., Hazzard, J. & Smith, T. (2002) *Non-intrusive investigation techniques for groundwater pollution studies*. R&D Technical Report P2-178/TR/1, Environment Agency, Rio House, Waterside Drive, Aztec West, Almondsbury, Bristol, BS32 4UD.

Mack, T.J. (1993) Detection of contaminant plumes by borehole geophysical logging. *Ground Water Monitoring and Remediation*, 13 (1), 107–114.

Matias, M.S., da Silva, M.M., Ferreira, P. & Ramalho, E. (1994) A geophysical and hydrogeological study of aquifers contamination by a landfill. *Journal of Applied Geophysics*, 32 (2 & 3), 155–162.

McNeill, J.D., Bosnar, M. & Snelgrove, F.B. (1990) *Resolution of an electromagnetic borehole conductivity logger for geotechnical and groundwater applications*. Technical Note TN-25, Geonics Ltd., Ontario, Canada.

Merkel, R.H. (1972) The use of resistivity techniques to delineate acid mine drainage in ground water. *Ground Water*, 10 (5), 38–42.

Metzger, L.F. & Izbicki, J.A. (2012) Electromagnetic-induction logging to monitor changing chloride concentrations. *Ground Water*, 50 (aop), 1–14.

Monier-Williams, M.E., Greenhouse, J.P, Mendes, J.M. & Ellert, N. (1990) Terrain conductivity mapping with topographic corrections at three waste disposal sites in Brazil. In: Ward, S.H. (ed.) *Geotechnical and Environmental Geophysics, Vol. II: Environmental and Groundwater.* Tulsa, Oklahoma, USA, Society of Exploration Geophysicists, pp. 41–56.

Muralidharan, D., Andrade, R. & Rangarajan, R. (2007) Evaluation of check-dam recharge through water-table response in ponding area. *Current Science*, 92 (10), 1350–1352.

Muralidharan D., Nair, A.P. & Sathyanarayana, U. (2002) Fluoride in shallow aquifers in Rajgarh Tehsil of Churu District, Rajasthan – an arid environment. *Current Science*, 83 (6), pp. 699–702.

National Academy Press (1984) *Groundwater Contamination (Studies in Geophysics)*, NRC, Washington, USA.

Nimmer, R.E., Osiensky, J.L., Binley, A.M., Sprenke, K.F.& Williams, B.C. (2007) Electrical resistivity imaging of conductive plume dilution in fractured rock. *Hydrogeology Journal*, 15 (5), 877–890.

Nobes, D.C. (1996) Troubled waters: environmental application of electrical and electromagnetic methods. *Surveys in Geophysics*, 17 (4), 393–454.

Nwankwo, C.N. & Emujakporue, G.O. (2012) Geophysical method of investigating groundwater and sub-soil contamination – a case study. *American Journal of Environmental Engineering*, 2 (3), 49–53.

Osiensky, J.L. & Donaldson, P.R. (1994) A modified mise-a-la-masse method for contaminant plume detection. *Groundwater*, 32 (3), 448–457.

Paterson, N. (1997) Remote mapping of mine wastes. In: Gubins, A.G. (ed.) *Proceedings of Exploration 97: Fourth Decennial International Conference on Mineral Exploration, 14–18 September 1997, Toronto, Canada,* GEO F/X Div. of AG Inf. Sys. Ltd. Canada, pp. 905–616.

Price, C.E. (1961) *Artificial recharge through a well tapping basalt aquifers, Walla Walla area, Washington.* U.S. Geological Survey Water-Supply Paper 1594-A, U.S.G.S.

Pujari, P.R., Padmakar, C., Sanam, R., Khandekar, V. & Labhasetwar, P.K. (2011) Near surface characterization near tailing. Presented at: *The Workshop on Recent Advances in Ground & Airborne Electromagnetic Methods, 27-28 September 2011, AMD-DAE Hyderabad, India,* Abstract Volume pp. 55–56.

Renganayaki, S.P. & Elango, L. (2013) A review on managed aquifer recharge by check dams: a case study near Chennai, India. *Int. Jour. Research in Engineering & Technology (IJRET)*, 2 (4), 416–423.

Rucker, D.F., Loke, M.H., Levitt, M.T. & Noonan, G.L. (2010) Electrical-resistivity characterization of an industrial site using long electrodes. *Geophysics*, 75 (4), WA95-WA104.

Sakthivadivel, R. (2007) The groundwater recharge movement in India. In: M., Giordano & Villholth, K.G. (eds.) *The Agricultural Groundwater Revolution: Opportunities and Threats to Development.* CABI, UK Publication in association with IWMI, Sri Lanka pp. 195–210.

Sara, M.N. (2003) *Site assessment and remediation handbook.* Florida, USA, CRC Press LLC.

Schüth, C., Kallioras, A., Piepenbrink, M., Pfletschinger, H., Al-Ajmi, H., Engelhardt, I., Rausch, R. & Al-Saud, M. (2011) New approaches to quantify groundwater recharge in arid areas. *International Journal of Water Resources and Arid Environments*, 1(1), 33–37.

Shevnin V., Delgado-Rodríguez, O., Mousatov, A., Nakamura-Labastida, E., & Mejía-Aguilar, A. (2003) Oil pollution detection using resistivity sounding. *Geofisica International*, 42 (4), 613–622.

Singh, H., Chandra P.C. & Tata, S. (1984) Application of geophysical surveys for artificial recharge – a case history. Presented at: *The AEG Decennial Convention and Seminar* on *Exploration Geophysics (Geophysics in Developing Countries), 26–28 October 1984, Osmania University, Hyderabad, India.*

Singha, K. & Gorelick, S.M. (2005) Saline tracer visualized with three-dimensional electrical resistivity tomography: Field-scale spatial moment analysis. *Water Resources Research*, 41 (5) W05023, doi:10.1029/2004WR003460.

Stewart, M.T. (1982) Evaluation of electromagnetic methods for rapid mapping of salt water interfaces in coastal aquifers. *Ground Water*, 20 (5), 538–545.

Stewart, M. & Gay, M.C. (1986) Evaluation of transient electromagnetic soundings for deep detection of conductive fluids. *Ground Water*, 24 (3), 351–356.

Suter, M. & Harmeson, R.H. (1960) *Artificial ground-water recharge at Peoria, Illinois.* Illinois State Water Survey Bulletin 48.

Tizro, A.T., Voudouris, K.S. & Akbari, K. (2011) Simulation of a groundwater artificial recharge in a semi-arid region of Iran. *Irrigation and Drainage*, 60 (3), 393–403.

Topp, G.C. (1980) Electromagnetic determination of soil water content: measurements in coaxial transmission lines. *Water Resources Research*, 16 (3), 574–582.

Trivedi, B.B., Srivastava, M.M., Adil, M. & Chandra, P.C. (2000) Artificial recharge of aquifers in saline groundwater environ of Mathura district, UP. In: Mohan, J. & Nusrat J. (eds.) *Proceedings of All India Seminar on Ground Water Recharge: Strategy for Sustainable Water World, 15–16 July, 2000, Lucknow, U.P., India.* The Institution of Engineers (India). pp. 194–198.

Verbist, K., Cornelis, W.M., McLaren, R., Gabriels, D. & Soto, G. (2010) A frame work for evaluating water harvesting efficiency. In: Verbist, K. & Gabriels, D. (eds.) *Proceedings of the International Conference "Arid and Semi-arid Development Through Water Augmentation" 13–16 December 2010, Valparaiso, Chile.* UNESCO Technical Document IHP-LAC N° 31, ISBN 978-92-9089-183-3, pp. 4–7.

Wang, W., Zhou, Y., Sun, X. & Wang, W. (2014) Development of managed aquifer recharge in China. *Boletín Geológico y Minero*, 125 (2), 227–233, ISSN: 0366-0176.

Warner, D.L. (1969) Preliminary field studies using earth resistivity measurements for delineating zones of contaminated groundwater. *Ground Water*, 7 (1), 9–16.

Weeks, E.P. (2002) A historical overview of hydrologic studies of artificial recharge in the U.S. Geological Survey. In: Aiken, G.R. & Kuniansky, E.L. (eds.) Proceedings of U.S. Geological Survey Artificial Recharge Workshop, 2–4 April 2002, Sacramento, California, USA. USGS Open-File Report 02-89.

White, P.A. (1988) Measurement of groundwater parameters using salt water injection and surface resistivity. *Ground Water*, 26 (2), 179–186.

Williams, J.H., Lapham, W.W. & Barringer, T.H. (1993) Application of electromagnetic logging to contamination investigation in glacial sand-and-gravel aquifers. *Groundwater Monitoring, & Remediation*, 13 (3), 129–138.

Yuval, & Oldenberg, D.W. (1996) DC resistivity and IP methods in acid mine drainage problems: results from the Copper Cliff mine tailings impoundments. *Journal of Applied Geophysics*, 34 (3), 187–198.

Subject index